Trends in Grain Processing for Food Industry

Trends in Grain Processing for Food Industry

Guest Editors

Georgiana Gabriela Codină
Adriana Dabija

Basel • Beijing • Wuhan • Barcelona • Belgrade • Novi Sad • Cluj • Manchester

Guest Editors

Georgiana Gabriela Codină	Adriana Dabija
Ștefan cel Mare University	Ștefan cel Mare University
Suceava	Suceava
Romania	Romania

Editorial Office
MDPI AG
Grosspeteranlage 5
4052 Basel, Switzerland

This is a reprint of the Special Issue, published open access by the journal *Applied Sciences* (ISSN 2076-3417), freely accessible at: https://www.mdpi.com/journal/applsci/special_issues/T1X16Q7LHR.

For citation purposes, cite each article independently as indicated on the article page online and as indicated below:

Lastname, A.A.; Lastname, B.B. Article Title. *Journal Name* **Year**, *Volume Number*, Page Range.

ISBN 978-3-7258-2895-1 (Hbk)
ISBN 978-3-7258-2896-8 (PDF)
https://doi.org/10.3390/books978-3-7258-2896-8

© 2025 by the authors. Articles in this book are Open Access and distributed under the Creative Commons Attribution (CC BY) license. The book as a whole is distributed by MDPI under the terms and conditions of the Creative Commons Attribution-NonCommercial-NoDerivs (CC BY-NC-ND) license (https://creativecommons.org/licenses/by-nc-nd/4.0/).

Contents

About the Editors vii

Preface ix

Georgiana Gabriela Codină and Adriana Dabija
Innovative Grain Processing: Trends and Technologies
Reprinted from: *Appl. Sci.* 2024, *14*, 10954, https://doi.org/10.3390/app142310954 1

Pavel Skřivan, Marcela Sluková, Andrej Sinica, Roman Bleha, Ivan Švec, Evžen Šárka and Veronika Pourová
Glycaemic Index of Bakery Products and Possibilities of Its Optimization
Reprinted from: *Appl. Sci.* 2024, *14*, 6070, https://doi.org/10.3390/app14146070 5

Alina Alexandra Dobre, Elena Mirela Cucu and Nastasia Belc
Influence of Technological Parameters on Sourdough Starter Obtained from Different Flours
Reprinted from: *Appl. Sci.* 2024, *14*, 4955, https://doi.org/10.3390/app14114955 26

Mihaela Brumă (Călin), Iuliana Banu, Ina Vasilean, Leontina Grigore-Gurgu, Loredana Dumitrașcu and Iuliana Aprodu
Influence of Soy Protein Hydrolysates on Thermo-Mechanical Properties of Gluten-Free Flour and Muffin Quality
Reprinted from: *Appl. Sci.* 2024, *14*, 3640, https://doi.org/10.3390/app14093640 37

Sylvestre Dossa, Christine Dragomir, Loredana Plustea, Cosmin Dinulescu, Ileana Cocan, Monica Negrea, et al.
Gluten-Free Cookies Enriched with Baobab Flour (*Adansonia digitata* L.) and Buckwheat Flour (*Fagopyrum esculentum*)
Reprinted from: *Appl. Sci.* 2023, *13*, 12908, https://doi.org/10.3390/app132312908 55

Ioana Stanciu, Elena Loredana Ungureanu, Elisabeta Elena Popa, Mihaela Geicu-Cristea, Mihaela Draghici, Amalia Carmen Mitelut, et al.
The Experimental Development of Bread with Enriched Nutritional Properties Using Organic Sea Buckthorn Pomace
Reprinted from: *Appl. Sci.* 2023, *13*, 6513, https://doi.org/10.3390/app13116513 71

Ionica Coțovanu, Costel Mironeasa and Silvia Mironeasa
Insights into the Potential of Buckwheat Flour Fractions in Wheat Bread Dough
Reprinted from: *Appl. Sci.* 2022, *12*, 2302, https://doi.org/10.3390/app12052302 97

Ana Batariuc, Mădălina Ungureanu-Iuga and Silvia Mironeasa
Characterization of Sorghum Processed through Dry Heat Treatment and Milling
Reprinted from: *Appl. Sci.* 2022, *12*, 7630, https://doi.org/10.3390/app12157630 116

Manee Saelee, Bhagavathi Sundaram Sivamaruthi, Periyanaina Kesika, Sartjin Peerajan, Chawin Tansrisook, Chaiyavat Chaiyasut and Phakkharawat Sittiprapaporn
Response-Surface-Methodology-Based Optimization of High-Quality *Salvia hispanica* L. Seed Oil Extraction: A Pilot Study
Reprinted from: *Appl. Sci.* 2023, *13*, 6600, https://doi.org/10.3390/app13116600 134

Bianca-Maria Tihăuan, Ioana-Cristina Marinaș, Marian Adascălului, Alina Dobre, Grațiela Grădișteanu Pîrcălăbioru, Mădălina Axinie, et al.
Nutritional Profiling and Cytotoxicity Assessment of Protein Rich Ingredients Used as Dietary Supplements
Reprinted from: *Appl. Sci.* 2023, *13*, 6829, https://doi.org/10.3390/app13116829 154

Vinko Krstanović, Kristina Habschied, Iztok Jože Košir and Krešimir Mastanjević
The Influence of Texture Type and Grain Milling Degree on the Attenuation Limit, Protein Content, and Degradation in Wheat Wort
Reprinted from: *Appl. Sci.* **2023**, *13*, 10626, https://doi.org/10.3390/app131910626 . . . . **171**

Joanna Le Thanh-Blicharz, Jacek Lewandowicz, Patrycja Jankowska, Przemysław Łukasz Kowalczewski, Katarzyna Zając, Miroslava Kačániová and Hanna Maria Baranowska
Pulses-Fortified Ketchup: Insight into Rheological, Textural and LF NMR-Measured Properties
Reprinted from: *Appl. Sci.* **2023**, *13*, 11270, https://doi.org/10.3390/app132011270 . . . . **179**

Lucie Jurkaninová, Ivan Švec, Iva Kučerová, Michaela Havrlentová, Matěj Božik, Pavel Klouček and Olga Leuner
The Use of Thyme and Lemongrass Essential Oils in Cereal Technology—Effect on Wheat Dough Behavior and Bread Properties
Reprinted from: *Appl. Sci.* **2024**, *14*, 4831, https://doi.org/10.3390/app14114831 . . . . **191**

Natalia Netreba, Elena Sergheeva, Angela Gurev, Veronica Dragancea, Georgiana Gabriela Codină, Rodica Sturza and Aliona Ghendov-Mosanu
The Influence of Pomace Powder of Musky Squash on the Characteristics of Foamy Confectionery Products during Storage
Reprinted from: *Appl. Sci.* **2024**, *14*, 6671, https://doi.org/10.3390/app14156671 . . . . **209**

About the Editors

Georgiana Gabriela Codină

Georgiana Gabriela CODINĂ (Professor Habilitate, PhD Eng.) joined the Ștefan cel Mare University, Faculty of Food Engineering (Suceava, Romania) in 2005, where she teaches different courses in the food science and technology field. She holds a PhD in Industrial Engineering obtained in 2009 and became a PhD supervisor in food engineering in 2017 when she sustained her habilitation thesis. Dr. Codină has expertise in food rheology, food sensory analysis, bread making, the beer industry, and food quality analysis through different rheological, textural, and sensory methods, as well as in the design of experiments and data analysis. Her research activities focus on improving the technology of food products and the quality of different foods such as baked goods, beer, dairy products, etc. She has been involved in more than 10 interdisciplinary research projects, has published more than 150 scientific papers, and has authored more than 35 patents under evaluation, of which 11 were published.

Adriana Dabija

Adriana DABIJA is a Professor in the Department of Food Technologies, Food and Environmental Safety, within the Faculty of Food Engineering at Stefan cel Mare University of Suceava, Romania. She also serves as a PhD supervisor at the Doctoral School of Applied Sciences and Engineering at the same university.

She has experience in production, teaching, and research in the field of Food Engineering (39 years) and is an Expert Evaluator for the Romanian Agency for Quality Assurance in Higher Education, Romania.

She has conducted research activities in the fields of Food Biotechnology, the Microbiology of Food Technology, Quality Control in the Food Industry, and Quality Management in the Food Industry and has published 60 ISI Thomson Reuters-indexed papers, 178 articles in international data-based journals and proceedings, 88 papers in non-indexed scientific journals and proceedings, 4 patents, 1 Euro patent, and 31 patent proposals.

Preface

In recent years, there has been an increasing trend in consumer preferences for healthier and more sustainable food products. This trend had a significant impact on the food industry, particularly in grain processing technologies. Grains, with their high nutritional values, may be incorporated into many food products, which can meet the evolving needs of health-conscious consumers.

This Special Issue explores the latest trends in grain processing for the food industry, showing how technological advancements, health considerations, and sustainability concerns are shaping the way grains are processed and used in food products. From the development of gluten-free and low-glycemic food products to the incorporation of functional ingredients such as probiotics and superfoods, the grain processing sector is witnessing significant innovation. At the same time, the industry is responding to the increasing demand for convenience, offering ready-to-eat and nutritionally enhanced food products that align with modern lifestyles.

As the world continues to grapple with health challenges such as diabetes, gluten intolerance, and other metabolic conditions, the role of grains in creating functional and therapeutic food products has become more crucial than ever. Moreover, sustainability concerns have prompted the exploration of more eco-friendly practices in grain sourcing and processing, ensuring that food production can meet the needs of a growing global population while minimizing environmental impact.

This Special Issue is primarily addressed to researchers, food technologists, nutritionists, and industry professionals involved in the grain processing, food production, and dietary health sectors. It will also be of interest to academics and students pursuing studies in food science, agriculture, and biotechnology. The text aims to provide these audiences with an in-depth understanding of the current trends in grain processing and the potential for innovation in developing health-conscious, functional food products.

This Special Issue aims to provide insight into these key trends, offering a comprehensive overview of how grain-based products are evolving in response to both consumer demand and scientific advancements. It also highlights the potential of grains as a cornerstone of the future food industry, leading innovation in product development and contributing to better public health and well-being. Through this exploration, the goal of this Special Issue is to inspire further research, innovation, and collaboration within the field of grain processing.

Georgiana Gabriela Codină and Adriana Dabija
Guest Editors

Editorial

Innovative Grain Processing: Trends and Technologies

Georgiana Gabriela Codină * and Adriana Dabija

Faculty of Food Engineering, Ștefan cel Mare University of Suceava, 13 Universitatii Street, 720229 Suceava, Romania; adriana.dabija@fia.usv.ro
* Correspondence: codina@fia.usv.ro; Tel.: +40-7-4546-072

1. Introduction

Consumers' desire for a rational and healthy diet has led to an increase in the consumption of grain-based products. The high interest in the use of grain products in food consumption is because they bring together, in a small volume, important amounts of nutrients (proteins, carbohydrates, mineral salts, vitamins) that provide high energy value [1,2]. Also, the use of grain as a raw material for obtaining food products presents economic advantages related to its short vegetation period and ease of transport and storage. Depending on the form in which they are consumed, cereals may be considered functional foods, with health benefits such as cardiovascular system protection and digestion improvement [3,4]. As consumer demand for functional and nutritious food increases, the role of grain in the diet, along with innovative processing techniques, continues to evolve, promising a future with food products that are both nutritious and tailored to specific health needs. This increasing demand is related to growing awareness of the importance of a balanced diet, particularly for addressing health conditions such as diabetes and celiac disease [5].

Diabetes is a chronic disease caused by inherited or acquired deficient or insufficient insulin production by the pancreas. Insulin deficiency causes high concentrations of glucose to be present in the blood, which affects many of the body's systems, particularly the circulatory and nervous systems [6]. Improving diet is the first measure taken in the treatment of diabetes; the bakery industry can contribute to this by providing various products that can be consumed as part of a hypoglycemic diet [7–10]. To reduce the glycemic indexes of bakery products, we must increase their resistant starch contents and decrease their total starch contents. Some processes, such as fermentation during leavening or pre-fermentation, the hydrothermal treatment of cereal suspensions, subsequent dough maturation, freezing or chilling, and storage and pre-consumption treatment, may reduce the glycemic indexes of bakery products [11]. Generally, whole grains have a lower glycemic index than refined grains because they contain more fiber and experience a slower digestion process.

Another type of product is gluten-free grain-based products, intended for people who cannot metabolize gluten [12,13]. Gluten is a protein found naturally in grains such as wheat, rye and barley [14]. Due to its varied symptoms, celiac disease is difficult to diagnose. The only treatment for celiac disease is consuming a strict diet that eliminates products that contain gluten [15]. Cereal products and floury foods intended for this population are obtained from gluten-free grains, typically using non-gluten flours. These products use alternative cereals or gluten-free flour sources such as rice, quinoa, buckwheat, corn, millet, amaranth, teff, etc. which have high contents of carbohydrates and essential nutrients but are gluten-free [16–18]. However, their use in bakery products is a real challenge due to the role of gluten in bread-making. It is formed during mixing using the gluten proteins of wheat flour. It also influences the rheological properties of the dough, which allow the dough to be modeled, maintain the form received during modeling, and retain gases [19–21].

The processing of food products inevitably transforms their physical–chemical characteristics and nutritional value. There are processes that allow for the better assimilation

Citation: Codină, G.G.; Dabija, A. Innovative Grain Processing: Trends and Technologies. *Appl. Sci.* **2024**, *14*, 10954. https://doi.org/10.3390/app142310954

Received: 6 November 2024
Accepted: 25 November 2024
Published: 26 November 2024

Copyright: © 2024 by the authors. Licensee MDPI, Basel, Switzerland. This article is an open access article distributed under the terms and conditions of the Creative Commons Attribution (CC BY) license (https://creativecommons.org/licenses/by/4.0/).

of such products by the human body; however, these process reduce the contents of key nutritional compounds. As a result, there is a need to restore products' nutritional value; an effective solution is their supplementation with affected nutrients such as vitamins, essential minerals, fibers, essential fatty acids, antioxidants, proteins, etc. [22]. Also, to limit products' nutritional value loss, different techniques may be used, such as minimal processing, grinding to optimal granulations sizes, extrusion, germination, fermentation, etc. [23].

2. An Overview of Published Articles

Thus, by grinding wheat grains, flour is obtained, traditionally used in the bakery and pasta industries. However, it may be used in other industries such as beer brewing. According to Krstanović et al. [24], different wheat varieties, which vary in terms of hardness and the degree of milling required, produce significantly different wort. A hybrid wheat variety may be needed to produce high-quality wheat beer. Adding different nutrients into wheat flour dough to improve bakery product quality from a nutritional point of view will significantly affect its rheological behavior and the quality of the finished products. The addition of different milling fractions of buckwheat will significantly affect the dough's rheological properties during mixing and extension and improve bread characteristics. The addition of buckwheat will increase α-amylase activity in composite flours, water absorption, the dough development time, protein weakening, starch gelatinization, storage and loss modulation, dough fermentation parameters obtained via a Rheofermentometer and dough tenacity. Adding rheologically optimum levels of different buckwheat milling fractions of will lead to new bread formulations with higher volume, elasticity and porosity values [25]. To improve bread quality, the possibility of using a sourdough starter obtained from different flours has been discussed. The use of different flours, recipes, rations and fermentation temperatures affect the microbial populations of sourdough dough. Sourdough can be successfully used in a preliminary phase of bread-making, involving increasing the bacteria from the starter culture. Sourdough typically creates semi-finished products with high acidity, which ferment under set time and temperature conditions. In bakery products, the use of sourdough can lead to products with superior volume, crumb elasticity, slicing capacity and flavor characteristics [26,27].

Grain products are one of the most important food products included in the diet food category. They are intended for people with various conditions (diabetes, hypertension, people with gluten intolerance, etc.), as well as for those who are healthy but have certain nutritional needs. The use of different non-gluten flours, such as baobab flour (*Adansonia digitata* L.), buckwheat flour (*Fagopyrum esculentum*), rice flour, and quinoa flour, may lead to high-quality bakery products. Also, the use of some ingredients, such as sea buckthorn, protein isolate, different types of legumes flours (lentils, lupine, chickpeas, peas), pea powder, yeast powder, almond powder, and spirulina powder, will increase the nutritional value of cereal food products. A product with an improved nutritional value was developed by Netreba et al. [28]. This product was a foamy confectionery product with pomace powder formed by musky squash and artichoke syrup. Adding up to 15% pomace powder from musky squash to the foam confectionary product's recipe gave it good sensory characteristics and texture parameters. It was concluded that the use of pumpkin pulp in the foamy confectionery product's recipe improved its biological value and sensory characteristics, as well as its shelf life.

These studies collectively highlight significant trends in grain processing, aiming to improve the quality of finished food products.

3. Conclusions

It can be concluded that in the last few decades, grain processing has experienced significant development due to scientific progress; this progress has included the development of biotechnologies, the improvement of technological processes used in grain processing, and the transfer of knowledge through fundamental and applied research. Future grain processing will be improved due to the need for healthier and more sustainable food products. Advances in food processing, especially through fermentation and heat treatments, have further improved the rheology and quality of grain-based products. By incorporating alternative flours and functional ingredients, the industry can create food products that not only meet nutritional needs but also have enhanced sensory characteristics and shelf lives.

Funding: This research received no external funding.

Conflicts of Interest: The authors declare no conflicts of interest.

References

1. Mesta Corral, M.; Gómez-García, R.; Balagurusamy, N.; Torres-León, C.; Hernández-Almanza, A.Y. Technological and Nutritional Aspects of Bread Production: An Overview of Current Status and Future Challenges. *Foods* **2024**, *13*, 2062. [CrossRef] [PubMed]
2. Zhang, W.; Boateng, I.D.; Xu, J.; Zhang, Y. Proteins from Legumes, Cereals, and Pseudo Cereals: Composition, Modification, Bioactivities, and Applications. *Foods* **2024**, *13*, 1974. [CrossRef]
3. Abdi, R.; Joye, I.J. Prebiotic Potential of Cereal Components. *Foods* **2021**, *10*, 2338. [CrossRef]
4. Grasgruber, P.; Sebera, M.; Hrazdira, E.; Hrebickova, S.; Cacek, J. Food consumption and the actual statistics of cardiovascular diseases: An epidemiological comparison of 42 European countries. *Food Nutr. Res.* **2016**, *60*, 31694. [CrossRef]
5. Le, T.H.; Disegna, M.; Lloyd, T. National food consumption patterns: Converging trends and the implications for health. *EuroChoices* **2023**, *22*, 66–73. [CrossRef]
6. Cho, Y.R.; Ann, S.H.; Won, K.B.; Park, G.M.; Kim, Y.G.; Yang, D.H.; Kang, J.W.; Lim, T.H.; Kim, H.K.; Choe, J.; et al. Association between insulin resistance, hyperglycemia, and coronary artery disease according to the presence of diabetes. *Sci. Rep.* **2019**, *9*, 6129. [CrossRef] [PubMed]
7. Romão, B.; Falcomer, A.L.; Palos, G.; Cavalcante, S.; Botelho, R.B.A.; Nakano, E.Y.; Raposo, A.; Shakeel, F.; Alshehri, S.; Mahdi, W.A.; et al. Glycemic Index of Gluten-Free Bread and Their Main Ingredients: A Systematic Review and Meta-Analysis. *Foods* **2021**, *10*, 506. [CrossRef] [PubMed]
8. Di Cairano, M.; Condelli, N.; Caruso, M.C.; Marti, A.; Cela, N.; Galgano, F. Functional properties and predicted glycemic index of gluten free cereal, pseudocereal and legume flours. *LWT-Food Sci. Technol.* **2020**, *133*, 109860. [CrossRef]
9. Punia Bangar, S.; Sharma, N.; Singh, A.; Phimolsiripol, Y.; Brennan, C.S. Glycaemic response of pseudocereal-based gluten-free food products: A review. *Int. J. Food Sci. Technol.* **2022**, *57*, 4936–4944. [CrossRef]
10. Atkinson, F.S.; Brand-Miller, J.C.; Foster-Powell, K.; Buyken, A.E.; Goletzke, J. International tables of glycemic index and glycemic load values 2021: A systematic review. *Am. J. Clin. Nutr.* **2021**, *114*, 1625–1632. [CrossRef]
11. Skřivan, P.; Sluková, M.; Sinica, A.; Bleha, R.; Švec, I.; Šárka, E.; Pourová, V. Glycaemic Index of Bakery Products and Possibilities of Its Optimization. *Appl. Sci.* **2024**, *14*, 6070. [CrossRef]
12. Dean, D.; Rombach, M.; Vriesekoop, F.; Mongondry, P.; Le Viet, H.; Laophetsakunchai, S.; Urbano, B.; Briz, T.; Xhakollari, V.; Atasoy, G.; et al. Against the Grain: Consumer's Purchase Habits and Satisfaction with Gluten-Free Product Offerings in European Food Retail. *Foods* **2024**, *13*, 3152. [CrossRef] [PubMed]
13. Matera, M.; Guandalini, S. How the Microbiota May Affect Celiac Disease and What We Can Do. *Nutrients* **2024**, *16*, 1882. [CrossRef] [PubMed]
14. Estévez, V.; Rodríguez, J.M.; Schlack, P.; Navarrete, P.; Bascuñán, K.A.; Núñez, V.; Oyarce, C.; Flores, C.; Ayala, J.; Araya, M. Persistent Barriers of the Gluten-Free Basic Food Basket: Availability, Cost, and Nutritional Composition Assessment. *Nutrients* **2024**, *16*, 885. [CrossRef] [PubMed]
15. Stanciu, D.; Staykov, H.; Dragomanova, S.; Tancheva, L.; Pop, R.S.; Ielciu, I.; Crișan, G. Gluten Unraveled: Latest Insights on Terminology, Diagnosis, Pathophysiology, Dietary Strategies, and Intestinal Microbiota Modulations—A Decade in Review. *Nutrients* **2024**, *16*, 3636. [CrossRef]
16. Bueno, C.; Thys, R.; Tischer, B. Potential Effects of the Different Matrices to Enhance the Polyphenolic Content and Antioxidant Activity in Gluten-Free Bread. *Foods* **2023**, *12*, 4415. [CrossRef]
17. Jaroszewska, A.; Jedrejek, D.; Sobolewska, M.; Kowalska, I.; Dzięcioł, M. Mineral, Nutritional, and Phytochemical Composition and Baking Properties of Teff and Watermelon Seed Flours. *Molecules* **2023**, *28*, 3255. [CrossRef]
18. Khairuddin, M.A.N.; Lasekan, O. Gluten-Free Cereal Products and Beverages: A Review of Their Health Benefits in the Last Five Years. *Foods* **2021**, *10*, 2523. [CrossRef]
19. Huertas-García, A.B.; Guzmán, C.; Ibba, M.I.; Rakszegi, M.; Sillero, J.C.; Alvarez, J.B. Processing and Bread-Making Quality Profile of Spanish Spelt Wheat. *Foods* **2023**, *12*, 2996. [CrossRef]
20. Al-Khayri, J.M.; Alshegaihi, R.M.; Mahgoub, E.I.; Mansour, E.; Atallah, O.O.; Sattar, M.N.; Al-Mssallem, M.Q.; Alessa, F.M.; Aldaej, M.I.; Hassanin, A.A. Association of High and Low Molecular Weight Glutenin Subunits with Gluten Strength in Tetraploid Durum Wheat (*Triticum turgidum* spp. *Durum* L.). *Plants* **2023**, *12*, 1416. [CrossRef]
21. Tóth, V.; Láng, L.; Vida, G.; Mikó, P.; Rakszegi, M. Characterization of the Protein and Carbohydrate Related Quality Traits of a Large Set of Spelt Wheat Genotypes. *Foods* **2022**, *11*, 2061. [CrossRef] [PubMed]

22. Pinto, D.; Castro, I.; Vicente, A.; Bourbon, A.I.; Cerqueira, M.A. Functional Bakery Products—An Overview and Future Perspectives. In *Bakery Products Science and Technology*, 2nd ed.; Zhou, W., Hui, H., De Leyn, I., Pagani, M.A., Rosell, C.M., Selman, D., Therdthai, N., Eds.; John Wiley & Sons: Hoboken, NJ, USA, 2014; pp. 431–452.
23. Dewettinck, K.; Van Bockstaele, F.; Kühne, B.; Van de Walle, D.; Courtens, T.M.; Gellynck, X. Nutritional value of bread: Influence of processing, food interaction and consumer perception. *J. Cereal Sci.* **2008**, *48*, 243–257. [CrossRef]
24. Krstanović, V.; Habschied, K.; Košir, I.J.; Mastanjević, K. The Influence of Texture Type and Grain Milling Degree on the Attenuation Limit, Protein Content, and Degradation in Wheat Wort. *Appl. Sci.* **2023**, *13*, 10626. [CrossRef]
25. Coțovanu, I.; Mironeasa, C.; Mironeasa, S. Insights into the Potential of Buckwheat Flour Fractions in Wheat Bread Dough. *Appl. Sci.* **2022**, *12*, 2302. [CrossRef]
26. Ma, S.; Wang, Z.; Guo, X.; Wang, F.; Huang, J.; Sun, B.; Wang, X. Sourdough improves the quality of whole-wheat flour products: Mechanisms and challenges—A review. *Food Chem.* **2021**, *360*, 130038. [CrossRef]
27. De Vuyst, L.; Comasio, A.; Kerrebroeck, S.V. Sourdough Production: Fermentation Strategies, Microbial Ecology, and Use of Non-Flour Ingredients. *Crit. Rev. Food Sci. Nutr.* **2022**, *63*, 2447–2479. [CrossRef]
28. Netreba, N.; Sergheeva, E.; Gurev, A.; Dragancea, V.; Codină, G.G.; Sturza, R.; Ghendov-Mosanu, A. The Influence of Pomace Powder of Musky Squash on the Characteristics of Foamy Confectionery Products during Storage. *Appl. Sci.* **2024**, *14*, 6671. [CrossRef]

Disclaimer/Publisher's Note: The statements, opinions and data contained in all publications are solely those of the individual author(s) and contributor(s) and not of MDPI and/or the editor(s). MDPI and/or the editor(s) disclaim responsibility for any injury to people or property resulting from any ideas, methods, instructions or products referred to in the content.

Review

Glycaemic Index of Bakery Products and Possibilities of Its Optimization

Pavel Skřivan, Marcela Sluková *, Andrej Sinica, Roman Bleha, Ivan Švec, Evžen Šárka and Veronika Pourová

Department of Carbohydrates and Cereals, University of Chemistry and Technology Prague, Technická 5, 166 28 Prague, Czech Republic; pavel.skrivan@vscht.cz (P.S.); andrej.sinica@vscht.cz (A.S.); roman.bleha@vscht.cz (R.B.); ivan.svec@vscht.cz (I.Š.); evzen.sarka@vscht.cz (E.Š.); werrily@seznam.cz (V.P.)
* Correspondence: marcela.slukova@vscht.cz

Abstract: Common bakery and many other cereal products are characterised by high glycaemic index values. Given the increasing number of people suffering from type 2 diabetes at a very young age, technological approaches to reduce the glycaemic index of cereal products are extremely important. In addition to increasing the dietary fibre content, either by using wholemeal flours or flours with added fibre from other sources, practices leading to an increase in resistant starch content are also of great interest. This review summarises the most important technological processes used to reduce the glycaemic index of bread and other bakery products. The summarization shows that the potential of various technological processes or their physical and physicochemical modifications to reduce the glycaemic index of common bakery products exists. At the same time, however, it has been shown that these processes have not been sufficiently explored, let alone applied in production practice.

Keywords: glycaemic index; resistant starch; bakery products; cereal processing; dietary fibre; cereal technology; diabetes

Citation: Skřivan, P.; Sluková, M.; Sinica, A.; Bleha, R.; Švec, I.; Šárka, E.; Pourová, V. Glycaemic Index of Bakery Products and Possibilities of Its Optimization. *Appl. Sci.* **2024**, *14*, 6070. https://doi.org/10.3390/app14146070

Academic Editors: Marco Iammarino, Georgiana Gabriela Codină and Adriana Dabija

Received: 23 April 2024
Revised: 9 July 2024
Accepted: 11 July 2024
Published: 11 July 2024

Copyright: © 2024 by the authors. Licensee MDPI, Basel, Switzerland. This article is an open access article distributed under the terms and conditions of the Creative Commons Attribution (CC BY) license (https://creativecommons.org/licenses/by/4.0/).

1. Introduction

Cereals are rightly considered one of the most important components of human nutrition. Cereal cultivation largely enabled the concentration of Neolithic populations, their numerical growth and the emergence of civilizations. Cereals are relatively easy to grow, and cereal grains can be stored safely and for a long time. They are therefore a ready and available source of food. Cereals and products made from them (especially bread) have thus enjoyed respect and popularity for millennia, becoming a truly staple food for large sections of the population of civilized regions, and their place in the diet seemed, until recently, to be completely unquestioned. In many languages, bread is synonymous with food as such. However, this situation has changed in recent decades, with cereals and their products moving from the base of the food pyramid to the higher rungs and being replaced in the base mainly by vegetables and legumes [1].

The main reason for this shift in the perception of the role of cereals in human nutrition is mainly due to their high starch content and, for some of them, their gluten content. While the gluten content of some cereals (wheat, barley, rye and, to a lesser extent, oats) poses a risk to a specific population group consisting of individuals suffering from coeliac disease or other forms of gluten intolerance, high starch content is a general problem. Starch is usually very well and above all quickly digested, so it is a ready and abundant source of energy, while at the same time, after its consumption, blood glucose concentration rises rapidly. Cereals and their products are therefore foods with a medium to high glycaemic index (GI) [1–4].

The glycaemic index is a dimensionless quantity that indicates the rate of increase in postprandial glycaemia from a particular food. Exactly, the GI is defined as the area under the glycaemia curve within two hours after ingestion of a given food, expressed as the proportion of the area under the curve after ingestion of the same amount of carbohydrate in the form of pure glucose. The glycaemic index is always a guideline number, it depends very much on the specific composition and processing of the food. Each organism reacts slightly differently and is sensitive to fluctuations in sugar levels. The glycaemic index related to pure glucose is sometimes confused with the Brot-index used in Germany, which refers to white wheat bread, which has a glycaemic index of 70. As a result, there is often confusion about the numbers in the various tables reported in the literature.

The reason why these properties of cereals, and especially their high content of rapidly digestible starch, have emerged as a serious problem in recent decades (especially since the mid-20th century) is not the cereals themselves; cereals have always had these properties, but the radical change in lifestyle and nutritional potential of people in the developed world. These are two factors that have completely changed our situation compared to that of hundreds of generations of our ancestors. For the first time in the history of mankind, we have reached a situation in which the majority of the population in the developed world, not only does not suffer from a lack of food but, on the contrary, lives in a state of surplus availability, with much of the food that is readily available being energy-rich. The second factor is the complete change in the way humans work. Whereas throughout history, i.e., throughout the historical development of human civilization, the vast majority of the population has necessarily needed to work physically for its livelihood, this situation has changed fundamentally in the last century. Modern man not only does not work physically for the most part, but in recent decades has not even been forced to move around actively thanks to various means of transport. Simply put, those who do not want to move and exert themselves physically voluntarily do not have to do so [5].

However, this leads to a complete change in the nutritional and energy balance. While energy intake can be in a high surplus, energy expenditure is quite low. In addition, many foods, not only cereals, contain rapidly digestible starch (RDS) or sugars, and thus have high GI values. The human metabolism, shaped for hundreds of generations by a completely different situation and thus genetically conditioned to cope with a relatively scarce food supply and the need for physical exercise, is unable to cope with the new situation. This leads in a large part of the population, including the very young, to excess weight, obesity, and the resulting health problems. Alongside other diseases of civilization, one of the most widespread and serious diseases is diabetes. In particular type 2 diabetes, which is clearly linked to an excess of high GI foods and lack of physical activity [5].

Cereals and their products (particularly breads and pastries), which are the focus of this paper, are, as noted, a typical and fundamentally important example of foods with higher to high GI values directly related to the prevalence of type 2 diabetes [6–14]. What for many generations was their benefit for hard-working communities of people, for whom they provided a rapid energy intake, represents a serious risk in today's technologically advanced society. It is the responsibility of those who develop and produce food in today's world to look for ways to optimize its energy value. This means to reduce its GI values and, in turn, exploit its full potential to increase its nutritional benefits, particularly its fibre content. This applies in full to cereals and their products. Therefore, we will focus not only on an evaluation of the most important cereals of the Central European region in terms of overall starch content and its fractions and its influence on the resulting GI.

The main focus of our work is on the used and potential ways to reduce GI values in bakery products, which in Central Europe are clearly the most important part of cereal-based foods. The main ways are, firstly, a relative reduction in the total starch content by increasing the fibre or protein content and, secondly, a reduction in the proportion of rapidly digestible starch (RDS) in favour of slowly digestible starch (SDS) and especially resistant starch (RS).

The contribution of our work and its novelty in terms of summarizing and profiling the available literature data lies in the focus on the relationship between technological processes and their influence on the glycaemic index of bread and bakery products. We focused in particular on the relationship between the transitions of starch types (modifications) in terms of its digestibility, i.e., RDS (rapid digestible starch), SDS (slow digestible starch) and RS (resistant starch) and the physical and physicochemical conditions of the production processes. These include hydrothermal treatment, fermentation and heat treatment. A similar summary from this point of view has not yet been published, according to all available information.

2. Cereals as an Important Source of Starch—Importance in Human Nutrition with Respect to Glycaemic Index

The Czech Republic has a population of just under 11 million (10,882,235 as of 30 September 2023, according to the Czech Statistical Office). According to the data of the Institute of Health Insurance, one in three people in the Czech Republic over the age of 65 is diabetic. The highest increase in the specific prevalence of diabetes mellitus is between the ages of 50 and 75, when the number of diabetics per 100,000 inhabitants increases more than fourfold from 8.700 to 38.000. However, the numbers of patients in lower age categories are also significant, and are on the order of hundreds to units of thousands, starting from the age of 15.

According to statistical predictions, there will be almost 1.3 million people with diabetes in the Czech Republic in 2030 [15]. It is therefore a real epidemic with widespread personal, social and economic consequences.

The situation is very similar in other Central European countries. The main (bread) cereals in Central Europe are wheat (*Triticum aestivum* L.), which accounts for 80–90% of bread and pastry production, and rye (*Secale cereale* L.), which accounts for 10–15% (historically, its share of bread and pastry production in Central Europe was much higher).

Other cereals (barley, oats, maize) and the pseudocereal buckwheat or other cereals contribute marginally to the production of bakery flours. The importance of buckwheat has increased slightly in recent decades in the context of gluten-free products [16].

In light of this brief summary of the situation, basic cereals should be judged according to their chemical composition and the resulting nutritional properties (Table 1). The basic and most abundant component of cereals is starch. Starch occurs in its native state in the form of starch grains composed of ordered amylose and amylopectin molecules.

Table 1. Basic chemical composition of cereals (g/100 g; average values for whole grain) [17].

Cereal	Water	Proteins	Lipids	Starch	Minerals
Wheat	13.2	11.7	2.2	59.2	1.5
Rye	13.7	11.6	1.7	52.4	1.9
Barley	11.7	10.6	2.1	52.2	2.3
Oat	13.0	12.6	5.7	40.1	2.9
Rice	13.1	7.4	2.4	70.4	1.2
Maize	12.5	9.2	3.8	62.6	1.3

Although the starch content itself is essential for the glycaemic index value (Table 2), the condition of the starch grains and their changes during food processing and production of the final products are equally essential [1,18].

Table 2. Starch content and glycaemic index of cereals, legumes, pseudocereals and selected cereal products based on data from National Institute of Health, Czech Republic) (https://szu.cz/wp-content/uploads/2023/12/Glykemicky-index-2003.pdf) (accessed on 23 April 2024) [18–22].

Crops-Whole Grains	Starch Content (% in d.m.)	Glycaemic Index
Wheat	65–75	50–60
Rye	65–75	50–60
Barley	65–75	50–60
Oat	55–65	40–50

Table 2. *Cont.*

Crops-Whole Grains	Starch Content (% in d.m.)	Glycaemic Index
Rice	70–80	55–60
Maize	60–70	50–60
Sorghum	65–75	55–65
Millet	70–80	55–65
Legumes	45–55	30–45
Buckwheat	45–55	35–45
Amaranth	55–60	30–40
Bread and bakery products		
White wheat bread	70–75	70
Wheat-rye bread (Central Europe type)	65–70	65–70
Dark rye bread	55–65	45–55
Whole grain bread	55–65	45–55
Bread type pumpernickel	55–65	45–55
Buckwheat bread	50–60	40–50
Wheat common bakery products	70–75	70
Croissants	50–60 *	60–70 *
Donuts	45–60 *	70–75 *
Muffins and other sweet pastry	45–60 *	60–75 *

* Starch content and glycaemic index decrease with higher fat content.

The processing of cereals for human consumption can be divided into two successive stages. Primary processing, the raw material of which is the sorted and decontaminated grain of the given cereal, generally consists of surface treatment of the grain, hydrothermal treatment and possible disintegration of the grain into flakes, groats, shreds and, in particular, flours and meal. After surface treatment and hydrothermal treatment, the grain may remain whole, usually for direct culinary processing or for use in multigrain breads and pastries and other cereal products. Soaking and germination (or malting) of the grain is also one of the primary processing methods. The most common primary processing method in Europe for wheat (and in Central Europe also for rye) is milling into bakery flours. Oats are mostly processed into oat flakes. Most barley is malted for the production of beer and other alcoholic beverages, and malt is used partly in the bakery industry as a recipe ingredient [23].

Secondary processing is the second stage in which the primary product (flour, meal, whole grain) is processed into final, usually directly consumable products—bread, pastries, other cereal products or pasta, which are cooked in water or steam before consumption. An overview of the basic technological steps is given in Figure 1. The first step in secondary processing is always the preparation of the primary mass (dough) or suspension by mixing the primary cereal product with water or fat and mechanical processing (kneading, whisking, etc.) The secondary processing of cereals always involves some form of heat treatment (baking, frying, drying, cooking, extrusion), which for bread and pastry is preceded by one or more stages of fermentation. The basic fermentation process used in cereal technology for leavening bread and pastry is ethanol fermentation, which produces sufficient carbon dioxide in addition to ethanol. In Central Europe, both traditional rye and wheat sourdoughs undergo homo- and heterofermentative lactic acid fermentation [23].

In all the steps summarized above, during both stages of cereal processing, gradual changes occur at all levels of starch structure. Damage and disintegration of native starch grains at the quaternary structure level, changes in the tertiary structure of amylose and amylopectin molecules, and the disintegration of secondary and primary structure due to hydrolysis. Starch grains swell to the extent that the proportion of water in the primary mass or suspension is high, and are subject to partial hydrolysis, mainly enzymatic via amylase, but also to acid hydrolysis during any lactic acid fermentation. The heat supplied results in the formation of a starch slurry and gel and subsequent retrogradation. The first significant damage to starch grains in the processing of cereals in mills occurs mechanically and thermally during the actual milling [1]. All of these physico-chemical and biochemical processes acting on starch during cereal processing have an impact on the final GI of the product [24,25].

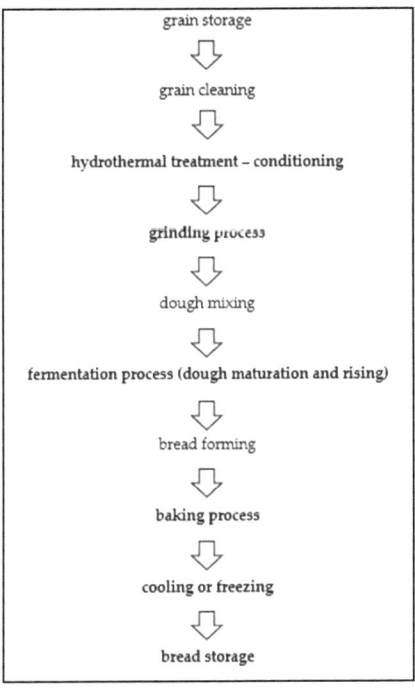

Figure 1. Diagram of basic operations in the processing of cereals for bread production and the sections where the GI value is significantly affected (bold).

In addition to starch, other important components of grain—proteins and fibre components (non-starch polysaccharides, oligosaccharides and lignin)—are also subject to changes during processing. Fibre is concentrated almost exclusively in the grain outer layers and seed coat layers [26–30]. However, these are removed during the standard processing of wheat (and rye) into common baking and pastry flours before milling. Only wholemeal flours contain all the fibre components. The removal of a significant part or almost all of the grain envelope leads to an increase in the relative starch content of flours and other products, which in itself has a major effect on the glycaemic index [31]. Although the fibre components that remain in the processed grain do not undergo such dramatic changes as the starch during primary and secondary processing, they also swell with water, change their tertiary and, in particular, quaternary structure (the native fibrous structures of the chaff are loosened) and undergo partial enzymatic hydrolysis due to enzymes added in the formulation or produced by certain lactic acid bacteria. These changes, which may potentiate some of the biological functions of the fibre components, may also have an indirect effect on the GI values of the final product [32].

2.1. Rapidly Digestive, Slowly Digestive and Resistant Starch

The starch contained in cereals and other starchy raw materials for food production in intermediate products and in the products, themselves can be divided, from the point of view of human nutrition, into rapidly digestible starch (RDS), slowly digestible starch (SDS), and resistant starch (RS). RDS and SDS are, as a result, completely metabolized. RDS represents the highest glycaemic load, and is the main factor responsible for the high glycaemic index of cereal products and the resulting consequences for human nutrition and consumer health [33–38].

A high GI value caused by a high proportion of RDS is the de facto cause of the increased potential for type 2 diabetes and related health problems and diseases. SDS is more favourable in this respect. Digestibility (rate of resorption) is influenced by a number of factors, including origin (source plant), starch grain size, the ratio of amylose to amylopectin, the extent of molecular associations between starch components, the type and degree of crystallinity, the length of the amylose chain, the molecular structure of amylopectin, and the presence of amylose-lipid complexes. Morphology and ultrastructure should also be taken into account, e.g., specific surface area, channels, and the porosity of starch grains [3,32,38,39].

Starch is hydrolysed in the small intestine by α-amylase from the pancreas (α-1,4-glucan-4-glucanohydrolase, EC 3.2.1.1), hydrolysing the α-D-(1$\rightarrow$4) linkage. In contrast to the hydrolysis of amylose, which is randomly cleaved into maltooligosaccharides, the action of α-amylase on amylopectin is systematic: its resulting products are maltose, maltotriose, and branched α-boundary dextrins containing all the original α-D-(1$\rightarrow$6) linkages and adjacent α-D-(1$\rightarrow$4) linkages. Amylose is resorbed more slowly and to a lesser extent than amylopectin. The glycaemic response of amylose has been shown to be less than that of the same amount of amylopectin [40]. The specific surface area of the grains is determined by their size. Native starch grains are divided according to size into fractions A and B, with the smaller starch grains (fraction B) having a larger specific surface area and (according to X-ray diffraction patterns [34]) having most of their branching points clustered in the amorphous region and therefore more being readily subject to enzymatic hydrolysis. The specific surface area is related to the particle size distribution that is characteristic of the crop, i.e., with cereals not only for the cereal species, but also for its variety [41,42].

Resistant starch (RS, which is subdivided into types 1 to 5) is starch that is not digested in the human small intestine and passes into the large intestine, where it is partially metabolized by the microorganisms present, and is therefore a component of fibre [41,42]. RS is therefore a heterogeneous group of starch-based compounds that are classified into 5 groups according to their resistance to digestion, i.e., according to the nature and properties of the starch grain. Resistant starches, depending on their origin and the way the food or food preparation is processed, exhibit health-promoting physiological effects on the human body similar to other fibre components with a probiotic function [41,42].

Resistant starch types RS1, RS2, RS3 and RS5 are naturally present in foods. Resistant starch type 1 (RS1) is a physically inaccessible (unavailable) starch, e.g., starch in intact whole or partially disturbed grains or seeds of legumes and raw cereal grains, where the starch is part of the cell wall or protein matrix and is not amenable to enzymatic hydrolysis by amylolytic enzymes. A physical barrier (in particular the seed or grain envelope) is the cause of the low glycaemic index of the food concerned [43,44].

Resistant starch type 2 (RS2) is a native starch contained in the starch grains of raw potatoes, green bananas, legumes and high-amylose corn starch. The reason for RS2 resistance is the native conformation of the starch grain, or the presence of a higher proportion of a tightly 'packed' crystalline amylose structure (either in native form or as ungreased parts of the starch grain). During the appropriate heat treatment of foods, RS1 and RS2 starch types undergo gelatinization and become partially digestible components [43,44].

Resistant starch type 3 (RS3) is formed as a result of certain technological processes, by the action of heat and moisture on the material being processed. It is a form of retrograded starch, retrograded amylose and amylopectin. During the gelatinization of the starch (suspension of the starch in water), the amylose chains are released into solution, and after the mixture has cooled, the amylose chains (ordered double helices stabilized by hydrogen bonds) are reconnected. This is the process of retrogradation of amylose (the retrogradation of amylose is much faster than the retrogradation of amylopectin). The resulting solid helices of amylose prevent amylolytic enzymes from accessing the glycosidic bonds, and this type of amylose is resistant to enzymatic hydrolysis. The retrogradation of amylopectin is much slower due to the complex structure of amylopectin. Type RS3 (retrograded starch) is found in large quantities in foods such as aged breads and pastries, chilled cooked potatoes, chilled cooked pasta, sterilized peas, sterilized beans, etc. One option for increasing the RS3 content in starch (in the laboratory or in industry) is to use starch gelatinization under various physical and physicochemical conditions of hydrothermal treatments, including extrusion [44–46].

Resistant starch type 4 (RS4) is a modified starch—a chemically or physically modified starch that does not occur naturally. The RS4 content of the starch can be increased by, for example, acid hydrolysis of the starch isolated from high-amylose barley followed by repeated heating and cooling of the mixture (suspension). Chemical modification of the starch consists of cross-linking the starch, e.g., after reacting the starch with phosphate and other reagents. Resistant starch type 5 (RS5) is a thermally stable starch fraction containing an inclusion of the hydrophobic part of the lipid into the helical cavity of the amylose helix. Fats and oils and monoacylglycerols used as emulsifiers form inclusion compounds with the amylose, thus retarding the swelling of the starch and reducing the extent of starch gelatinization [44,47–49].

It should also be mentioned that other factors that hinder starch digestion are the presence of α-amylase inhibitors, a high content of non-starch polysaccharides (fibre) in the food, higher viscosity of the food, etc. In a general context, it is the slowing down of the access of enzymes to the substrate and the creation of a certain resistance, a barrier, that limits the digestion of starch [43,50,51].

During the processing of cereals and other raw materials, the digestibility of starch varies depending on the technological processes used; in addition, RS2, RS3 and RS4 types can be added to food as a functional ingredient (additive). An important source of resistant starch is cooked foods/feeds made from legumes, potatoes and cereals. Cooked and cooled (or stored) foods/feeds have higher RS contents than freshly cooked ones [40,51].

2.2. Cereal Fibre, Its Components and Its Effect on Postprandial Glycaemia and GI

Cereal fibre is composed of non-starch polysaccharides and oligosaccharides (as well as resistant starch) and lignin. The presence of fibre in its entirety (total fibre) in cereal products has a primary role in terms of GI, in that its proportion correspondingly reduces the relative proportions of digestible starch (RDS and SDS). The higher the proportion of fibre in a given product, the lower the proportion of starch. However, the effect of fibre on the GI value is not limited to reducing the relative starch concentration. The polysaccharides and oligosaccharides of dietary fibre play a physiological role in this respect, and this role varies from one fibre component to another [52–56].

Cellulose is composed of long unbranched chains of D-glucose units linked by a β-1,4 bond. Cellulose fibres together with other non-starch polysaccharides (hemicelluloses, see below) form cell walls and are the basic building material in plants that fixes plant tissues.

In cereal grains, cellulose is mainly present in the outer layers (mechanical, protective function of the tissues). Cellulose is insoluble in water and does not swell significantly at normal temperatures. It is a component of insoluble fibre. The function of cellulose in the human body is to accelerate intestinal transit, improve intestinal peristalsis and increase stool volume. A small portion of cellulose is fermented by bacteria in the colon to form so-called SCFAs (Short-Chain Fatty Acids), short aliphatic chain acids (such as acetic, propionic, butyric). The formation of these acids lowers the pH in the colon, which may have a preventive effect against cancerous growth (prevention of colorectal cancer) [57].

Hemicelluloses are another important group of non-starch polysaccharides. These heteropolysaccharides have a lower molecular weight than cellulose and are composed of monosaccharides such as D-xylose, D-galactose, L-arabinose, D-glucose, and uronic acids may also be bound. Hemicelluloses fill the spaces between cellulose fibres and are divided into two main groups of polysaccharides: heteroglucans and heteroxylans. Heteroglucans are further subdivided into xyloglucans and β-glucans. The group of heteromannans (galactomannans and glucomannans) contained in the coatings of cereal grains is also somewhat important [57].

Cereal β-glucans, β-(1→3), (1→4)-D-glucans or β-glucans with mixed linkages, are structural polysaccharides composed of β-D-glucose molecules. Unlike cellulose, they typically contain about 70% β-(1→4) units and about 30% β-(1→3) units. On average, the ratio of β-(1→4) to β-(1→3) bonds is reported to be about 3:1. For barley β-glucans, the ratio is reported to be 2.8–3.3 and for oat β-glucans 2.1–2.4. These β-glucans are capable of forming highly viscous gels. This is also linked to a number of technological and nutritional aspects in the production and consumption of products with a barley or oat component. They are found in all cereals, and to a greater extent in barley and oat grains. The physicochemical properties of cereal β-glucans depend on their primary structure, the type (or ratio) of linkages in the molecule, and their molecular weight. The β-glucans of cereals are partly soluble (extractable), part of them is a component of the so-called insoluble fibre. The solubility of β-glucans depends on their structure and origin, which decreases in the order oats (most soluble β-glucans), barley, wheat (least soluble β-glucans). The solubility of β-glucans depends on the number of (1→4) bonds in the chain, with a higher number of these bonds the solubility of β-glucans decreases. The solubility of β-glucans increases with increasing temperature. Protein-bound β-glucans are mostly insoluble in water [57].

High-viscosity, high-molecular-weight β-glucans increase viscosity in the human intestine (effect on satiety and satiety, reduced resorption of some nutrients and enzymes); this property of cereal β-glucans is important for reducing the rate of starch resorption [57,58]. Some studies have reported that β-glucans with lower viscosity and lower molecular weight are preferable, due to easier and more rapid utilization by bacteria present in the human colon. Thus, these β-glucans have a prebiotic function [59].

Arabinoxylans, like β-glucans, are structural non-starchy cereal polysaccharides, but are classified as heteroxylans. They are a diverse group of substances that can be divided into water-insoluble arabinoxylans, which accompany cellulose in cell walls, and water-soluble arabinoxylans, which form gels and mucilages. Often arabinoxylans are called by the older name pentosans (polysaccharides containing pentoses in the molecule). They consist of a β-(1,4) xylose skeleton with arabinose attached to either the second or third carbon. In addition to xylose and arabinose, arabinoxylans contain D-glucose and sometimes other minor building units (D-galactose, D-glucuronic acid, etc.). In different cereals, arabinoxylans differ in the manner of substitution of their xylan chain and in their arabinose content, or the ratio of the two sugars, arabinose and xylose. The average xylose content is 52–60%, and the arabinose content 36–46% [57].

The average relative molecular weight of wheat arabinoxylans ranges from about 220 to 260 kDa, that of rye arabinoxylans from 520 to 770 kDa. The differences in solubility depend on the degree of branching, with more branched arabinoxylan molecules being more soluble. Soluble arabinoxylans have a high water-binding capacity, even at room temperature. Most arabinoxylans derived from the endosperm of both rye and wheat grains are water soluble, whereas arabinoxylans from the aleurone layer and pericarp are insoluble in water. Arabinoxylans form the main component of rye fibre (8–12%) and are mainly found in the aleurone layer of the rye grain. Rye flours contain approximately 4–7% arabinoxylans, whereas wheat flours contain only 1–3% (depending, of course, on the degree of milling and the type of flour) [57,60,61].

Arabinoxylans are a fibre component that affects the nutritional and technological value of food (in particular the viscosity of dough, the softness of bread and pastry crumbs) and have positive health effects. From the point of view of the effect on GI, it is significant that rye arabinoxylans in particular exhibit a higher viscosity in the gut than barley or oat β-glucans [57].

Other non-starchy cereal carbohydrates are fructooligosaccharides and fructans. These are non-digestible storage oligo- and polysaccharides with β-(1→2) linkages. They differ in their structure, degree of polymerization and molecular weight, and can be grouped into inulins, levans and branched structures. They act as prebiotics. They serve as a substrate for bifidobacteria in the colon and promote their proliferation. These bifidobacteria ferment fructooligosaccharides and fructans into short-chain fatty acids, which have a positive effect on, for example, lipid metabolism, lowering the pH in the colon, etc. [57].

Among cereals, wheat is an important source of fructans. Fructans can also be found in rye, in which fructans accounted for around 2% of the total fibre content (15%). Short-chain fructans isolated from plants have a sweet taste and form the ingredients of natural low-calorie sweeteners. Long-chain fructans are taste neutral and form emulsions with a fat-like structure (they can serve as fat replacements) [62].

3. Overview of Technological Routes to Modify the Glycaemic Index of Bakery Products and Their Evaluation

3.1. Increasing the Fibre Content

Increasing the fibre content primarily results in a relative reduction in the digestible (RDS and SDS) starch content of the final product. However, this alone does not exhaust the full potential effect of fibre addition. While the effect of some fibre components on slowing down the resorption of starch is low and their addition does indeed act almost exclusively as a partial substitution of starch in the total dry weight of a given product, intermediate or raw material (composite flour), other components do indeed act to slow down its resorption and thus slow down the increase in postprandial glycaemia. In particular, it is the fibre components that cause an increase in the viscosity of the intestinal contents [63,64].

Fibre is one of the factors that can influence the glycaemic index of foods, along with fat and protein. It has the ability to reduce the rate of glucose absorption after eating high glycemic index foods containing carbohydrates: the blood glucose response curve decreases and the need for insulin decreases. Soluble and insoluble fibre may contribute to glycaemic control by reducing gastric emptying, modifying gastrointestinal hormone release, inhibiting amylase activity, interacting with the mucosa to form a barrier layer, and delaying starch hydrolysis, thereby reducing the rate of diffusion of amylolytic products into the small intestine [65].

In this case we can speak of an effective substitution of digestible starch. Among the components of cereal fibre, the main ones are arabinoxylans and β-glucans, whereas with fibre from other plant sources, apart from β-glucans, the main ones are pectin, guar or gum Arabic or tragacanth, or natural polysaccharides used as hydrocolloid additives, originating from seaweed or of microbial origin (agar, alginate, carrageenan, xanthan); β-glucans originating from certain higher fungi (Pleurotus, Ganoderma species) may also be important in this respect. However, the addition of flours from other sources, in particular certain pseudocereals and legumes, which contain high proportions of resistant starch in their native state, is also becoming increasingly important. Of the pseudocereals, buckwheat [63,66] is particularly important in this respect.

However, hydrocolloid polysaccharides, also used as food additives, which are the result of the physicochemical modification of cellulose or starch, also have a similar effect. This category also includes resistant starch type 4 (RS4) or starch type RS3 obtained by physical modification (HTM) from starch from another source and used as an additive. A great deal of research has been devoted to the physical modification of starch in order to reduce its digestibility or to the production of RS3 starch from various sources (rice, cassava, maize, potato). In particular, heat treatment, often for several hours at temperatures mainly between 80 and 140 °C with controlled water content (usually less than 40% w/w), is one of the many possible types of HTS, and the extrusion of starchy materials is another, so the source of resistant RS3 starch may thus include, for example, flours produced by milling extruded cereals and semi-finished products [56,67].

Whichever route is used to increase the fibre content or reduce the GI by the effective substitution of digestible (RDS and SDS) starch in the production of bread or other bakery products, the quality of the resulting product, particularly in terms of its sensory properties, must always be kept in mind. This is where the biggest problem lies in the approaches to reduce the GI of bakery products [68–70].

The point is that most of the substances, fibre components, through which effective substitution of digestible starch can be achieved, significantly alter the consistency of doughs and consequently the structure and properties of bread and pastry crumb. These are generally substances of a hydrocolloidal nature which are capable of binding significant quantities of water and thus altering the ratio of water to solid cereal material (and other formulation ingredients) in the dough. This has, of course, on the one hand, significant positive effects—firstly, an increase in the water binding capacity of the dough and thus an increase in the yield of the dough. (Note: the yield of the dough—expressed as a percentage of the weight of the dough relative to the weight of all the solid cereal fraction (flour)—is an important indicator of the efficiency and economy of bakery production.) Dough yields can rise from the normal range of 160–170% to values approaching 200% with higher additions of hydrocolloid substances. Another important effect is a significant increase in the fixation of water in the dough microstructure and consequently a slowing down of its release from the microstructures of the bread and pastry crumb, a slowing down of the retrogradation of starch and thus a slowing down of the ageing of the product. This results in an increase in the content of most of the above-mentioned fibre components. They include cereal β-glucans and arabinoxylans as well as cellulose, natural polysaccharides from other sources (other plants, fungi or micro-organisms), and in particular chemically and physically modified polysaccharide derivatives. It is these substances that are often used in bakery practice (in low doses) as additives to improve the properties of doughs and to retard the ageing of bread and pastry [70,71].

However, these positives are balanced by many negatives. Particularly for wheat dough and bakery products, increasing the natural content, or adding more of these substances in concentrations that actually lead to the required effective substitution of digestible starch and thus a significant reduction in GI, significantly alters the properties of doughs and bread and pastry crumbs. This is a significant increase in the dough denseness, reduction in softness and fluffiness that we generally expect from wheat bakery products, and some of these substances may also adversely affect the taste and aroma of bakery products, especially if they are not sufficiently isolated and purified from their original sources (which invariably increases the cost of their production) [68,70,71].

Higher additions of buckwheat and, in particular, legume flours often also cause an impairment, or at least an unexpected change in taste and aroma. The addition of substances of a hydrocolloidal nature often also causes technological problems in the processing of doughs (e.g., by increasing their stickiness), and has the significant consequence of increasing baking times. Extended baking times can lead to an increase in the production of process contaminants (e.g., acrylamide) on the surface (crust) of bread and bakery products. Longer baking times at high concentrations of hydrocolloids still do not lead to a sufficient reduction in the moisture content of the crumb, which not only has a negative impact on the sensory properties, but may also increase the risk of microbial contamination. The considerable amount of water bound in the microstructure of the crumb in the presence of high concentrations of hydrocolloids is gradually released, the water migrates to the surface, and there is an increase in water activity, which always poses a risk of the growth of micro-organisms, in this case in particular moulds [23].

All these facts must be kept in mind when considering increasing the fibre content of bread and bakery products. The most natural and easiest way to increase fibre content is to use wholemeal flours. This will significantly increase the content of all the fibre components naturally present in cereal grains. That is to say, both those substances which have a greater influence in retarding the resorption of starch (in particular the extractable arabinoxylans and β-glucans) and those substances (cellulose and lignin) which make a less significant contribution to what we have called effective substitution, and whose importance (particularly for lignin) lies rather in the relative, quantitative substitution of part of the starch in the total weight of the final product. In addition, both cellulose and lignin in particular, due to their physical properties, complicate both the processability and subsequent structure of the dough (small, tough, flake-like or shavings-like particles significantly disturb the fibrous microstructure of wheat doughs) and the sensory properties of the product. It therefore seems preferable, from this point of view, not to work with entirely wholemeal flours, but with flours which, thanks to the previous surface treatment of the grain by exfoliation, are free of hull particles. It is the pericarp that contains the highest proportion of cellulose and lignin [72].

If the GI value is addressed by the addition of fibre from other sources as additives or formulation ingredients, for the reasons given above, it is necessary to carefully weigh how important the properties of the substance or mixture of substances in question are in slowing down resorption in terms of what we call effective substitution [63,66], in addition to the simple substitution of digestible starch.

In central northern and eastern Europe, rye is traditionally used as a bread cereal alongside wheat. Rye flours (not only wholemeal and dark flours with a high degree of milling) contain significantly higher levels of arabinoxylans than wheat. The arabinoxylan content of rye bread flours reaches values of several percent. Bread and other products made from rye flours and rye-wheat mixtures are therefore nutritionally more advantageous in terms of fibre content, and to some extent GI, than pure wheat products. Rye breads and pastries generally have a different, denser texture than pure wheat products, which is another advantage. In fact, the increase in the hydrocolloidal fibre content of rye or rye-wheat bakery products is not as pronounced in their sensory characteristics as in pure wheat bakery products [1].

3.2. Hydrothermal Treatment of Raw Materials

The hydrothermal treatment of grain or its products (meal, flour or flakes, for example) has traditionally been used in bakery technology, and is nowadays used in many sophisticated processes. The hydrothermal treatment of cereals (including pseudocereals and legumes, or other grains and seeds) represents a very wide range of different processes involving all the anatomical components of the grain and based on interactions between biopolymers (polysaccharides and proteins) of a hydrocolloidal nature and water, often in combination with tempering or various intensive heating processes, i.e., with various modes and technical designs of interactions with heat [72,73].

Hydrothermal treatment, in its simpler forms (slicing and dehulling of the grain, slicing and tempering of the grain during dehulling—so-called conditioning), is part of the preparation of the grain for primary processing, particularly milling [72].

With processes that are mainly intended to prepare grain for flaking or milling, these are usually less vigorous, i.e., often involving less intensive wetting or heating. The simplest is sprinkling (moistening by a few per cent) and dehulling, which is used in preparation for the standard milling of wheat or rye. More intensive tempering combined with wetting (conditioning) or steaming of the grains is mainly used before flaking oats and other cereals [72].

However, the possibilities of hydrothermal treatment are incomparably broader, and it is applied to whole grains as well as to the products of their disintegration—hulled and otherwise surface-treated grains (hail), broken grains, cracked grains, shreds, flakes and flours. The grains and other raw materials are soaked in excess water (yield 200–300%), and then the resulting suspension is heated intensively for several hours to temperatures ranging from 40 to 80 °C, with temperatures of around 60 °C being the most common. The grains are 'cooked' in this way, whole or disintegrated, to form scrap or flour, often very rich mixtures of grains and parts thereof. Enzyme-active malt preparations (malt flours or extracts) are often added to the mixtures [74–77].

During the preparation of mash or similar products, many parallel physico-chemical and biochemical processes occur in the tempered suspension (the lubrication and partial enzymatic hydrolysis of starch, partial hydrolysis of proteins, a non-enzymatic browning reaction—the Maillard reaction). The result is the formation of a very wide range of sensory active substances, both aromatic and gustatory. An important nutritional benefit and therefore a reason for the increase in the frequency of use of these technological processes in the production of bread and bakery products is the demonstrable effect on increasing the bioavailability of nutritionally important fibre components and accompanying substances (phenolic compounds, etc.) [75,77].

However, the effect of these procedures on the structure and properties of starch, and thus on the GI, is problematic. In contrast to the hydrothermal treatment processes used for the preparation of RS3 starch, the processes used in the baking industry for the above purposes differ, firstly, by a significantly (and even several times) higher addition of water, usually working in excess in suspension, by a lower temperature (often below 80 °C) and in some cases, as mentioned, by the presence of amylolytic enzymes. It is therefore reasonable to assume, and in some cases has been demonstrated, that these procedures not only do not have a similar effect to the hydrothermal procedures for modifying starch in terms of resistance, but may on the contrary increase the digestibility of the starch present. The hydrothermal treatment of grain and its products (meal, flour), particularly when carried out in excess water at lower temperatures and in the presence of amylolytic enzymes, is ambiguous in terms of its effect on GI, and depends very much on the raw material used and its previous processing and on the very specific conditions under which the actual hydrothermal procedures is carried out [77,78].

3.3. Fermentation

Fermentation processes play an essential role in the most common types of secondary processing of cereals. In the production of bread, and of common and most types of fine bakery products, they serve to leaven the dough and, more generally, create the conditions for the final structure, consistency and texture of the product crumb, which is their primary and original role. Fermentation processes have accompanied bread-making since its origins in antiquity. For many centuries, they have been skilfully exploited and modified on the basis of purely empirical knowledge. If we look at the basic types of fermentation (detailed in the following chapter), the most massive production of CO_2 occurs during ethanol fermentation, to a lesser extent this gas is produced during heterofermentative lactic acid fermentation, and practically no CO_2 is produced during homofermentative lactic acid fermentation [23,79].

However, lactic fermentation is, together with ethanol fermentation, the principal fermentation process used in fermentations, including traditional Central European rye sourdoughs. In the modern industrial production of rye and rye-wheat sourdough bread, strongly acidic sourdoughs are often used with only lactic fermentation initiated by starter cultures, and the ethanol fermentation necessary for sufficient CO_2 production is only achieved by the addition of yeast to the bread dough [23].

Lactic acid sourdoughs are increasingly used in bakery technology for many reasons. In the production of traditional sourdough bread, the sensory effects are significant, resulting in the typical taste and aroma of these breads. However, this is by no means the only reason why the use of sourdough is expanding. The products of lactic fermentation have favourable nutritional properties, not only the acids themselves (especially lactic acid), but also for example the exopolysaccharides produced by certain strains of lactic acid bacteria (LAB). Similar to the above-mentioned hydrothermal processes used in baking, lactic acid fermentation results in an increase in the bioavailability of fibre components and co-formulants in sourdoughs and doughs, especially with longer maturation times. However, its effect on starch and its structure and properties, and thus on GI, is similarly ambiguous. On the one hand, the production of exopolysaccharides as well as the release of some fibre components from native fibre complexes may cause a partial effective substitution of digestible starch. However, the effect on pre-existing forms of MS is problematic [69,74].

3.4. Heat Treatment and Finishing of Bread and Other Cereal Products

The conversion of some starch to the resistant type (especially RS3) is generally caused by a combination of temperature and the presence of water. Baking is the basic method of heat treatment that is applied to bread and virtually all types of bakery products. In some special cases (doughnuts, donuts, linguine), frying is also used. Cooking has a double meaning. Firstly, it gives the products their final form and characteristics. During the heat treatment process, the appearance, shape, volume and internal structure of the products, as well as the taste and aroma, are completed. Most products are only made edible by heat treatment. The second fundamental importance of heat treatment is the stabilization of the product. Most doughs and masses are physico-chemically and biochemically highly unstable before heat treatment. The heat treatment fixes the structure, in particular by forming a starch slurry which, when cooled, becomes a flexible gel that binds water in its structure, coagulates proteins, inactivates enzymes, and stops the biochemical processes associated, in particular, with enzymatic hydrolysis and fermentation processes. The product is also fundamentally stabilized microbiologically [23].

Dough is a poor conductor of heat, so the temperature on the surface is significantly higher than inside, and the temperature gradient from the surface to the centre is not linear. In most cases, a large proportion of the heat transferred by radiation is reflected away from the surface of the baked dough piece and does not penetrate deeper. The heat transferred by the flow of air also primarily heats the surface and penetrates the interior of the dough piece slowly through conduction, corresponding to its low conductivity, and the same applies to the heat shared by conduction to the surface when baking in moulds. As a result, the surface and only a thin layer of a few millimetres underneath quickly reach high temperatures (around 150–170 °C on the surface), while the temperature inside slowly rises, and at the end of baking is between 95–98 °C in the centre of the baked item. For safe completion in terms of microbiological stability, the temperature in the centre (core) needs to reach 96 °C [23].

So what effect does baking bread and bakery products have on starch and its structure in terms of resistance to digestion and thus a reduction in GI? During the heat treatment of food and storage, RS content can be reduced under some conditions, but also increased under other conditions. It depends very much on the temperature, the time of heat treatment and also the method of cooling and storage [26,66,67,73,80].

A range of baking processes are used in bakery technology today, involving different types of heating and heat transfer. Convection, convection and radiation are applied to very different degrees in different types of ovens and in different baking regimes. This mostly affects the surface of the baked piece of dough and the relatively thin layer of dough just below the surface. However, the processes inside the crumb, where heat sharing is dominated by conduction through the crumb mass, are similar. The effect of baking technique on the interior of the baked piece is greater in inverse proportion to its weight, shape and volume. This means that the more massive and bulky the baked piece of dough, typically bread, the less the influence of what happens above the surface on the processes taking place inside [23].

Longer baking times for wheat breads, whose crumb dries out more quickly during baking and especially at lower weights and volumes, may have a positive effect on the increase in RS content, whereas for rye breads, rye-wheat breads or doughs with a higher content of hydrocolloidal water-binding components, the RS content may not only not increase, but may also decrease. In addition, in the first stages of baking, the rate of some reactions, including the enzymatic hydrolysis of starch, increases due to increasing temperature until inhibitory values are reached [23].

The significant influence of the cooling processes and the method of storage of the given finished product has been mentioned above. In general, the RS content increases with cooling and further increases with storage time due to progressive retrogradation. A large part of the range of bakery products in supermarket chains in the Czech Republic and throughout Central Europe consists of products baked directly in shops from pre-baked and frozen semi-finished products. The technique and speed of shock freezing and storage conditions and time, as well as the thawing and baking process, also have a significant influence on the RS content of the product before consumption [23].

The various options for modifying and assessing the GI of breads and bakery products are shown in Table 3.

Table 3. Review of studies on GI modification of bread and bakery products.

Type of GI Modification of Baked Goods (Addition/Replacement/Food Processing)	References
Rye flour with larger starch granules particles	[81,82]
Blends of wheat and buckwheat/oat/teff flours	[83]
High-amylose wheat flour	[84,85]
Mixture of oat-buckwheat flours	[86]
Wholemeal cereal flours	[87–89]
Ancient, colored and non-traditional wheat varieties	[90,91]
Cereal β-glucans	[58,92–94]
Pseudocereal and legume flours	[95–98]
Chickpea flour and psyllium	[99]
Acorn and chickpea flour blend	[100]
Lupine flour and resistant starch	[101]
Addition of pea protein	[102]
Effect of soluble dietary fibres	[103,104]
Prebiotic components (inulin, oligofructose, polydextrose, etc.)	[105,106]
Addition of guar gum	[107]
Addition of psyllium	[108]
Effects of xanthan gum, carrageenan and psyllium husk	[109]
Mucilage polysaccharides	[63,110]
Pomelo (*Citrus maxima*) fruit segments	[111]
Pomegranate peel powder	[112]
Resistant starch	[113]
Potato, cassava, sweet potato, banana and lentil starches modified with citric acid	[114]
RS4 enriched octenyl succinylated sweet potato, banana and lentil starches	[115]
Prebiotic dietary insoluble fibre from sweet potato peel and haricot bean flours	[116]
Aqueous extract of *Camellia sinensis* (green tea)	[117]
Addition of mushroom (*Pleurotus eryngii* and *Cantharellus cibarius*) powder	[118,119]
Addition of defatted melon seeds	[120]
Addition of Chinese chestnut flour	[121]
Wheat bran, oat bran, and oat β-glucan	[122,123]
Superfine (micronized) wheat bran	[123]
Pearl millet starch germ complex	[124]
Hydrothermal treatment of whole wheat grains and pulses	[125,126]
Fermented wheat bran and wheat germ	[127]
Addition of sourdoughs	[128–134]
Physically treated sugarcane fibre	[135]
Thermal and non-thermal approaches on the physical and chemical modification of starch	[136]

4. Future Research Trends

The impact of technological processes in the production of bread and bakery products on the resulting GI value, in terms of their selection and the selection of specific conditions for their different stages, includes many uncertainties which should be investigated and defined in detail. The fate of starch during the hydrothermal treatment of cereal suspensions under different conditions (water content, temperature, tempering time and enzyme activity of amylases) should be studied in detail, and procedures should be sought which do not decrease or increase the proportions of RS and SDS. The same applies to the maturation conditions of the yeasts. For sourdoughs, particular attention should be paid to the effect of pH and titratable acidity on the transitions between the different forms and types of starch (RDS, SDS and RS types). A third important area, which is also not sufficiently investigated, is the influence of thermal processes (baking, cooling, storage, freezing of pre-baked semi-finished products and baking). These processes certainly change the ratios of the different forms of starch in terms of their digestibility, and it is certainly possible to optimize them and to look for processes that can effectively stabilize or increase the RS content.

Bread and pastry will continue to be an important part of our diet for the foreseeable future. It is essential that their contribution to the prevalence of type 2 diabetes in our population is kept to a manageable minimum.

5. Conclusions

The problem of diabetes, especially type 2 diabetes, is an extremely serious global problem and affects most of the parts of the world described as developed. It is, of course, also fully applicable to the Czech Republic and the whole of Central Europe, whose population in many respects shares similar dietary habits. Bread and bakery products, or pasta, are the most important source of starch in the diet of the population, with the vast majority of the total starch content of these products occurring in digestible form, mainly as RDS and partly as SDS. The contribution of starch to the disease itself and its complications is undisputed. The GI of standard breads and pastries is most often in the range 60–75, with 70 being the threshold for labelling a product as a high GI food. Conversely, the threshold below which foods are labelled as low GI is 55.

Traditional Central European bread is a mixed rye-wheat or wheat-wheat bread leavened with rye sourdough spontaneously developed from the natural microflora of rye bread flour. In modern industrial production, spontaneous vital sourdoughs are being replaced by more stable, acidic vital yeasts initiated by BMK starter cultures, and the dough is leavened by the addition of yeast [137]. The emphasis on the use of yeasts is increasing with the growing awareness of their nutritional benefits. The use of rye, but also wheat sourdoughs, sourdoughs from gluten-free cereals and pseudocereals (rice, sorghum or buckwheat flours) and legume flours is expanding into the production of non-traditional bakery products [138].

Similar to the use of leavening, the use of many types of hydrothermal treatment of raw materials in aqueous suspensions is expanding, both for whole grains for the production of multi-grain types of bread and for bakery products, cereal meal and flours. As with leavening, these processes are rightly considered important not only as a means of improving the sensory properties of bread and pastry but also, and in particular, for their positive nutritional effects.

To summarize the nutritional effects demonstrated for the leavening and hydrothermal treatment of raw materials, these are mainly the production of certain nutritionally important substances (e.g., organic acids and exopolysaccharides in leavening) or the effect of both technological processes on increasing the bioavailability of fibre components and accompanying substances (phenolic compounds, minerals, vitamins, etc. [139].

The effect of the use of sourdoughs and hydrothermally treated suspensions not only occurs via their addition to the dough itself, but also by their action at other stages of the technological process of bread and pastry production, i.e., during the maturation of the dough and during heat treatment (baking), cooling and storage.

What remains unanswered, however, is the influence of these desirable technological processes on the final GI values of bread and pastry. The pathways leading to a reduction in the GI of a bakery product are firstly to reduce the relative starch content, and secondly to increase the proportion of some type of resistant RS starch in the total starch content.

The hydrothermal treatment of cereal suspensions, fermentation at the leavening or pre-fermentation stage, and the subsequent processes of dough maturation, mode of heat treatment, chilling or freezing, storage, finishing, and pre-consumption treatment also have an impact on the fate of starch and changes in its digestibility, and on the transitions between RDS, SDS and RS, and thus have a significant impact on the final GI of the bakery product. The impact of many of these is ambiguous and varies substantially depending on many specific conditions.

Author Contributions: Conceptualization, P.S., M.S., A.S., R.B., I.Š., E.Š. and V.P.; writing—original draft, P.S., M.S., A.S., R.B., I.Š., E.Š. and V.P.; writing—review and editing, P.S., M.S., I.Š., E.Š. and V.P. All authors have read and agreed to the published version of the manuscript.

Funding: This research received no external funding.

Data Availability Statement: The results are securely maintained by the authors and can be provided.

Conflicts of Interest: The authors declare no conflict of interest.

References

1. Delcour, J.A.; Hoseney, R.C. *Principles of Cereal Science and Technology*, 3rd ed.; AACC Inc.: St. Paul, MN, USA, 2010; pp. 1–22, 23–52.
2. Lindsay, D. The nutritional enhancement of plant foods. In *The Nutrition Handbook for Food Processors*; Woodhead Publishing Limited: Cambridge, UK, 2002; pp. 195–208.
3. Englyst, K.N.; Vinoy, S.; Englyst, H.N.; Lang, V. Glycaemic index of cereal products explained by their content of rapidly and slowly available glucose. *Br. J. Nutr.* **2003**, *89*, 329. [CrossRef] [PubMed]
4. Sobotka, L. *Basics in Clinical Nutrition*, 5th ed.; Galén: Prague, Czech Republic, 2011; pp. 445–448.
5. Ludwig, D.S. Dietary glycemic index and obesity. *J. Nutr.* **2000**, *130*, 280. [CrossRef] [PubMed]
6. Augustin, L.S.; Kendall, C.W.; Jenkins, D.J.; Willett, W.C.; Astrup, A.; Barclay, A.W.; Björck, I.; Brand-Miller, J.C.; Brighenti, F.; Buyken, A.E.; et al. Glycemic index, glycemic load and glycemic response: An International Scientific Consensus Summit from the International Carbohydrate Quality Consortium (ICQC). *Nutr. Metab. Cardiovasc. Dis.* **2015**, *25*, 795–815. [CrossRef]
7. Insel, P.; Ross, D.; McMahon, K.; Bernstein, M. *Nutrition*, 4th ed.; Jones & Bartlett Learning: Burlington, MA, USA, 2010; pp. 122–126.
8. ISO 26642:2010; Food Products—Determination of the Glycaemic Index (GI) and Recommendation for Food Classification. International Standards Organisation: Geneva, Switzerland, 2010.
9. Ijarotimi, O.S.; Fakayejo, D.A.; Oluwajuyitan, T.D. Nutritional characteristics, glycaemic index and blood glucose lowering property of gluten-free composite flour from wheat (*Triticum aestivum*), soybean (*Glycine max*), oat-bran (*Avena sativa*) and rice-bran (*Oryza sativa*). *Appl. Food Res.* **2021**, *1*, 100022. [CrossRef]
10. Bajka, B.H.; Pinto, A.M.; Ahn-Jarvis, J.; Ryden, P.; Perez-Moral, N.; Van Der Schoot, A.; Stocchi, C.; Bland, C.; Berry, S.E. The impact of replacing wheat flour with cellular legume powder on starch bioaccessibility, glycaemic response and bread roll quality: A double-blind randomised controlled trial in healthy participants. *Food Hydrocoll.* **2021**, *114*, 106565. [CrossRef] [PubMed]
11. Englyst, H.N.; Veenstra, J.; Hudson, G.J. Measurement of rapidly available glucose (RAG) in plant foods: A potential in vitro predictor of the glycemic response. *Brit. J. Nutr.* **1996**, *67*, 327–337. [CrossRef] [PubMed]
12. Englyst, K.N.; Englyst, H.N.; Hudson, G.J.; Cole, T.J.; Cummings, J.H. Rapidly available glucose in foods: An in vitro measurement that reflects the glycemic response. *Am. J. Clin. Nutr.* **1999**, *69*, 448–454. [CrossRef] [PubMed]
13. Englyst, H.N.; Cummings, J.H. Digestion of the polysaccharides of some cereal foods in the human small intestine. *Am. J. Clin. Nutr.* **1985**, *42*, 778–787. [CrossRef] [PubMed]
14. Sulung, N.K.; Aziss, N.A.S.M.; Kutbi, N.F.; Ahadaali, A.A.; Zairi, N.A.; Mahmod, I.I.; Sajak, A.A.B.; Sultana, S.; Azlan, A. Validation of in vitro glycaemic index (eGI) and glycaemic load (eGL) based on selected baked products, beverages, and canned foods. *Food Chem. Adv.* **2023**, *3*, 100502. [CrossRef]
15. Institute of Health Information and Statistics of the Czech Republic. Statistical Outcomes 2024. *Diabetology, Diabetes Care*. Available online: https://www.uzis.cz/index.php?pg=vystupy--statistika-vybranych-oboru-lekarske-pece--diabetologie (accessed on 23 April 2024).
16. Cereal Consumption in the Czech Republic—Increase in Consumption of Marginal Cereals and Buckwheat. Available online: https://www.fao.org/faostat/en/#data (accessed on 23 April 2024).
17. Velíšek, J. *The Chemistry of Food*; John Wiley & Sons: New York, NY, USA, 2013; p. 51.
18. Strassner, C. Chapter 7—Food, nutrition and health in Germany. In *Nutritional and Health Aspects of Food in Western Europe*; Academic Press: Cambridge, Massachusetts, USA, 2020; p. 147.
19. The glycaemic index of different types of food. Data from National Institute of Health of the Czech Republic. Available online: https://szu.cz/wp-content/uploads/2023/12/Glykemicky-index-2003.pdf (accessed on 23 April 2024). (only in Czech language).
20. Potkule, J.; Punia, S.; Kumar, M. Buckwheat: Nutritional Composition, Health Benefits, and Applications. In *Handbook of Cereals, Pulses, Roots, and Tubers: Functionality, Health Benefits, and Applications*; CRC Press Taylor & Francis Group: Boca Raton, FL, USA, 2022; p. 263.

21. Taylor, J.R.N.; Awika, J. Amaranth: Its Unique Nutritional and Health-promoting Attributes. In *Gluten-Free Ancient GrainsCereals, Pseudocereals, and Legumes: Sustainable, Nutritious, and Health-Promoting Foods for the 21st Century*; Woodhead Publishing Limited: Cambridge, UK, 2017; p. 133.
22. Villemejane, C.; Denis, S.; Marsset-Baglieri, A.; Alric, M.; Aymard, P.; Michon, C. In vitro digestion of short-dough biscuits enriched in proteins and/or fibres using a multi-compartmental and dynamic system (2): Protein and starch hydrolyses. *Food Chem.* **2016**, *190*, 164–172. [CrossRef]
23. Sluimer, P. *Principles of Breadmaking*, 2nd ed.; AACC Inc.: St. Paul, MN, USA, 2007; pp. 4–6, 113–138, 139–164, 183–202.
24. Ekström, L.M.N.K.; Björck, I.M.E.; Östman, E.M. On the possibility to affect the course of glycaemia, insulinaemia, and perceived hunger/satiety to bread meals in healthy volunteers. *Food Funct.* **2013**, *4*, 522–529. [CrossRef] [PubMed]
25. Ekström, L.M.N.K.; Björck, I.M.E.; Östman, E.M. An improved course of glycaemia after a bread based breakfast is associated with beneficial effects on acute and semi-acute markers of appetite. *Food Funct.* **2016**, *7*, 1040–1047. [CrossRef] [PubMed]
26. Jones, J.M. CODEX-aligned dietary fiber definitions help to bridge the 'fiber gap'. *Nutr. J.* **2014**, *13*, 1–10. [CrossRef] [PubMed]
27. Cummings, J.H.; Mann, J.I.; Nishida, C.; Vorster, H.H. Dietary fibre: An agreed definition. *Lancet* **2009**, *373*, 365–366. [CrossRef] [PubMed]
28. EC. Regulation (EC) No 1924/2006 of the European Parliament and of the Council of 20 December 2006 on Nutrition and Health Claims Made on Foods. 2006. Available online: https://eur-lex.europa.eu/legal-content/EN/ALL/?uri=celex:32006R1924 (accessed on 15 March 2024).
29. EFSA. Scientific Opinion on the substantiation of a health claim related to oat beta glucan and lowering blood cholesterol and reduced risk of (coronary) heart disease pursuant to Article 14 of Regulation (EC) No 1924/2006. *EFSA J.* **2010**, *8*, 1885–1899.
30. Sibakov, J.; Lehtinen, P.; Poutanen, K. Cereal brans as dietary fibre ingredients. In *Fibre-Rich and Wholegrain Foods: Improving Quality*; Woodhead Publishing Limited: Cambridge, UK, 2013; pp. 170–192.
31. Goff, H.D.; Repin, N.; Fabek, H.; El Khoury, D.; Gidley, M.J. Dietary fibre for glycaemia control: Towards a mechanistic understanding. *Bioact. Carbohydr. Diet. Fibre* **2018**, *14*, 39–53. [CrossRef]
32. Kahraman, K.; Aktas-Akyildiz, E.; Ozturk, S.; Köksel, H. Effect of different resistant starch sources and wheat bran on dietary fibre content and in vitro glycaemic index values of cookies. *J. Cereal Sci.* **2019**, *90*, 102851. [CrossRef]
33. Shin, S.I.; Choi, H.J.; Chung, K.M.; Hamaker, B.; Park, K.H.; Moon, T.W. Slowly digestible starch from debranched waxy sorghum starch: Preparation and properties. *Cereal Chem.* **2004**, *81*, 404. [CrossRef]
34. Jane, J.-L.; Wong, K.-S.; McPherson, A.E. Branch-structure difference in starches of A-and B-type X-ray patterns revealed by their Naegeli dextrins. *Carbohydr. Res.* **1997**, *300*, 219. [CrossRef]
35. Patterson, M.A.; Maiya, M.; Stewart, L. Resistant Starch Content in Foods Commonly Consumed in the United States: A narrative Review. *J. Acad. Nutr. Diet.* **2020**, *120*, 230–244. [CrossRef]
36. Elmståhl, H.L. Resistant starch content in a selection of starchy foods on the Swedish market. *Eur. J. Clin. Nutr.* **2002**, *56*, 500–505. [CrossRef] [PubMed]
37. Roberts, J.; Jones, G.P.; Rutishauser, I.H.E.; Birkett, A.; Gibbons, C. Resistant starch in Australian diet. *Nutr. Diet.* **2004**, *61*, 98–104.
38. Liu, X.; Qiao, L.; Kong, Y.; Wang, H.; Yang, B. Characterization of the starch molecular structure of wheat varying in the content of resistant starch. *Food Chem.* **2024**, *21*, 101103. [CrossRef] [PubMed]
39. Guo, J.; Gutierrez, A.; Tan, L.; Kong, L. Considerations and strategies for optimizing health benefits of resistant starch. *Curr. Opin. Food Sci.* **2023**, *51*, 101008. [CrossRef]
40. Hoover, R.; Zhou, Y. In vitro and in vivo hydrolysis of legume starches by α-amylase and resistant starch formation in legumes—A review. *Carbohydr. Polym.* **2003**, *54*, 401–417. [CrossRef]
41. Figueroa-González, I.; Quijano, G.; Ramírez, G.; Cruz-Guerrero, A. Probiotics and prebiotics—Perspectives and challenges. *J. Sci. Food Agric.* **2011**, *91*, 1341. [CrossRef] [PubMed]
42. Yonekura, L.; Suzuki, H. Effects of dietary zinc levels, phytic acid and resistant starch on zinc bioavailability in rats. *Eur. J. Nutr.* **2005**, *44*, 384. [CrossRef] [PubMed]
43. Englyst, H.N.; Kingman, S.M.; Cummings, J.H. Classification and measurement of nutritionally important starch fractions. *Eur. J. Clin. Nutr.* **1992**, *46*, S33–S50. [PubMed]
44. Harris, K. An introductory review of resistant starch type 2 from high-amylose cereal grains and its effect on glucose and insulin homeostasis. *Nutr. Rev.* **2019**, *77*, 748–764. [CrossRef]
45. Yi, M.; Tang, X.; Liang, S.; He, R.; Huang, T.; Lin, Q.; Zhang, R. Effect of microwave alone and microwave-assisted modification on the physicochemical properties of starch and its application in food. *Food Chem.* **2024**, *446*, 138841. [CrossRef]
46. Martins Fonseca, L.; Mello El Halal, S.L.; Guerra Dias, A.R.; da Rosa Zavareze, E. Physical modification of starch by heat-moisture treatment and annealing and their applications: A review. *Carbohydr. Polym.* **2021**, *274*, 11866.
47. Ma, R.; Cai, C.; Wang, F.; Zhan, J.; Tian, Y. Improvement of resistant starch content and thermal-stability of starch-linoleic acid complex: An attempt application in extruded recombinant rice. *Food Chem.* **2024**, *445*, 138768. [CrossRef] [PubMed]
48. Wang, Z.; Wang, S.; Xu, Q.; Kong, Q.; Li, F.; Lu, L.; Xu, Y.; Wei, Y. Synthesis and functions of resistant starch. *Adv. Nutr.* **2023**, *14*, 1131–1144. [CrossRef] [PubMed]
49. Oyeyinka, S.A.; Singh, S.; Amonsou, E.O. A review on structural, digestibility and physicochemical properties of legume starch-lipid complexes. *Food Chem.* **2021**, *349*, 129165. [CrossRef] [PubMed]

50. Trumbo, P.; Schlicker, S.; Yates, A.A.; Poos, M. Dietary reference intakes for energy, carbohydrate, fiber, fat, fatty acids, cholesterol, protein and amino acids. *J. Am. Diet. Assoc.* **2002**, *102*, 1621–1630. [CrossRef] [PubMed]
51. Ajala, A.; Kaur, L.; Lee, S.J.; Singh, J. Native and processed legume seed microstructure and its influence on starch digestion and glycaemic features: A review. *Trends Food Sci. Technol.* **2023**, *133*, 65–74. [CrossRef]
52. Marangoni, F.; Poli, A. The glycemic index of bread and biscuits is markedly reduced by the addition of a proprietary fiber mixture to the ingredients. *Nutr. Metab Cardiovasc. Dis.* **2008**, *18*, 602–625. [CrossRef]
53. Brennan, C.S.; Blake, D.E.; Ellis, P.R.; Schofield, J.D. Effects of guar galactomannan on wheat bread microstructure and on the in vitro and in vivo digestibility of starch in bread. *J. Cereal Sci.* **1996**, *24*, 151–160. [CrossRef]
54. McRorie, J.W., Jr.; McKeown, N.M. Understanding the physics of functional fibers in the gastrointestinal tract: An evidence-based approach to resolving enduring misconceptions about insoluble and soluble fiber. *J. Acad. Nutr. Diet.* **2017**, *117*, 251–264. [CrossRef]
55. Kendall, C.W.; Esfahani, A.; Jenkins, D.J. The link between dietary fibre and human health. *Food Hydrocoll.* **2010**, *24*, 42–48. [CrossRef]
56. Ma, M.; Gu, Z.; Cheng, L.; Li, Z.; Li, C.; Hong, Y. Effect of hydrocolloids on starch digestion: A review. *Food Chem.* **2024**, *444*, 138636. [CrossRef] [PubMed]
57. Plaami, S.P. Content of dietary fiber in foods and its physiological effects. *Food Rev. Int.* **1997**, *13*, 29–76. [CrossRef]
58. Ekström, L.M.N.K.; Henningsson Bok, E.A.E.; Sjöö, M.E.; Östman, E.M. Oat β-glucan containing bread increases the glycemic profile. *J. Funct. Foods* **2017**, *32*, 106–111. [CrossRef]
59. Brennan, C.S.; Cleary, L.J. Utilisation Glucagel® in the β-glucan enrichment of breads: A Physicochemical and nutritional evaluation. *Food Res. Int.* **2007**, *40*, 291–296. [CrossRef]
60. Damen, B.; Pollet, A.; Dornez, E.; Broekaert, W.F.; Van Haesendonck, I.; Trogh, I.; Arnaut, F.; Delcour, J.A.; Courtin, C.M. Xylanase-mediated in situ production of arabinoxylan oligosaccharides with prebiotic potential in whole meal breads enriched with arabinoxylan rich materials. *Food Chem.* **2011**, *131*, 111–118. [CrossRef]
61. Noort, M.W.J.; van Haaster, D.; Hemery, Y.; Schols, H.A.; Hamer, R.J. The effect of particle size of wheat bran fractions on bread quality—Evidence for fibre-protein interactions. *J. Cereal Sci.* **2010**, *52*, 59–64. [CrossRef]
62. Diez-Sánchez, E.; Llorca, E.; Quiles, A.; Hernando, I. Using different fibers to replace fat in sponge cakes: In vitro starch digestion and physico-structural studies. *Food Sci. Technol. Int.* **2018**, *24*, 533–543. [CrossRef] [PubMed]
63. Salgado-Cruz, M.d.l.P.; Ramírez-Miranda, M.; Díaz-Ramírez, M.; Alamilla-Beltran, L.; Calderón-Domínguez, G. Microstructural characterization and glycemic index evaluation of pita bread enriched with chia mucilage. *Food Hydrocoll.* **2017**, *69*, 141–149. [CrossRef]
64. Ibrügger, S.; Kristensen, M.; Mikkelsen, M.S.; Astrup, A. Flaxseed dietary fiber supplements for suppression of appetite and food intake. *Appetite* **2012**, *58*, 490–495. [CrossRef]
65. Kristensen, M.; Jensen, M.G.; Riboldi, G.; Petronio, M.; Bügel, S.; Toubro, S.; Tetens, I.; Astrup, A. Wholegrain vs. refined wheat bread and pasta. Effect on postprandial glycemia, appetite, and subsequent ad libitum energy intake in young healthy adults. *Appetite* **2010**, *54*, 163–169. [CrossRef]
66. Hernández-Nava, R.G.; Berrios, J.D.J.; Pan, J.; Osorio-Díaz, P.; Bello-Perez, L.A. Development and characterization of spaghetti with high resistant starch content supplemented with banana starch. *Food Sci. Technol. Int.* **2009**, *15*, 73. [CrossRef]
67. van Hung, P.; Yamamori, M.; Morita, N. Formation of enzyme-resistant starch in bread as affected by high-amylose wheat flour substitutions. *Cereal Chem.* **2005**, *82*, 690. [CrossRef]
68. Sozer, N.; Cicerelli, L.; Heiniö, R.-L.; Poutanen, K. Effect of wheat bran addition on in vitro starch digestibility, physico-mechanical and sensory properties of biscuits. *J. Cereal Sci.* **2014**, *60*, 105–113. [CrossRef]
69. Rakha, A.; Åman, P. Fibre-enriched and wholegrain breads. In *Fibre-Rich and Wholegrain Foods: Improving Quality*; Woodhead Publishing Limited: Cambridge, UK, 2013; pp. 211–235.
70. Sozer, N.; Poutanen, K. Fibre in extruded products. In Fibre-Rich and Wholegrain Foods: Improving Quality; Woodhead Publishing Limited: Cambridge, UK, 2013; pp. 256–272.
71. Chen, H.; Rubenthaler, G.L.; Leung, H.K.; Baranowski, J.D. Chemical, physical, and baking properties of apple fiber compared with wheat and oat bran. *Cereal Chem.* **1988**, *65*, 244–247.
72. Ulmer, K. *Technology & Equipment Grain Milling*; Bühler AG: Uzwill, Switzerland, 2011; pp. 323–354.
73. Chang, Y.K.; Wang, S.S. *Advances in Extrusion Technology*; Technomic Publishing Company: Lancaster, PA, USA, 1999; p. 266.
74. Rahmadani, M.; Tryas, A.A.; Susanto, I.; Nahrowi, L.; Khotijah, L.; Jayanegara, A. Enhancing resistant starch in foods through organic acid intervention: A meta-analysis on thermal properties, nutrient composition, and in vitro starch digestibility. *J. Agri. Food Res.* **2024**, *15*, 101037. [CrossRef]
75. Santos Dorneles, M.; Silva de Azevedo, E.; Zapata Norena, C.P. Effect of heat treatment at low moisture on the increase of resistant starch content in Araucaria angustifolia seed starch. *Food Hydrocoll.* **2024**, *150*, 109639. [CrossRef]
76. Bao, J.; Zhou, X.; Hu, Y.; Zhang, Z. Resistant starch content and physicochemical properties of non-waxy rice starches modified by pullulanase, heat-moisture treatment, and citric acid. *J. Cereal Sci.* **2022**, *105*, 103472. [CrossRef]
77. Van Hung, P.; Lam Vien, N.; Lan Phi, N.T. Resistant starch improvement of rice starches under a combination of acid and heat-moisture treatments. *Food Chem.* **2016**, *191*, 67–73. [CrossRef] [PubMed]

78. Gao, M.; Jia, J.; Zhang, C.; Liu, Y.; Dou, B.; Zhang, N. Structure, properties, and resistant starch content of modified rice flour prepared using dual hydrothermal treatment. *Int. J. Biol. Macromol.* **2024**, *262*, 130050. [CrossRef]
79. Crittenden, R.G.; Morris, L.F.; Harvey, M.L.; Tran, L.T.; Mitchell, H.L.; Playne, M.J. Selection of a Bifidobacterium strain to complement resistant starch in a synbiotic yoghurt. *J. Appl. Microbiol.* **2001**, *90*, 268. [CrossRef]
80. Bayomy, H.M.; Alamri, E.S.; Albalawi, A.N.; Alharbi, R.; Ozaybi, N.A.; Rozan, M.A.; Shamsia, S.M. Production of extruded functional snacks based on resistant starch using waste rice and whey milk. *LWT—Food Sci. Technol.* **2024**, *197*, 115871. [CrossRef]
81. Akila, S.R.V.; Mishra, S.; Hardacre, A.; Matia-Merino, L.; Goh, K.; Warren, F.; Monro, J. Kernel structure in breads reduces in vitro starch digestion rate and estimated glycaemic potency only at high grain inclusion rates. *Food Structure* **2019**, *21*, 100109.
82. Tagliasco, M.; Fon, G.; Renzetti, S.; Capuano, E.; Pellegrini, N. Role of particle size in modulating starch digestibility and textural properties in a rye bread model system. *Food Res. Int.* **2024**, *190*, 114565. [CrossRef] [PubMed]
83. Wolter, A.; Hager, A.S.; Zannini, E.; Arendt, E.K. In vitro starch digestibility and predicted glycaemic indexes of buckwheat, oat, quinoa, sorghum, teff and commercial gluten-free bread. *J. Cereal Sci.* **2013**, *58*, 431–436. [CrossRef]
84. Li, C.; Dluital, S.; Gidley, M.J. High-amylose wheat bread with reduced in vitro digestion rate and enhanced resistant starch content. *Food Hydrocoll.* **2022**, *123*, 107181. [CrossRef]
85. Di Rosa, C.; De Arcangelis, E.; Vitelli, V.; Crucillà, S.; Angelicola, M.; Trivisonno, M.C.; Sestili, F.; Blasi, E.; Cicatiello, C.; Lafiandra, D.; et al. Effect of three bakery products formulated with high-amylose wheat flour on post-prandial glycaemia in healthy volunteers. *Foods* **2023**, *12*, 319. [CrossRef] [PubMed]
86. Bączek, N.; Jarmułowicz, A.; Wronkowska, M.; Haros, C.M. Assessment of the glycaemic index, content of bioactive compounds, and their in vitro bioaccessibility in oat-buckwheat breads. *Food Chem.* **2020**, *330*, 127199. [CrossRef] [PubMed]
87. Moretton, M.; Alongi, M.; Melchior, S.; Anese, M. Adult and elderly in vitro digestibility patterns of proteins and carbohydrates as affected by different commercial bread types. *Food Res. Int.* **2023**, *167*, 112732. [CrossRef]
88. Lužnik, I.A.; Polak, M.L.; Demšar, L.; Gašperlin, L.; Polak, T. Does type of bread ingested for breakfast contribute to lowering of glycaemic index? *J. Nutr. Intermed. Metab.* **2019**, *16*, 100097. [CrossRef]
89. Dong, H.; Pineda, D.G.; Li, N.; Xu, Y. Investigation of the postprandial glycaemic response to white bread and wholemeal bread consumption among healthy young adults. *Proceedings* **2023**, *91*, 194.
90. Koksel, H.; Cetiner, B.; Shamanin, V.P.; Tekin-Cakmak, Z.H.; Pototskaya, I.V.; Kahraman, K.; Sagdic, O.; Morgounov, A.I. Quality, nutritional properties, and glycemic index of colored whole wheat breads. *Foods* **2023**, *12*, 3376. [CrossRef] [PubMed]
91. Păucean, A.; Șerban, L.R.; Chiș, M.S.; Mureșan, V.; Pușcaș, A.; Man, S.M.; Pop, C.R.; Socaci, S.A.; Igual, M.; Ranga, F.; et al. Nutritional composition, in vitro carbohydrates digestibility, textural and sensory characteristics of bread as affected by ancient wheat flour type and sourdough fermentation time. *Food Chem.* **2024**, *22*, 101298. [CrossRef] [PubMed]
92. Lante, A.; Canazza, E.; Tessari, P. Beta-glucans of cereals: Functional and technological properties. *Nutrients* **2023**, *15*, 2124. [CrossRef] [PubMed]
93. Bozbulut, R.; Sanlier, N. Promising effects of β-glucans on glyceamic control in diabetes. *Trends Food Sci. Technol.* **2019**, *83*, 159–166. [CrossRef]
94. Hu, H.; Lin, H.; Xiao, L.; Guo, M.; Yan, X.; Su, X.; Liu, L.; Sang, S. Impact of native form oat β-glucan on the physical and starch digestive properties of whole oat bread. *Foods* **2022**, *11*, 2622. [CrossRef] [PubMed]
95. Di Cairano, M.; Condelli, N.; Caruso, M.C.; Marti, A.; Cela, N.; Galgano, F. Functional properties and predicted glycemic index of gluten free cereal, pseudocereal and legume flours. *LWT—Food Sci. Technol.* **2020**, *133*, 109860. [CrossRef]
96. Vinod, B.R.; Asrey, R.; Rudra, S.G.; Urhe, S.B.; Mishra, S. Chickpea as a promising ingredient substitute in gluten-free bread making: An overview of technological and nutritional benefits. *Food Chem. Adv.* **2023**, *3*, 100473. [CrossRef]
97. Gallo, V.; Romano, A.; Ferranti, P.; D'Auria, G.; Masi, P. Properties and in vitro digestibility of a bread enriched with lentil flour at different leavening times. *Food Structure* **2022**, *33*, 100284. [CrossRef]
98. Sowdhanya, D.; Singh, J.; Rasane, P.; Kaur, S.; Kaur, J.; Ercisli, S.; Verma, H. Nutritional significance of velvet bean (*Mucuna pruriens*) and opportunities for its processing into value-added products. *J. Agri. Food Res.* **2024**, *15*, 100921. [CrossRef]
99. Santos, F.G.; Aguiar, E.V.; Rosell, C.M.; Capriles, V.D. Potential of chickpea and psyllium in gluten-free breadmaking: Assessing bread's quality, sensory acceptability, and glycemic and satiety indexes. *Food Hydrocoll.* **2021**, *113*, 106487. [CrossRef]
100. Gkountenoudi-Eskitzi, I.; Kotsiou, K.; Irakli, M.N.; Lazaridis, A.; Biliaderis, C.G.; Lazaridou, A. In vitro and in vivo glycemic responses and antioxidant potency of acorn and chickpea fortified gluten-free breads. *Food Res. Int.* **2023**, *166*, 112579. [CrossRef] [PubMed]
101. Yaver, E.; Bilgiçli, N. Ultrasound-treated lupin (*Lupinus albus* L.) flour: Protein- and fiber-rich ingredient to improve physical and textural quality of bread with a reduced glycemic index. *LWT—Food Sci. Technol.* **2021**, *148*, 111767. [CrossRef]
102. Moretton, M.; Alongi, M.; Renoldi, N.; Anese, M. Steering protein and carbohydrate digestibility by food design to address elderly needs: The case of pea protein enriched bread. *LWT—Food Sci. Technol.* **2023**, *190*, 115530. [CrossRef]
103. Zhou, Z.; Ye, F.; Lei, L.; Zhou, S.; Zhao, G. Fabricating low glycemic index foods: Enlightened by the impacts of soluble dietary fibre on starch digestibility. *Trends Food Sci. Technol.* **2022**, *122*, 110–122. [CrossRef]
104. Torres, J.D.; Dueik, V.; Carré, D.; Contardo, I.; Bouchon, P. Non-invasive microstructural characterization and in vivo glycemic response of white bread formulated with soluble dietary fiber. *Food Bioscience* **2024**, *61*, 104505. [CrossRef]
105. Pimentel, T.C.; de Assis, B.B.T.; dos Santos Rocha, C.; Marcolino, V.A.; Rosset, M.; Magnani, M. Prebiotics in non-dairy products: Technological and physiological functionality, challenges, and perspectives. *Food Bioscience* **2022**, *46*, 101585. [CrossRef]

106. Devi, R.; Sharma, E.; Thakur, R.; Lal, P.; Kumar, A.; Altaf, M.A.; Singh, B.; Tiwari, R.K.; Lal, M.K.; Kumar, R. Non-dairy prebiotics: Conceptual relevance with nutrigenomics and mechanistic understanding of the effects on human health. *Food Res. Int.* **2023**, *170*, 112980. [CrossRef] [PubMed]
107. Mæhre, H.K.; Weisensee, S.; Ballance, S.; Rieder, A. Guar gum fortified white breads for prospective postprandial glycaemic control—Effects on bread quality and galactomannan molecular weight. *LWT—Food Sci. Technol.* **2021**, *152*, 112354. [CrossRef]
108. Mallikarjunan, N.; Rajalakshmi, D.; Maurya, D.K.; Jamdar, S.N. Modifying rheological properties of psyllium by gamma irradiation enables development of low glycaemic index food with a predicted gastrointestinal tolerance. *Int. J. Biol. Macromol.* **2024**, *257*, 128625.
109. Yassin, Z.; Tan, Y.L.; Srv, A.; Monro, J.; Matia-Merino, L.; Lim, K.; Hardacre, A.; Mishra, S.; Goh, K.K.T. Effects of xanthan gum, lambda-carrageenan and psyllium husk on the physical characteristics and glycaemic potency of white bread. *Foods* **2022**, *11*, 1513. [CrossRef]
110. Goksen, G.; Demir, D.; Dhama, K.; Kumar, M.; Shao, P.; Xie, F.; Echegaray, N.; Lorenzo, J.M. Mucilage polysaccharide as a plant secretion: Potential trends in food and biomedical applications. *Int. J. Biol. Macromol.* **2023**, *230*, 123146. [CrossRef] [PubMed]
111. Reshmi, S.K.; Sudha, M.L.; Shashirekha, M.N. Starch digestibility and predicted glycemic index in the bread fortified with pomelo (*Citrus maxima*) fruit segments. *Food Chem.* **2017**, *237*, 957–965. [CrossRef] [PubMed]
112. García, P.; Bustamante, A.; Echeverría, F.; Encina, C.; Palma, M. A Feasible approach to developing fiber-enriched bread using pomegranate peel powder: Assessing its nutritional composition and glycemic index. *Foods* **2023**, *12*, 2798. [CrossRef]
113. Amaral, O.; Guerreiro, C.; Almeida, A.; Cravo, M. Bread with a high level of resistant starch influenced the digestibility of the available starch fraction. *Bioact. Carbohydr. Diet. Fibre* **2022**, *28*, 100318. [CrossRef]
114. Remya, R.; Jyothi, A.N.; Sreekumar, J. Effect of chemical modification with citric acid on the physicochemical properties and resistant starch formation in different starches. *Carb. Polym.* **2018**, *202*, 29–38. [CrossRef] [PubMed]
115. Remya, R.; Jyothi, A.N.; Sreekumar, J. Morphological, structural and digestibility properties of RS4 enriched octenyl succinylated sweet potato, banana and lentil starches. *Food Hydrocoll.* **2018**, *24*, 219–229. [CrossRef]
116. Admasu, F.; Fentie, E.G.; Admassu, H.; Shin, J.-H. Functionalization of wheat bread with prebiotic dietary insoluble fiber from orange-fleshed sweet potato peel and haricot bean flours. *LWT—Food Sci. Technol.* **2024**, *200*, 116182. [CrossRef]
117. Lawal, T.A.; Ononamadu, C.J.; Okonkwo, E.K.; Adedoyin, H.J.; Muhammad Liman Shettima, M.L.; Muhammad, I.U.; Alhassan, A.J. In vitro and in vivo hypoglycaemic effect of Camellia sinensis on alpha glucosidase activity and glycaemic index of white bread. *Appl. Food Res.* **2022**, *2*, 100037. [CrossRef]
118. Liu, Y.; Zhang, H.; Brennan, M.; Brennan, C.; Qin, Y.; Cheng, G.; Liu, Y. Physical, chemical, sensorial properties and in vitro digestibility of wheat bread enriched with yunnan commercial and wild edible mushrooms. *LWT—Food Sci. Technol.* **2022**, *169*, 113923. [CrossRef]
119. Zhao, H.; Wang, L.; Brennan, M.; Brennan, C. How does the addition of mushrooms and their dietary fibre affect starchy foods. *J. Future Foods* **2022**, *2*, 18–24. [CrossRef]
120. Zhang, G.; Chatzifragkou, A.; Charalampopoulos, D.; Rodriguez-Garcia, J. Effect of defatted melon seed residue on dough development and bread quality. *LWT—Food Sci. Technol.* **2023**, *183*, 114892. [CrossRef]
121. Wang, L.; Shi, D.; Chen, J.; Dong, H.; Chen, L. Effects of Chinese chestnut powder on starch digestion, texture properties, and staling characteristics of bread. *GOST* **2023**, *6*, 82–90. [CrossRef]
122. Lin, S.; Jin, X.; Gao, J.; Qiu, Z.; Ying, J.; Wang, Y.; Dong, Z.; Zhou, W. Impact of wheat bran micronization on dough properties and bread quality: Part II—Quality, antioxidant and nutritional properties of bread. *Food Chem.* **2022**, *396*, 133631. [CrossRef] [PubMed]
123. Duță, D.E.; Culețu, A.; Mohan, G. 10—Reutilization of cereal processing by-products in bread making. In *Sustainable Recovery and Reutilization of Cereal Processing By-Products*; Woodhead Publishing Limited: Cambridge, UK, 2018; pp. 279–317.
124. Vidhyalakshmi, R.; Prabhasankar, P.; Muthukumar, S.P.; Prathima, C.; Meera, M.S. The impact of addition of pearl millet starch-germ complex in white bread on nutritional, textural, structural, and glycaemic response: Single blinded randomized controlled trial in healthy and pre-diabetic participants. *Food Res. Int.* **2024**, *183*, 114186. [CrossRef]
125. King, J.; Leong, S.Y.; Alpos, M.; Johnson, C.; McLeod, S.; Peng, M.; Sutton, K.; Oey, I. Role of food processing and incorporating legumes in food products to increase protein intake and enhance satiety. *Trends Food Sci. Technol.* **2024**, *147*, 104466. [CrossRef]
126. Abhilasha, A.; Kaur, L.; Monro, J.; Hardacre, A.; Singh, J. Effects of hydrothermal treatment and low-temperature storage of whole wheat grains on in vitro starch hydrolysis and flour properties. *Food Chem.* **2022**, *395*, 133516. [CrossRef]
127. Pontonio, A.; Lorusso, A.; Gobbetti, M.; Rizzello, C.G. Use of fermented milling by-products as functional ingredient to develop a low-glycaemic index bread. *J. Cereal Sci.* **2017**, *77*, 235–242. [CrossRef]
128. Das, S.; Pegu, K.; Arya, S.S. Functional sourdough millet bread rich in dietary fibre—An optimization study using fuzzy logic analysis. *Bioact. Carbohydr. Diet. Fibre* **2021**, *26*, 100279. [CrossRef]
129. Chatonidi, G.; Poppe, J.; Verbeke, K. Plant-based fermented foods and the satiety cascade: A systematic review of randomized controlled trials. *Trends Food Sci. Technol.* **2023**, *133*, 127–137. [CrossRef]
130. Wang, Z.; Wang, L. Impact of sourdough fermentation on nutrient transformations in cereal-based foods: Mechanisms, practical applications, and health implications. *GOST* **2024**, *7*, 124–132. [CrossRef]

131. Sun, X.; Wu, S.; Li, W.; Koksel, F.; Du, Y.; Sun, L.; Fang, Y.; Hu, Q.; Pei, F. The effects of cooperative fermentation by yeast and lactic acid bacteria on the dough rheology, retention and stabilization of gas cells in a whole wheat flour dough system—A review. *Food Hydrocoll.* **2023**, *135*, 108212. [CrossRef]
132. Islam, M.d.A.; Islam, S. Sourdough bread quality: Facts and Factors. *Foods* **2024**, *13*, 2132. [CrossRef]
133. Demirkesen-Bicak, H.; Arici, M.; Yaman, M.; Karasu, S.; Sagdic, O. Effect of different fermentation condition on estimated glycemic index, in vitro starch digestibility, and textural and sensory properties of sourdough bread. *Foods* **2021**, *10*, 514. [CrossRef] [PubMed]
134. Taras, M.A.; Cherchi, S.; Campesi, I.; Margarita, V.; Carboni, G.; Rappelli, P.; Tonolo, G. Utility of flash glucose monitoring to determine glucose variation induced by different doughs in persons with type 2 diabetes. *Diabetology* **2024**, *5*, 129–140. [CrossRef]
135. Yassin, Z.A.R.; Halim, F.N.B.A.; Taheri, A.; Goh, K.K.T.; Du, J. Effects of microwave, ultrasound, and high-pressure homogenization on the physicochemical properties of sugarcane fibre and its application in white bread. *LWT—Food Sci. Technol.* **2023**, *184*, 115008. [CrossRef]
136. Jeevarathinam, G.; Ramniwas, S.; Singh, P.; Rustagi, S.; Asdaq, S.M.B.; Pandiselvam, R. Macromolecular, thermal, and nonthermal technologies for reduction of glycemic index in food-A review. *Food Chem.* **2024**, *445*, 138742. [CrossRef] [PubMed]
137. Sluková, M.; Jurkaninová, L.; Gillarová, S.; Horáčková, Š.; Skřivan, P. Rye—Nutritional and technological evaluation in Czech cereal technology—A review: Sourdoughs and bread. *Czech J. Food Sci.* **2021**, *39*, 65–70. [CrossRef]
138. Skřivan, P.; Chrpová, D.; Klitschová, B.; Švec, I.; Sluková, M. Buckwheat flour (*Fagopyrum esculentum* Moench)—A contemporary view on the problems of its production for human nutrition. *Foods* **2023**, *12*, 3055. [CrossRef]
139. Skřivan, P.; Sluková, M.; Švec, I.; Čížková, H.; Horsáková, I.; Rezková, E. The use of modern fermentation techniques in the production of traditional wheat bread. *Czech J. Food Sci.* **2023**, *41*, 173–181. [CrossRef]

Disclaimer/Publisher's Note: The statements, opinions and data contained in all publications are solely those of the individual author(s) and contributor(s) and not of MDPI and/or the editor(s). MDPI and/or the editor(s) disclaim responsibility for any injury to people or property resulting from any ideas, methods, instructions or products referred to in the content.

Article

Influence of Technological Parameters on Sourdough Starter Obtained from Different Flours

Alina Alexandra Dobre, Elena Mirela Cucu and Nastasia Belc *

National Research & Development Institute for Food Bioresources—IBA Bucharest, 6 Dinu Vintilă Street, District 2, 021102 Bucharest, Romania; alina.dobre@bioresurse.ro (A.A.D.)
* Correspondence: nastasia.belc@bioresurse.ro

Abstract: One of the oldest biotechnological processes used in bread manufacture is sourdough production which relies on wild yeast and lactobacillus cultures naturally present in flour. The aim of this paper was to evaluate the influence of selected flours of different cereal grains (ancient wheat, corn, and rye), different dough variations, and temperature of fermentation on the quality of spontaneous sourdough. Two values of fermentation temperatures were tested (25 °C and 35 °C), and for each temperature analyzed, three backslopping steps were carried out to obtain mature doughs according to the traditional type I sourdough scheme. In total, 14 different sourdoughs were produced, and microbiology, pH, and total titration acidity for 96 h were determined. Optimal pH values for the samples determined that the optimal fermentation period was 48 h. The acidification rate of the dough was faster at 35 °C than at 25 °C. This fact became evident via the pH values obtained in the first 24 h. However, from this point, the pH values were lower in the samples kept at 25 °C, showing that a cooler fermentation temperature allows the acidification activity of the microorganisms to be prolonged for a longer time. In the study carried out, the ideal fermentation time for the population of LAB and yeasts is 72 h at a temperature of 25 °C, and the most productive sourdoughs were the dough with 100% Einkorn wheat flour and the dough obtained from the 1:1 combination of flour rye and corn flours.

Keywords: different flours; sourdough starter; spontaneous fermentation; technological factors; LAB; yeast

Citation: Dobre, A.A.; Cucu, E.M.; Belc, N. Influence of Technological Parameters on Sourdough Starter Obtained from Different Flours. *Appl. Sci.* **2024**, *14*, 4955. https://doi.org/10.3390/app14114955

Academic Editor: Marco Iammarino

Received: 19 April 2024
Revised: 4 June 2024
Accepted: 4 June 2024
Published: 6 June 2024

Copyright: © 2024 by the authors. Licensee MDPI, Basel, Switzerland. This article is an open access article distributed under the terms and conditions of the Creative Commons Attribution (CC BY) license (https://creativecommons.org/licenses/by/4.0/).

1. Introduction

The food industry is in continuous transition and is constantly evolving from its traditional status based on raw materials to adopting an innovative system and a market-oriented position. This transformation is accompanied by increasing demand for high value-added products that meet consumer demands—for taste, convenience, health, food safety and well-being. In many cases, innovations in biotechnology research succeed in meeting these requirements.

Bakery products provide over 50% of humanity's food sources, and therefore, worldwide, a growing number of specialists and institutions undertake studies and carry out extensive research in order to finalize in the industry the processes and technologies that ensure obtaining quality products and varied assortments, adapted to local specifics and consumer taste.

To improve the quality of bakery products in terms of texture, shelf life, and flavor, an important step is dough fermentation, largely attributed to the metabolic interaction of microorganisms [1]. Lactic bacteria represent, along with yeasts, the predominant microflora. Most of the microorganisms isolated from doughs are represented by the Lactobacillus genus, and among the yeast species, Candida and Saccharomyces are the most common [2]. In the study by Dan Xu [3], mixed starter cultures of yeast and lactobacilli were used to evaluate their ability to improve bread quality and enhance its flavor. The survival of fermentation microorganisms depends on the competitiveness against certain microbial strains present in the conditions of the dough ecosystem. Their survival and growth are also determined by the presence of suitable substrates. However, a thorough understanding of the behavior of microorganisms and the impact of certain substrates in the ingredients used can lead to obtaining new, stable doughs and implicitly to innovative bread recipes with the addition of sourdough [4]. It was shown that dough that was fermented with both lactic bacteria and yeast had an extended shelf life, and its sensory properties were also improved due to the presence of organic acids, amino acids, and a group of B vitamins produced by the lactic starter culture [5]. The results showed that the synergistic activity between lactic acid bacteria (*Lactobacillus bulgaricus*) and baker's yeast (*Saccharomyces cerevisiae*) improved the sensory properties of bread and also extended its shelf life.

Studies and specialized literature have highlighted the importance of biotechnological applications of lactic bacteria in grain-based products, creating new opportunities to improve the functional and nutritional quality and texture of flour products using sourdoughs. Bread obtained by the addition of sourdough obtained from a combination of lactobacilli and yeast had a more complex profile of volatile substances. In the study by Noriko K. [6], different effects of lactic acid bacteria and yeast on sourdough bread were examined, and the development of a new sourdough bread made with a yeast isolated from fruit and lactic acid bacteria, namely *Lactobacillus paracasei* NFRI 7415 isolated from a traditional Japanese fermented fish (funa-sushi). An increased content of organic acids and free amino acids was found in this dough, which favored the flavor of the final product. Also, in 2021, a study presented new hypotheses for the successful management of sourdough and proposed different directions for research and their application [7].

Although taste and convenience are important factors for consumers, most products that meet these consumer demands are line extensions or packaging innovations that involve limited biotechnological research. In any case, there is a continuous search for new products and processes that offer ingredients with new functionalities and cost-effective processing. Food fermentations offer these advantages, and there is continuous improvement in the microbial bioconversions that are at the heart of the production of functional metabolites. Microbial fermentation with the help of lactic acid bacteria strains and yeasts is of real interest to the bakery industry as a result of their significant antifungal activity and the ability to contribute to the extension of the shelf life. New dimensions have been introduced to improve fermentations by applying post-genomic approaches to almost all microorganisms involved in major food fermentations or used to produce different value-added compounds.

Lactic acid bacteria (LAB), with a tradition in industrial food fermentation, are used as starters for the fermentation of raw materials of plant and animal origin. Using lactic acid bacteria in food fermentation has an important role in the quality of food products. Fermented food products with lactic acid bacteria are less perishable, and their nutritional value may be enhanced, but one of the most important aspects is the safety of these products, which may be improved due to the inhibition of pathogenic bacteria and spoilage microorganisms by the low pH and the presence of organic acids and antimicrobial compounds. Bio-preservation is one of the many attributes of lactic acid bacteria under the scope of food safety.

The microbial ecosystem in sourdough obtained from different flours may present distinct lactic acid bacteria communities that can contribute to many variations of sourdough flavors and antimicrobial compounds. In this study, different flours (wheat, corn, and rye), variations, and technological parameters were used in order to obtain sourdough with different microbial ecosystems and characteristics. Therefore, the aim of this study is to evaluate the influence of different cereal grains (ancient wheat, corn, and rye), different dough variations, and temperature of fermentation on the quality of the spontaneous sourdough in order to obtain active food ingredients with a role in optimizing baking technology, improving nutritional quality and extending the shelf life of food products.

2. Materials and Methods

2.1. Raw Materials

The raw materials used in experiments were procured from Romanian manufacturers. Corn flour from SC Paradisul Verde SRL, Brasov, Romania, obtained from 100% corn ecologic crop produced in Romania, Einkorn wheat flour (*Triticum monococum*), whole wheat flour, rye flour from Biofarmland Manufactura SRL, Arad, Romania, 100% organic flours produced in Romania, packed in 5 kg paper bags.

The flour samples are freshly ground at the time of product purchase, using a mill that does not heat the flour. The purpose of this practice is to ensure the highest nutritional value of the obtained flours. The flours were used immediately after milling in order to obtain different variations of sourdough.

2.2. Sourdough Fermentation

Spontaneous dough was formed by fermenting a mixture of flour and water without the addition of an external starter culture. The four types of flour, together with plain Borsec water, were mixed to prepare seven different doughs. Mixing was carried out using a dough blender in a ratio of 1:1 to form a dough. The dough was kneaded for 4 min for homogenization and left to ferment at 25 °C and 35 °C, respectively, for 24 h. Every 24 h, the refreshing procedure (backslopping) was carried out by adding 100 g of fermented dough, 100 g of white wheat flour 650, and 100 mL of plain Borsec water (ratio 1:1:1). Four refreshment steps were carried out to obtain mature doughs according to the traditional type I acid dough scheme.

2.3. Assessment of the Nutritional Profile of Dry Raw Materials

2.3.1. The Moisture Content

The moisture content was achieved via a gravimetric method that involved drying the samples at 130 °C for 90 min until a constant mass was reached, the method described by ISO 712:2010 [8]. Approximately 5 g of the sample was weighed into a vial, which was dried to a constant mass in an oven (MRC DK-500WT, MRC Ltd., Holon, Israel). The analysis was performed in duplicate. The moisture content (M, %) was determined using the following formula:

$$M (\%) = (M_0 - M_1)/M_0 \times 100 \quad (1)$$

where M_1—mass of vial with cap (g); M_0—sample mass.

2.3.2. Acidity

The method consisted of titration of the aqueous flour extract with a 0.1 N sodium hydroxide solution in the presence of phenolphthalein as an indicator. Two parallel determinations were performed, the result being the arithmetic mean of the two determinations, if the difference between them does not exceed 0.2 degrees of acidity per 100 g sample. The analysis was performed in duplicate.

Acidity (degrees) is calculated using the following formula:

$$\text{Acidity} = (V \times 0.1)/m \times 100 \quad (2)$$

where
V—volume of 0.1N NaOH solution (mL);
m—mass of the working sample (g);
0.1—the normality of the NaOH solution.

2.3.3. Protein Content

Total protein content was determined according to ISO 20483:2013 [9]. The protein content was analyzed by the Kjeldahl method with a FOSS Kjeltec 2300 analyzer (FOSS Group, Hillerød, Denmark) after acid hydrolysis in an auto-digester (Behrotest InKjel, 450 P, Behr—Labor Technik GmbH, Dusseldorf, Germany). According to the classical

Kjeldahl method, samples were digested using concentrated sulfuric acid. The ammonium sulfate salt and alkali generated ammonia, which was trapped in boric acid via steam distillation. Titration was performed using a hydrochloric acid 0.2 N solution. The analysis was performed in duplicate.

2.3.4. Determination of Ash

The sample to be analyzed (2–3 g) is precalcined at a temperature of 750 °C by adding ethanol, and the calcination is continued for another 4 h. After the calcination is completed, the crucible with the sample is removed from the oven, placed in the desiccator until it reaches room temperature and weighed quickly with an accuracy of 0.1 mg.

The ash content is calculated as follows:

$$w_{a,d} = (m_2 - m_1) \times 100/m_0 \times 100/(100 - W_m) \tag{3}$$

where

m_0—mass of the working sample (g);
m_1—mass of the calcination crucible (g);
m_2—mass of calcination crucible and calcined residue (g);
W_m—sample moisture in mass percent.

As a result, the arithmetic mean between two determinations is taken. The results are expressed with an accuracy of 0.01%. The analysis was performed in duplicate.

2.3.5. Fat Content Determination

The determination is carried out with the Soxtec equipment from FOSS, in the presence of petroleum ether. The working steps are boiling step (40 min), rinsing (150 min), solvent recovery (15 min), solvent evaporation (10 min at 105 °C), and weighing of the glass vials.

The fat content is calculated according to the formula

$$M\,(\%) = (M_2 - M_1)/M_0 \times 100 \tag{4}$$

where

M_1—mass of the empty glass bottle (g);
M_2—mass of glass bottle with fat (g);
M_0—mass of the working sample (g).

The results are expressed with an accuracy of 0.01%. The analysis was performed in duplicate.

2.3.6. Determination of Fiber Content

The determination is carried out with the Fibertech equipment by treating the sample, placed in the FiberBag, with boiling sulfuric acid (acid digestion), rinsing with water, which is followed by alkaline digestion with sodium hydroxide and again rinsing with water. The FiberBag is wrapped and placed in the crucible brought to constant mass (in the oven at 750 °C for 1 h). The crucible with the FiberBag inside is dried in an oven at 105° ± 1 °C for a minimum of 4 h and then left to cool in a desiccator for 30 min and weighed. The sample is then calcined at 600 °C for 4 h, after which it is weighed.

The crude fiber content is calculated as follows:

$$w_f = (m_3 - m_1 - m_4 - m_5)/m_2 \times 100 \tag{5}$$

$$m_5 = m_7 - m_6$$

where

m_1—FiberBag table, g;
m_2—initial mass of the sample, g;
m_3—mass of the calcination crucible with the dry FiberBag, g;
m_4—the mass of the calcination crucible and the residue obtained after calcination, g;

m_5—mass of the empty FiberBag blank, g;
m_6—mass of the calcination crucible, g;
m_7—the mass of the crucible and the ash content of the empty FiberBag, g.

The result is rounded to the nearest 1 g/kg and expressed as a percentage. The analysis was performed in duplicate.

2.4. Assessment of the Physico-Chemical Properties and Microbiological Status for Spontaneous Sourdough Obtained from Different Flours

The maturity of the starter dough is evaluated by the stability of the pH, the acidity value, and its microbiota. Measurements of pH and total titratable acidity (TTA) were performed according to the Romanian standard methods 90/2007 [10]. The TTA value is defined as the amount of 0.1 N NaOH solution (mL) used to neutralize the 10 g sample weight.

Cell counts of LAB and yeast were performed using national and international food microbiology standards by using the viable cell count method on MRS agar and Potato Dextrose agar in duplicates using appropriate dilution of dough. The plates were anaerobically incubated at 30 °C for 72 h (LAB) and at 25 °C for 5 days (yeast). Incubation steps were performed using Panasonic and Memmert incubators. Results were expressed as CFU/g (colony forming units per gram dough). All the analyses were performed in duplicate.

2.4.1. Determination of Lactic Acid Bacteria (LAB) (According to ISO 15214/2001 [11])

Cell counts (expressed as CFU per gram of dough) were determined by mixing 10 g of sourdough with 90 mL of peptone physiological solution (made in a laboratory from ingredients—bacteriological peptone and NaCl (Oxoid, Ltd., Basingstoke, UK). Appropriate dilutions were made, and 1 mL inoculum was plated on MRS agar (Oxoid, Ltd., Basingstoke, UK), followed by incubation at 30 °C for 72 h. The colonies were numbered, and the interpretation of results was performed using the following Formula (6):

$$N = (\sum C)/(n1 + 0.1\, n2) \times d \tag{6}$$

where N = number of CFU from two serial dilutions; $\sum C$ = sum of colonies counted in all retained plates; n1 = number of plates retained at first dilution; n2 = number of plates retained at the second dilution; d = first retained.

2.4.2. Determination of Yeast Count (According to ISO 21527-1:2009 [12])

The method implies dispersing 0.1 mL of sample inoculum onto the surface of DG-18 Agar (Dicloran Glycerol Agar, Oxoid, Ltd., Basingstoke, UK) using a Drigalski spatula followed by incubation at 25 °C for 5 days. After the incubation period, the colonies were counted and analyzed according to Formula (4).

3. Results

3.1. Assessment of the Nutritional Profile of Dry Raw Materials

The fermentation of spontaneous sourdoughs depends on the endogenous parameters, mainly represented by the chemical and microbiological composition/quality (microbiota) of the flour. In order to characterize the flour samples, the value of the parameters related to 100% product will be taken into account. Results for all tested samples are presented in Table 1.

The acidity of the flour expresses the degree of freshness, the wheat flour samples being freshly ground, and the flour being more acidic due in particular to acid phosphates. The acidity of normal wheat flours depends on the degree of extraction. The higher it is, the higher the acidity. The white flours of low extraction, which come from the endosperm, contain little mineral salts and fatty substances and therefore have low acidity (2–2.8 degrees).

Table 1. Quality and nutritional profile of raw materials (flours). (Results are presented as mean of two determinations ± SD).

Samples	Moisture	Acidity	Fat	Protein	Ash	Fiber
	%	Grade	%	%	%	%
Einkorn wheat flour	10.04 ± 0.02	8.33 ± 0.29	2.59 ± 0.03	19.18 ± 0.18	2.45 ± 0.00	2.93 ± 0.00
Whole wheat flour	10.87 ± 0.01	3.65 ± 0.14	1.59 ± 0.02	9.40 ± 0.00	1.73 ± 0.02	3.66 ± 0.00
Corn flour	13.70 ± 0.02	6.25 ± 0.29	1.98 ± 0.01	6.11 ± 0.05	0.68 ± 0.01	1.70 ± 0.02
Rye flour	10.05 ± 0.00	6.46 ± 0.29	1.28 ± 0.01	7.91 ± 0.12	1.74 ± 0.01	3.08 ± 0.01

Some of the most important nutritional profiles of flours are their protein and fiber content. Wheat and rye are rich sources of fiber. Cereal fibers are rich in non-cellulosic polysaccharides, which have an extremely high capacity to bind water and quickly provide a feeling of satiety.

Einkorn wheat flour presented the highest protein value (19.18% ± 0.18) while also having a fairly high fiber content. Whole wheat flour has the highest fiber concentration, 3.66% ± 0.00, followed by rye flour and Einkorn wheat flour. Rye flour gliadin and gluten do not differ significantly in terms of structure and molecular mass compared to wheat proteins, but they have different colloidal properties (they do not form gluten, they do not form a continuous protein network in the dough, a structure which in the case of wheat flour obtain even for poor quality flour).

Regarding the content of mineral substances (included in the ash content), the two types of wheat flour but also rye flour presented the highest values, representing a rich growth substrate for lactic acid bacteria.

3.2. Assessment of the Physico-Chemical Properties and Microbiological Status for Spontaneous Sourdough Obtained from Different Flours

3.2.1. pH and Total Titratable Acidity

All sourdough variants showed, during the fermentation period at 25 °C and 35 °C, respectively, a decrease in pH values and an increase in acidity as a result of the production of lactic and acetic acid by the lactic bacteria from the spontaneous microflora of the tested flours. In the case of both variants of the fermentation temperature, the maximum acidity values were identified at 48 h of fermentation. The sourdough made from 100% Einkorn wheat flour presented the highest TTA value (13.4 ± 0.05 degrees) at 48 h of fermentation at 25 °C, which decreased considerably until the last day of testing, reaching 10.7 ± 0.12-degree acidity. This course was also observed in the case of the other variants of sourdough, less so in the case of the sourdough made of 100% corn flour and the mixture of corn flour 50% + rye flour 50%, the acidity constantly increasing until the last day of fermentation, reaching 11.7 ± 0.01 degrees and 12.4 ± 0.34 degrees, respectively.

Figure 1 shows that the acidity variations of different variations of spontaneous sourdough were influenced by the fermentation temperature used. A dough should be warm enough, at the optimum temperature for yeast growth, usually 29–32 °C, to encourage massive spontaneous fermentation and flavor creation. The highest values of TTA were determined at 48 h of fermentation at 35 °C, with the 100% Einkorn wheat dough having the highest value (27.4 ± 0.0 degrees), followed by the 50% Einkorn + rye flour dough 50%, respectively 24.8 ± 0.29 degrees.

One of the major characteristics of sourdough fermentation is a decrease in pH proportional to the maturation of the LAB community that produces lactic and acetic acid, eventually reaching a pH of about 4.0. The pH of the dough changes depending on the stage of fermentation it is in; the values determined in the case of the two fermentation temperatures are presented in Table 2. Since changes in pH can induce stress on cultures of lactic acid bacteria, evaluation of pH conditions is necessary to understand and control the evolution of microorganisms in dough. In addition, pH can influence the degree of lactic acid fermentation in the dough. The optimum pH of the dough should be between 4.2 and 4.5 [2], and therefore, the fermentation time required at 25 to 35 °C will be somewhere between 6 and 24 h.

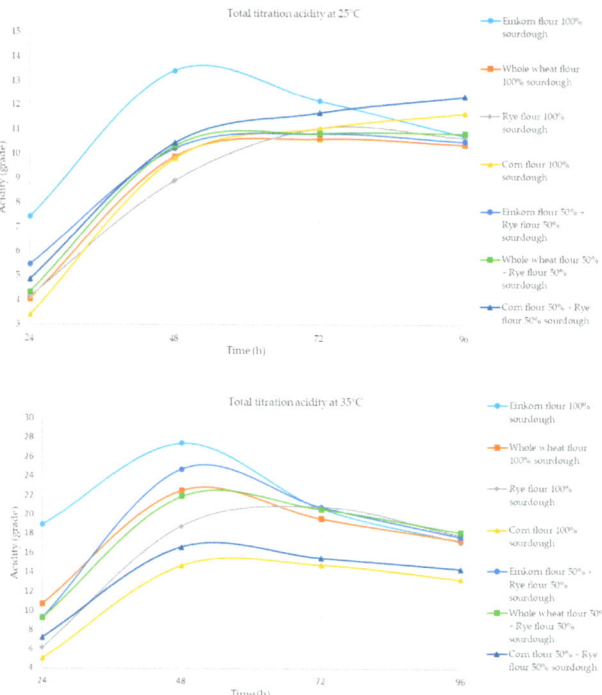

Figure 1. Evolution of total titration acidity values (TTA) for sourdough samples fermented at 25 °C and 35 °C (Results are presented as the mean of two determinations ± SD).

Table 2. Evolution of pH values during the fermentation period at temperatures of 25 °C and 35 °C (Results are presented as mean of two determinations ± SD).

Samples	24 h		48 h		72 h		96 h	
	25 °C	35 °C	25 °C	35 °C	25 °C	35 °C	25 °C	35 °C
Einkorn flour 100%	6.36 ± 0.11	4.98 ± 0.11	4.16 ± 0.21	3.52 ± 0.08	4.0 ± 0.00	3.46 ± 0.11	3.92 ± 0.10	3.47 ± 0.02
Whole wheat flour 100%	6.35 ± 0.14	5.05 ± 0.10	3.98 ± 0.10	3.47 ± 0.11	3.85 ± 0.21	3.41 ± 0.21	3.78 ± 0.10	3.43 ± 0.10
Rye flour 100%	6.48 ± 0.12	6.1 ± 0.15	4.15 ± 0.15	3.67 ± 0,21	3.86 ± 0.25	3.39 ± 0.05	3.78 ± 0.21	3.39 ± 0.16
Corn flour 100%	5.97 ± 0.10	5.14 ± 0.21	3.98 ± 0.10	3.59 ± 0.10	3.82 ± 0.08	3.55 ± 0.15	3.74 ± 0.12	3.56 ± 0.18
Einkorn flour 50% + Rye flour 50%	6.44 ± 0.12	5.89 ± 0.20	4.2 ± 0.13	3.54 ± 0.16	3.98 ± 0.21	3.43 ± 0.19	3.89 ± 0.09	3.44 ± 0.10
Whole wheat flour 50% + Rye flour 50%	6.38 ± 0.15	5.16 ± 0.18	4.1 ± 0.18	3.57 ± 0.18	4.0 ± 0.10	3.41 ± 0.20	3.93 ± 0.02	3.41 ± 0.21
Corn flour 50% + Rye flour 50%	6.21 ± 0.12	5.33 ± 0.10	3.93 ± 0.20	3.69 ± 0.21	3.84 ± 0.05	3.52 ± 0.31	3.72 ± 0.10	3.54 ± 0.11

Surveying the changes in the pH until it reached a value of about 4.3 allowed us to discover the optimal fermentation period for each of the two temperatures. It was observed that samples fermented at 35 °C reached a pH of 4.2 after 24 h, while samples maintained at 25 °C took 48 h to reach a pH of 4.1. It was also possible to find an increase in the volume of dough held at 35 °C that was higher than that occurring in dough fermented at 25 °C. This can be attributed primarily to a higher level of fermentation activity, producing more gas, which in turn decreases the density of the dough. From 48 h, the volume increase was stabilized.

3.2.2. Evolution of Lactic Acid Bacteria and Yeast Cultures of Spontaneous Fermentation Sourdough

The population of lactic acid bacteria for the sourdough fermented 24 h at 25°C had an average of 6.29 log UFC/g. These values increased to 8.97 log CFU/g after 72 h of fermentation, respectively, 8.91 log CFU/g at 96 h. These values increased to over 8 log CFU/g after 48 h of fermentation at 25 °C, with value reached in 24 h of fermentation at 35 °C for all doughs tested. The lactic acid bacteria population was much more productive at 24 h of fermentation at 35 °C for all types of sourdough, but the growth during the fermentation period was very low (Figure 2). The population of lactic acid bacteria present in the sourdough obtained with whole wheat flour showed an increase throughout the fermentation period up to 96 h for the dough fermented at 25°C, while for the dough kept at 35 °C, the maximum number was reached after 24 h, after which there was a slow decline until 96 h. The same pattern was also identified in the case of rye and corn flour, but also for the combinations of whole wheat flour 50% with rye flour 50% and corn flour 50% with rye flour 50%.

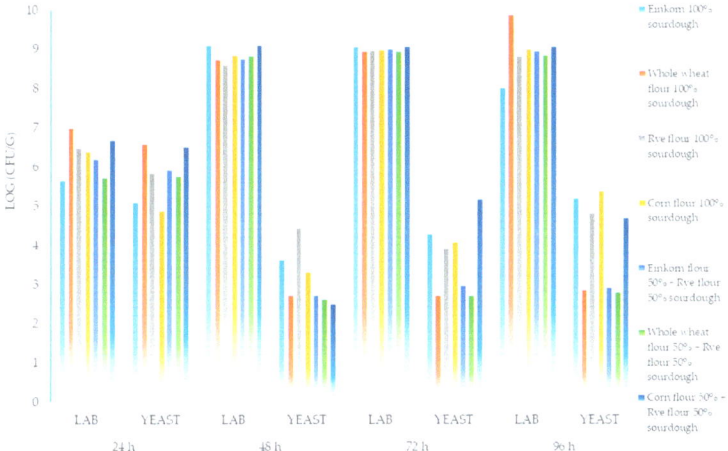

Figure 2. Evolution of lactic acid bacteria and yeast populations compared to pH values in the case of doughs fermented at 25 °C (Results are presented as mean of two determinations ± SD).

Lactobacilli constitute a well-adapted population to the physical and chemical conditions in the dough, competing with the remaining microbial life until they become the dominant microbiota [13]. This allows the dough to spread over long periods if water and flour are added at appropriate intervals. Therefore, it can be said that the LAB population increases as the pH decreases and finally both reach a plateau when the dough ripens.

The dynamics of the yeast population of the fermented doughs at the two established temperatures generally revealed a stable growth of approximately log 5.78 CFU/g (25 °C) and log 7.21 CFU/g (35 °C) at 24 h of fermentation. After 24 h of fermentation, the yeast population showed a constant decrease until the end of the fermentation period, less in the case of sourdoughs from 100% corn flour and 50% rye flour with 50% corn flour. At 48 h of fermentation, the yeast population decreased dramatically in all doughs tested and for both fermentation temperatures.

4. Discussion

The baking process is a complex process based on a large number of physical, chemical, colloidal, biochemical, and microbiological processes. Some of these cannot occur without others, some follow each other, and often, they condition each other so that they cannot be treated individually.

The complexity of the baking process is largely due to the development and fermentation of the dough, which is a viscoelastic semi-finished product with a certain consistency imposed by the technological process and which can hardly be acted upon from the outside. Consumer interest in the last 20 years has led to the re-invention of the technological process by increasing the production of sourdough, a fermented semi-finished product, and also the production of selected microbial strains for the preparation of sourdough [14].

The use of the sourdough-based technological process as the basis of fermentation and leavening is one of the oldest biotechnological processes in the production of cereal foods [15]. Each sourdough is a different natural ecosystem that can produce a different quality of bread, the microbial metabolism being specific to the strain, the intermediate metabolites formed, especially flavor compounds but also antibacterial compounds, conferring a sensory and hygienic quality to bakery products based on yeast in comparison to conventional bread. Type I sourdough can be maintained for years by continuous refreshment using the previous batch as inoculum.

The type of flour that is used during starter maintenance can affect the culture's ecology, technical efficiency as a leavening agent, and unique sensory qualities. Flour helps introduce microorganisms into the dough environment [16]. Also, flour provides different nutrients (carbohydrates and amino acids) and non-nutrients (phenolic acids, amylase, ash), whose presence and concentrations can influence the survival of bacteria and yeast species [17–20]. Phenolic compounds and ash are naturally present in flour in varying amounts and have been shown to affect the acidification rate of starter cultures, thereby influencing microbial succession. Similarly, the presence or absence of the bran portion of cereals may also influence microbial ecology in ways that are not yet clear [21].

The study followed the chemical and microbiological characteristics of different variants of spontaneous sourdoughs obtained by combinations of wheat flour, corn flour, and rye flour, fermented at different temperatures, 25 °C and 35 °C, over a similar period of time (96 h). Thus, seven variants of spontaneous fermentation sourdoughs were tested. The flours used to obtain them were purchased from local producers and were from organically certified crops. Sourdough production is relatively simple and does not need complicated installations and can obtain different products with different characteristics just by varying the ratio between raw materials. By changing the rations and recipe, different sourdough can be obtained to promote diversity and different microbial ecosystems.

At the beginning of fermentation, the dough fermented at 25 °C had higher pH values than those observed in the batches of dough fermented at 35 °C and a slower increase in acidity. In parallel, the number of LAB was much higher in the dough fermented at 35 °C at the beginning of fermentation, but after 24 h, the batches of dough fermented at 25 °C had a higher number, which remained more or less constant until the end of fermentation. In the case of sourdough samples fermented at 25 °C, the pH values started from an average of 6.31 and decreased to an average of 3.82, and in the case of sourdough fermented at 35 °C, the pH value varied from an average of 5.37 to 3.46. The optimal pH values for the dough samples fermented at the two experimental temperatures determined that the optimal fermentation period was 48 h. The acidification rate of the dough was faster at 35 °C than at 25 °C. This fact became evident via the pH values obtained in the first 24 h. However, from this point, the pH values were lower in the samples kept at 25 °C, showing that a cooler fermentation temperature allows the acidification activity of the microorganisms to be prolonged for a longer time.

An optimal frequency of backslopping promotes great diversity and fermentative activity. Intermediate fermentation times achieve a balance between pH and acid-tolerant microorganisms. The longer the fermentation time, the more acid-tolerant species of lactic acid bacteria (LAB) and yeasts will predominate. The ideal fermentation time allows both lactic acid bacteria and yeast to reach an optimal growth rate and cell density. In the study carried out, the ideal fermentation time for the population of LAB and yeasts is 72 h at a temperature of 25 °C, and the most productive sourdoughs were the dough with 100% Einkorn wheat flour and the dough obtained from the 1:1 combination of rye and corn flours. A stable number of LAB was obtained in all sourdough variations during the tested period, but the yeast population presented a decrease after 24 h backslopping, possibly because of the competitiveness of the LAB compared to the yeast. LAB numbers increased in all sourdough samples at the end of fermentation for 96 h. The results obtained based on the LAB population are interesting due to the effect of different flours used. For example, the flour of Einkorn wheat and corn in combination with rye flour showed important stimulating proprieties in the development of the LAB population during the tested period (96 h), while a different pattern was observed in the case of rye and corn flour was the maximum number was reached after 72 h, after which there was a slow decline until 96 h.

Sourdoughs fermented at 25 °C showed an increase in volume characteristic of a good fermentation, all samples showed some level of bubbling, and all smelled distinct from each other. The characteristic smell was that of acid and earth and was pungent. Only the sample of 100% corn flour in combination with rye flour 50/50 had a pleasant, sweet smell. These two sourdoughs and the technological parameters (fermentation temperature of 25 °C for a duration of 72 h) presented the best results and were selected for future experiments that imply phenotypic identification of isolated strains of lactic acid bacteria from selected sourdoughs and highlighting of strains with biotechnological potential for the bakery industry.

5. Conclusions

Using sourdough in bread making brings value to the final products, and one of the main advantages is the rich microbiota that grows spontaneously during fermentation time and daily backslopping. The microbial populations of a sourdough during the initial propagation of a starter and during continued maintenance are unique based on the flour types used and technological factors. Based on the results obtained in this study, future research implies phenotypic identification of isolated strains of lactic acid bacteria from selected sourdoughs and highlighting of strains with biotechnological potential for the bakery industry, but also to connect microbial ecology to the sensory and technical qualities of sourdough products.

In conclusion, this study showed that the use of different flours, rations, recipes, and fermentation temperatures has an important influence on microbial population, thus obtaining an active food ingredient with high potential to improve the technological and nutritional properties of bakery products.

Author Contributions: Conceptualization, A.A.D. and E.M.C.; methodology, A.A.D. and N.B.; formal analysis, A.A.D.; investigation, A.A.D. and E.M.C.; data curation, A.A.D.; writing—original draft preparation, A.A.D. and E.M.C.; writing—review and editing, A.A.D. and N.B. All authors have read and agreed to the published version of the manuscript.

Funding: This research was achieved through the Core Program, with the support of the Ministry of Research, Innovation and Digitization (MCID), contract 39N/2023, project PN 23 01 02 02.

Institutional Review Board Statement: Not applicable.

Informed Consent Statement: Not applicable.

Data Availability Statement: The original contributions presented in the study are included in the article, further inquiries can be directed to the corresponding author.

Conflicts of Interest: The authors declare no conflicts of interest. The funders had no role in the design of the study, in the collection, analyses, or interpretation of data, in the writing of the manuscript, or in the decision to publish the results.

References

1. Gobetti, M.; De Angelis, M.; Di Cagno, R.; Calasso, M.; Archetti, G.; Rizzello, C.G. Novel insights on the functional/nutritional features of the sourdough fermentation. *Int. J. Food Microbiol.* **2019**, *302*, 103–113. [CrossRef] [PubMed]
2. Xu, D.; Zhang, H.; Xi, J.; Jin, Y.; Chen, Y.; Guo, L.; Jin, Z.; Xu, X. Improving bread aroma using low-temperature sourdough fermentation. *Food Biosci.* **2020**, *37*, 100704. [CrossRef]

3. De Vuyst, L.; Neysens, P. The sourdough microflora: Biodiversity and metabolic interactions. *Trends Food Sci. Technol.* **2005**, *16*, 43–56. [CrossRef]
4. De Vuyst, L.; Comasio, A.; Van Kerrebroeck, S. Sourdough production: Fermentation strategies, microbial ecology, and use of non-flour ingredients. *Crit. Rev. Food Sci. Nutr.* **2021**, *63*, 2447–2479. [CrossRef] [PubMed]
5. Edeghor, U.; Lennox, J.; Etta-Abgo, B.; Aminadokiari, D. Bread fermentation using synergistic activity between lactic acid bacteria (*Lactobacillus bulgaricus*) and baker's yeast (*Sacchromyces cerevisae*). *Pak. J. Food Sci.* **2016**, *26*, 46–53. Available online: https://psfst.org/paper_files/6019_105651_265.pdf (accessed on 15 August 2023).
6. Komatsuzaki, N.; Izawa, M.; Suzumori, M.; Fujihara, S.; Shima, J. Characteristics of New Sourdough using Lactic Acid Bacteria and Wild Yeast. *J. Food Sci. Nutr. Res.* **2019**, *2*, 1–12. [CrossRef]
7. Calvert, M.D.; Madden, A.A.; Nichols, L.M.; Haddad, N.M.; Lahne, J.; Dunn, R.R.; McKenney, E.A. A review of sourdough starters: Ecology, practices, and sensory quality with applications for baking and recommendations for future research. *PeerJ* **2021**, *9*, e11389. [CrossRef] [PubMed]
8. *ISO 712:2010 (En)*; Cereals and Cereal Products—Determination of Moisture Content. International Organisation for Standardization (ISO): Geneva, Switzerland, 2010. Available online: https://www.iso.org/obp/ui/#iso:std:iso:711:ed-2:v1:en (accessed on 10 July 2023).
9. *ISO 20483:2013 (En)*; Cereals and Pulses—Determination of the Nitrogen Content and Calculation of the Crude Protein Content—Kjeldahl Method. International Organisation for Standardization (ISO): Geneva, Switzerland, 2013. Available online: https://www.iso.org (accessed on 10 July 2023).
10. *SR 90:2007*; Wheat Flour. Analysis Methods. Standardization Association of Romania (ASRO): Bucharest, Romania, 2007. Available online: https://www.asro.ro (accessed on 10 July 2023).
11. *ISO 15214:1998 (En)*; Microbiology of Food Animal Feeding Stuffs Horizontal Method for the Enumeration of Mesophilic Lactic Acid Bacteria Colony-Count Technique at 30 Degrees, C. International Organisation for Standardization (ISO): Geneva, Switzerland, 1998. Available online: https://www.iso.org/standard/26853.html (accessed on 10 July 2023).
12. *ISO 21527-1:2009 (En)*; Microbiology of Food and Animal Feeding Stuffs—Horizontal Method for the Enumeration of Yeasts and Moulds—Part 1: Colony Count Technique in Products with Water Activity Greater Than 0.95. International Organisation for Standardization (ISO): Geneva, Switzerland, 2009. Available online: https://www.iso.org/standard/38275.html (accessed on 10 July 2023).
13. Messens, W.; De, V.L. Inhibitory substances produced by Lactobacilli isolated from sourdoughs—A review. *Int. J. Food Microbiol.* **2002**, *72*, 31–43. [CrossRef] [PubMed]
14. Catzeddu, P. Chapter 14—Sourdough Breads. In *Flour and Breads and Their Fortification in Health and Disease Prevention (Second Edition)*; Preedy, V.R., Watson, R.R., Eds.; Academic Press: Cambridge, MA, USA, 2019; pp. 177–188, ISBN 9780128146392. [CrossRef]
15. Chavan, R.S.; Chavan, S.R. Sourdough Technology—A Traditional Way for Wholesome Foods: A Review. *Compr. Rev. Food Sci. Food Saf.* **2011**, *10*, 169–182. [CrossRef]
16. Reese, A.T.; Madden, A.A.; Marie, J.; Guylaine, L.; Dunn, R.R. Influences of ingredients and bakers on the bacteria and fungi in sourdough starters and bread. *mSphere* **2020**, *5*, e00950-19. [CrossRef] [PubMed]
17. Vogelmann, S.A.; Hertel, C. Impact of ecological factors on the stability of microbial associations in sourdough fermentation. *Food Microbiol.* **2011**, *28*, 583–589. [CrossRef] [PubMed]
18. Minervini, F.; De Angelis, M.; Di Cagno, R.; Gobbetti, M. Ecological parameters influencing microbial diversity and stability of traditional sourdough. *Int. J. Food Microbiol.* **2014**, *171*, 136–146. [CrossRef] [PubMed]
19. Minervini, F.; Dinardo, F.R.; Celano, G.; De Angelis, M.; Gobbetti, M. Lactic acid bacterium population dynamics in artisan sourdoughs over one year of daily propagations is mainly driven by flour microbiota and nutrients. *Front. Microbiol.* **2018**, *9*, 1984. [CrossRef] [PubMed]
20. Gänzle, M.G. Enzymatic and bacterial conversions during sourdough fermentation. *Food Microbiol.* **2014**, *37*, 2–10. [CrossRef] [PubMed]
21. Corsetti, A.; Settanni, L.; Chaves López, C.; Felis, G.E.; Mastrangelo, M.; Suzzi, G. A taxonomic survey of lactic acid bacteria isolated from wheat (*Triticum durum*) kernels and non-conventional flours. *Syst. Appl. Microbiol.* **2007**, *30*, 561–571. [CrossRef] [PubMed]

Disclaimer/Publisher's Note: The statements, opinions and data contained in all publications are solely those of the individual author(s) and contributor(s) and not of MDPI and/or the editor(s). MDPI and/or the editor(s) disclaim responsibility for any injury to people or property resulting from any ideas, methods, instructions or products referred to in the content.

Article

Influence of Soy Protein Hydrolysates on Thermo-Mechanical Properties of Gluten-Free Flour and Muffin Quality

Mihaela Brumă (Călin), Iuliana Banu, Ina Vasilean, Leontina Grigore-Gurgu, Loredana Dumitrașcu and Iuliana Aprodu *

Faculty of Food Science and Engineering, Dunarea de Jos University of Galati, 111 Domneasca St., 800201 Galati, Romania; mihaela.calin@ugal.ro (M.B.); iuliana.banu@ugal.ro (I.B.); ina.vasilean@ugal.ro (I.V.); leontina.gurgu@ugal.ro (L.G.-G.); loredana.dumitrascu@ugal.ro (L.D.)
* Correspondence: iuliana.aprodu@ugal.ro

Abstract: The influence of protease-assisted hydrolysis on the impact exerted by the soy protein isolate on the thermo-mechanical behavior and baking performance of the gluten-free composite flour, consisting of a mixture of rice and quinoa flours, was investigated. The mPAGE analysis revealed that soluble fractions of the hydrolysates, obtained with bromelain, Neutrase or trypsin, concentrated the peptides with a molecular weight lower than 20 kDa, whereas the insoluble ones retained higher molecular weight fragments. The influence of the separate and cumulative addition of the soluble and insoluble soy peptide fractions on the thermo-mechanical properties of dough was tested by means of a Mixolab device. Regardless of the enzyme used for hydrolysis, the addition of the soluble peptide fraction to the gluten-free composite flour resulted in delayed starch gelatinization, whereas the insoluble one caused a considerable increase in the dough consistency. The most important improvements in the dough behavior were observed when supplementing the gluten-free flour with 10% soy protein hydrolysates obtained with bromelain and trypsin. The gluten-free muffins enriched in soy protein hydrolysate exhibited important differences in terms of moisture, height and specific volume, compared to the control. Moreover, the ABTS- and DPPH-based methods indicated that protein hydrolysate addition caused a significant improvement in the antioxidant activity (by at least 38% and 23%, respectively) compared to the control. In conclusion, soy protein hydrolysate might be successfully used for increasing both the protein content and the antioxidant activity of the muffin samples.

Keywords: gluten-free composite flour; protein hydrolysis; rheological properties; muffins

Citation: Brumă, M.; Banu, I.; Vasilean, I.; Grigore-Gurgu, L.; Dumitrașcu, L.; Aprodu, I. Influence of Soy Protein Hydrolysates on Thermo-Mechanical Properties of Gluten-Free Flour and Muffin Quality. *Appl. Sci.* **2024**, *14*, 3640. https://doi.org/10.3390/app14093640

Academic Editor: Wojciech Kolanowski

Received: 16 April 2024
Revised: 23 April 2024
Accepted: 24 April 2024
Published: 25 April 2024

Copyright: © 2024 by the authors. Licensee MDPI, Basel, Switzerland. This article is an open access article distributed under the terms and conditions of the Creative Commons Attribution (CC BY) license (https://creativecommons.org/licenses/by/4.0/).

1. Introduction

The popularity of the gluten-free diet increased significantly in the last years, gluten-free products being popular not only among people who are intolerant or allergic to gluten but also among those who chose to have a healthy lifestyle [1]. The gluten-free diet has an important role in alleviating undesired or life-threatening symptoms in celiac or gluten-allergic patients, but careful attention should be paid to the long-term consequences caused by the nutritional limitations of the gluten-free diet. Considering the rather unique functionality of gluten in the breadmaking process and the quality of the final products, identifying a suitable gluten-free formulation for mimicking regular food products' quality, such as to meet the market demands, is challenging for food specialists. A widely used strategy for developing gluten-free products involves the use, as main ingredients, of gluten-free flours from cereals or pseudocereals and eventually of starches, proteins and hydrocolloids [1]. Do Nascimento et al. [2] pointed out that most commercially available gluten-free products are based on cheap ingredients, like rice and or corn flours, eventually combined with starch of different origins such as cassava and potato. Gluten-free products have higher amounts of fats, for improving the mouthfeel, carbohydrates and sodium, and are generally poor in high-quality proteins or fibers compared to regular products [1,3].

Because of the high digestibility, hypoallergenic properties, mild taste and white color [4], rice flour was used in the present study for developing gluten-free muffins. In order to balance the nutritional profile of the final product, quinoa flour was chosen to be used in ad-mixture with the rice flour. The quinoa contents of protein (14.12%), lipid (6.07%), ash (2.7%) and fiber (7.0%) are significantly higher compared to rice (6.81, 0.55, 0.19 and 2.8%, respectively) and most of the grains [5]. In particular, quinoa is acknowledged as a good source of high-quality proteins which are rich in amino acids deficient in cereals, such as lysine, methionine, histidine and threonine [6], of unsaturated fatty acids, like alpha-linoleic, oleic and linolenic acids which represent 87–88% of the total fatty acids [7] and dietary fibers [5]. Moreover, quinoa is rich in vitamins, mainly pyridoxine, folic acid and vitamin E, and minerals like calcium, iron, magnesium and potassium, in amounts considered sufficient for a balanced diet [5,6].

In order to compensate the low protein content in gluten-free mixtures, which do not usually meet the required daily dietary amounts and affect the quality of the backed products [3], different studies focused on identifying suitable sources of proteins for fortifying the gluten-free product. Among protein sources, legumes attained the interest of researchers, and soy is of particular importance because of the good availability, low cost and multiple benefits associated with the high biological value and good functional properties [3]. Moreover, different studies highlighted the possibility of enhancing the technological functionality of these proteins by using exogenous enzymes, in addition to improving the physiological properties because of the release of the encrypted peptides with antioxidant activity, antihypertensive and anticancer properties, hypocholesterolemic effects, etc. [8–10]. Unlike other physiologically active compounds, the use of peptides for providing potential health benefits to the host, through diet, is highly desired, because of the low costs, good absorption, avoidance of safety issues and contribution to nutritional value [10,11].

The aim of the study was to investigate the impact of soy protein hydrolysates' addition on the rheological properties of the gluten-free dough and on the properties of the muffins. The soy protein hydrolysates prepared using three different exogenous enzymes, namely bromelain, Neutrase and trypsin, were freeze-dried and were further added to the gluten-free composite flour consisting of whole rice (RF) and quinoa flour (QF). In addition to the influence of the total protein hydrolysates, the contribution of the soluble and insoluble fractions of soy protein hydrolysate to the rheological behavior of the gluten-free dough was investigated.

2. Materials and Methods

2.1. Materials

The commercial whole rice flour (RF; origin Greece) (moisture 11.2%, proteins 7.1%, fats 2.8%, fibers 4.6%) and quinoa flour (QF; origin Peru) (moisture 9.2%, proteins 14.0%, fats 6.1%, fibers 7.0%) distributed by Solaris Plant S.R.L. (Bucharest, Romania) and the soy protein isolate (SPI; Supro® XT 221D IP, distributed by KUK Romania, Voluntari, Romania) (moisture 4.8%, proteins 87.1%) were used in the study. Enzymes of various origins were selected for preparing the protein hydrolysates: bromelain of vegetal origin (Carl Roth, Karlsruhe, Germany), Neutrase 5.0 BG of microbial origin (Novo Nordisk, Bagsværd, Denmark) and trypsin of animal origin (Merck, Darmstadt, Germany).

1,1-diphenyl-2-picrylhydrazyl (DPPH), 2,2-azino-bis(3-ethylbenzothiazoline-6-sulfonic acid) diammonium salt (ABTS), 6-hydroxy-2,5,7,8-tetramethylchroman-2-carboxylic acid (Trolox) and sodium dodecyl sulphate (SDS) were purchased from Sigma-Aldrich Chemie GmbH (Taufkirchen, Germany).

2.2. Composition Analysis

The analysis of the gluten-free flours' composition was conducted using the following methods: SR ISO 712:2005 for moisture content [12], the semimicro-Kjeldahl method (Raypa Trade, R Espinar, SL, Barcelona, Spain) for protein content (nitrogen-to-protein conversion factor of 5.95 for rice flours and 6.25 for quinoa flour) and the Soxhlet method (SER-148; VELP Scientifica, Usmate Velate, Italy) for fat content.

2.3. The Hydrolysis of the Soy Protein Isolate

The hydrolysis of the soy protein suspensions (12% w/v) with bromelain (0.5 g/100 g proteins), Neutrase (1 g/100 g proteins) and trypsin (1 g/100 g proteins) was carried out for 68 h at 50 °C while continuously shaking (SI300R; Jeio Tech, Chalgrove, UK) at 100 rpm, as indicated in [10]. After enzymes' inactivation through heating for 5 min at 90 °C, half of each protein hydrolysate was directly subjected to freeze drying (CHRIST Alpha 1–4 LD plus, Osterode am Harz, Germany), and the resulting hydrolyzed SPIs (hSPIs) were coded hSPI-B, hSPI-N and hSPI-T; the other half was first centrifuged at 14,000 rpm for 10 min followed by freeze drying the soluble fractions concentrated in the supernatants (coded sSPI-B, sSPI-N and sSPI-T) and the insoluble ones separated in the pellets (coded rSPI-B, rSPI-N and rSPI-T).

2.4. mPAGE Analysis

The soy protein isolate and hydrolysates (hSPI, sSPI and hSPI) were diluted with distilled water, homogenized with mPAGE 4× LDS Sample Buffer (Merck, KGaA, Darmstadt, Germany), treated at 95 °C for 5 min and run in the denaturing mPAGE® Lux Casting Bis-Tris polyacrylamide gels (Merck, KGaA, Darmstadt, Germany). The concentrations of the separation and stacking gels were 12% and 5%, respectively. A volume of 10 µL (concentration of 4 µg/µL) of each sample was loaded on the stacking gel and run at 90 V for 80 minutes in the running buffer (Bio-Rad, Hercules, CA, USA). The protein bands were fixed with 50% methanol/10% acid acetic (v/v), stained in 0.1% w/v Coomassie Brilliant Blue R-250 (Sigma-Aldrich, St. Louis, MI, USA) and then de-stained in methanol/acetic acid/water (40:20:140). The Precision Plus Proteins Dual Xtra Prestained marker (Bio-Rad, Hercules, CA, USA) was migrated in parallel with the samples.

2.5. The Preparation of the Gluten-Free Flour Mixtures

The gluten-free composite flour consisting of equal parts of RF and QF was used as the basis for preparing dough. In order to increase the total protein content of the gluten-free composite flour, the SPI hydrolysates were further used to replace 10% of RF in the mixture. A total of nine protein-enriched gluten-free flour mixtures were prepared by incorporating the whole SPI hydrolysates (hSPI-B, hSPI-N and hSPI-T), the soluble fractions (sSPI-B, sSPI-N and sSPI-T) or the insoluble ones (rSPI-B, rSPI-N and rSPI-T) and were further used for testing the thermo-mechanical behavior of the dough. The composite flours supplemented with hSPI-B, hSPI-N and hSPI-T, which exhibited the most promising thermo-mechanical behavior, were further used for preparing muffins. The supplementation level of the gluten-free composite flour with soy proteins or peptides was established upon running preliminary tests, so as to ensure that about half of the total protein content of the final mixtures originate from soy. The composite flour consisting of equal parts of RF and QF was used as the control.

2.6. The Thermo-Mechanical Behavior of the Gluten-Free Flour Mixtures

The Mixolab device (Chopin Technology, Villeneuve La Garenne, France) was used to monitor the rheological properties of the gluten-free dough. The Mixolab curves recorded using the Chopin+ protocol include five distinct phases (Ph) describing the behavior of the dough subjected to double kneading and a temperature constraint. In the first phase (Ph I), ranging from 0 to 480 s, the dough is kneaded at a constant temperature of 30 °C; in the next 900 s, defining the second phase (Ph II), the dough is heated from 30 to 90 °C; in the third phase (Ph III), which lasts 420 s, the dough is kept at a constant temperature of 90 °C; the dough is afterwards cooled down from 90 to 50 °C over 600 s in the fourth phase (Ph IV); and finally, in the fifth phase (Ph V), is kept for 300 s at 50 °C. A specific parameter is marked in each phase on the Mixolab curve, as follows: C1 is the maximum consistency of the dough registered during Ph I; C2 is the minimum dough consistency associated with the weakening of the proteins, measured during Ph II; C3 is the maximum consistency which measures gelatinization that occurred during Ph III; C4 is the minimum dough consistency indicating the stability over heating in Ph IV; and C5 is the maximum consistency providing an indication on starch retrogradation during Ph V. The dough consistency values registered at the end of Ph I (after 8 min), Ph II (after 23 min), Ph III (after 30 min), Ph IV (after 40 min) and Ph V (after 45 min) were termed C-I, C-II, C-III, C-IV and C-V, respectively. All experiments were conducted using a modified protocol which uses a dough weight of 90 g, and a water absorption (WA) of 65% was selected for preparing all samples. The measurements were performed at least in duplicate.

2.7. Muffin Preparation

The gluten-free mixtures supplemented with hSPI, which presented the best thermo-mechanical behavior, as indicated by the Mixolab measurements, were further used to prepare muffins. In agreement with Singh et al. [13] with slight modification, the batter formulations used to prepare the muffin samples were obtained by mixing the following ingredients: gluten-free mixtures (100 g), refined white sugar (65 g), sunflower oil (40 g), baking powder (5 g), salt (1g) and xanthan (0.5 g), as indicated in Table 1. The ingredients were well mixed using a Braun mixer (De'Longhi, Neu Isenburg, Hessen, Germany), as indicated by Banu et al. [14], and after pouring 48 g of each batter into the paper cups placed in the muffin baking trays, the backing was performed at 180 °C for 20 min using an electric oven (Electrolux, Stockholm, Sweden). The samples were stored in the paper cups at room temperature until further characterization.

Table 1. Formulation of batter used to obtain gluten-free muffins (M—control muffin; M-SPI—muffins enriched with soy protein isolate; M-hSPI-B—muffins enriched with soy protein hydrolysate produced with bromelain; M-hSPI-N—muffins enriched with soy protein hydrolysate produced with Neutrase; M-hSPI-T—muffins enriched with soy protein hydrolysate produced with trypsin).

Ingredients	Muffin Samples				
	M	M-SPI	M-hSPI-B	M-hSPI-N	M-hSPI-T
Rice flour, g	50	40	40	40	40
Quinoa flour, g	50	50	50	50	50
SPI, g	-	10	-	-	-
hSPI-B, g	-	-	10	-	-
hSPI-N, g	-	-	-	10	-
hSPI-T, g	-	-	-	-	10
Sugar, g	65	65	65	65	65
Oil, mL	40	40	40	40	40
Baking powder, g	5	5	5	5	5
NaCl, g	1	1	1	1	1
Xanthan gum, g	0.5	0.5	0.5	0.5	0.5
Water, g	85	85	85	85	85

The weight loss during baking (WLB) was calculated as follows:

$$WLB = \frac{m_B - m_F}{m_B} 100 \quad (1)$$

where m_B is the mass of the batter used to obtain the muffin, and m_F is the mass of the sample upon cooling to room temperature.

2.8. Muffin Characterization

The muffin samples were characterized by assessing the main physico-chemical characteristics within 24 h after samples cooling to room temperature.

The moisture content was determined using the method SR 91:2007 [12]. The muffin height was estimated by measuring, with a caliper, the distance between the flat bottom surface to the highest point of each sample. The specific volume (cm^3/g) of the muffins was determined by the rapeseed displacement method (SR 91:2007, [12]).

The texture of the muffin samples was assessed by measuring the firmness of the crumb with the MLFTA apparatus (Guss, Strand, South Africa) and a probe of 7.9 mm in diameter. For each tested formulation, two samples were penetrated for 25 mm in three different points, at a penetration speed of 5 mm/s and a trigger threshold force of 20 g [15].

The CIElab color parameters of the muffin crust in crumb, in terms of the brightness/darkness (L*), redness/greenness (a*) and yellowness/blueness (b*), were measured using the Chroma Meter CR-410 (Konica Minolta Business Solutions Europe GmbH) colorimeter. In agreement with Matos et al. [16], in order to evaluate the effect of soy proteins' or peptides' addition on the color appearance parameter of the muffin samples, the chroma (C*) and hue angle (H) were calculated using Equations (2) and (3):

$$C^* = \sqrt{a^{*2} + b^{*2}} \qquad (2)$$

$$H = \tan^{-1}(b^*/a^*) \text{ for quadrant I } (+a^*, +b^*) \qquad (3)$$

The antioxidant activity of the muffin samples was determined, as described by [17], using the ABTS$^+$ and DPPH radical scavenging activity (ABTS-RSA and DPPH-RSA, respectively) methods. In short, the muffin samples were subjected to extraction in 80% aqueous methanol solution, for 2 h at 23 ± 2 °C. The extracts obtained upon centrifugation for 15 min at 10,000 rpm were further used for measuring the antioxidant activity. ABTS-RSA was determined by measuring the absorbance of a mixture consisting of a 40 µL extract and 2.96 mL ABTS$^{·+}$ solution, at a wavelength of 734 nm, immediately and after 6 min. DPPH-RSA was determined by measuring the decrease, over 20 min, of the absorbance, at a wavelength of 515 nm, of a mixture consisting of a 100 µL extract, 250 µL DPPH solution and 2.1 mL of 80% aqueous methanol solution. In the case of each method, a Trolox standard curve was prepared, and the antioxidant activity was expressed as µmol Trolox/g d.w.

The baking experiment was carried out in duplicate, and the measurements were performed at least in triplicate.

2.9. Statistical Analysis

The statistical analysis was carried out using the Minitab 19 (Minitab LLC, State College, PA, USA) software. The ANOVA method and the post hoc test, based on the Tukey method, when $p < 0.05$ were considered to identify significant differences between results.

3. Results and Discussion

3.1. mPAGE Analysis of Soy Protein Hydrolysates

In agreement with our previous study, Brumă et al. [10], bromelain, Neutrase and trypsin ensured different hydrolysis degrees (HDs) of soy protein isolate. Among all the tested enzymes, bromelain was the most efficient in hydrolyzing the soy proteins (HD of 10.3%), being followed by Neutrase (HD of 9.7%) and trypsin (HD of 1.9%). The electrophoretic analysis was employed to check the molecular weight distribution of the resulting peptides, separated through centrifugation into soluble and insoluble fractions. The mPAGE analysis revealed that the SPI used in the study has high amounts of 7S (β-conglycinin) and 11S (glycinin), which are the major globulins found in soybeans [18]. Figure 1 (lane SPI) emphasizes the specific subunits of the main globulins found in soybeans: α' (~75 kDa), α (~72 kDa) and β (~53 kDa) subunits of β-conglycinin, together with the acidic (33–42 kDa) and basic subunits (~22 kDa) of glycinin [19,20].

Analyzing the results presented in Figure 1, one can observe that the soluble fraction of soy protein hydrolysates retained only the peptides with molecular weights lower than 20 kDa (lanes sSPB, sSPN, sSPT). On the other hand, the insoluble fractions (residuals) of soy protein hydrolysates retained higher molecular weight fragments (Figure 1, lanes rSPB, rSPN, rSPT), probably resulting from the partial digestion of the main soy proteins (lane SPI in Figure 1). The smearing in the rSPB, rSPN and rSPT lanes (Figure 1) might also suggest the rather low specificity of exogenous proteases used in this study. In agreement with our observations, Yang et al. [21] reported the reducing of the number of protein bands as a consequence of the Alcalase-assisted hydrolysis of the α′ and α subunits of β-conglycinin, together with the decrease in the antigenic properties. Moreover, Wen et al. [22] emphasized that soy protein hydrolysis with Alcalase and Neutrase resulted in 84 peptides with immunoregulatory activities and other positive effects on human health. In addition to modulating the functional properties of the soy protein isolate, enzyme-assisted hydrolysis might also allow for the reduction in allergenic properties. Various studies assigned the allergenic properties of soybeans to various 7S and 11S globulins [23,24] or to proteins located in soybean hulls [25,26]. The major soybean protein allergens are considered to be Gly m 4 (17 kDa), Gly m 5 (β-conglycinin) and Gly m 6 (with five subunits having polypeptides of 20 kDa and 40 kDa) [26,27]. Some other proteins responsible for allergenic activities were described as Gly m1, Gly m 2, Gly m 3 or 2S-globulin fraction, having lower molecular weights of 7 kDa, 7.5 kDa, 8 kDa and 12–15 kDa, respectively [19,20]. To reduce or eliminate the allergenicity of foods based on soybean proteins, different thermal and non-thermal treatments were applied. Among these treatments, enzyme-assisted hydrolysis appears very efficient for altering the structure of allergenic proteins. This effect of allergenic potential reduction associated with hydrolysis with exogenous enzymes is of particular interest in the case of those proteins which are resistant to in vivo digestion, as is the case of soy proteins [10]. The results presented in Figure 1 suggest that the separation of the soluble fraction of the soy protein hydrolysates obtained with bromelain or Neutrase might allow for obtaining soy protein ingredients devoid of such allergens.

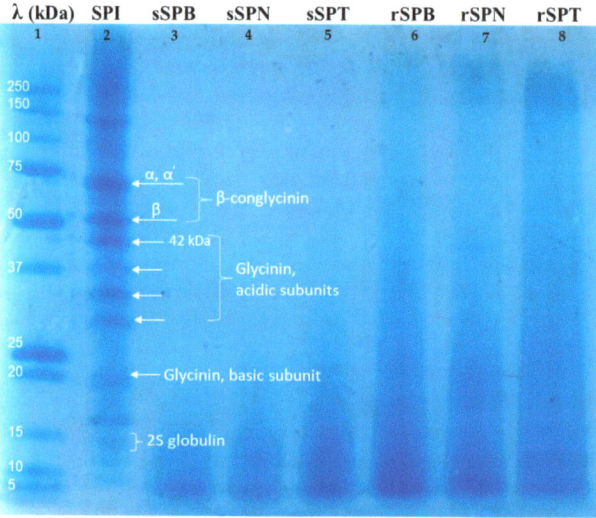

Figure 1. The mPAGE profile of the soy protein isolate (SPI, lane 2) and of the soluble and insoluble fractions of the hydrolysates obtained with bromelain (sSPI-B and rSPI-B, lanes 3 and 6), Neutrase (sSPI-N and rSPI-N, lanes 4 and 7) and trypsin (sSPI-T and rSPI-T, lanes 5 and 8). Lane 1—Dual Xtra marker (Bio-Rad, Hercules, CA, USA). The arrows on lane 2 indicate the subunits of β-conglycinin [20], glycinin [19] and 2S-globulin [20].

3.2. The Influence of SPI Addition on the Thermo-Mechanical Properties of the Gluten-Free Mixtures

In order to investigate the influence of the addition of soy protein isolate and hydrolysates on the rheology of the dough based on gluten-free composite flour consisting of whole rice flour and quinoa flour, the evolution of the doughs' consistency over the entire Mixolab curves was considered (Figures 2 and 3). Particular attention was paid to the consistency values at the end of each phase of the Mixolab curves (Table 2), as well as to the specific parameters highlighted by the Mixolab software (version 4.1.2.10) (Table 3).

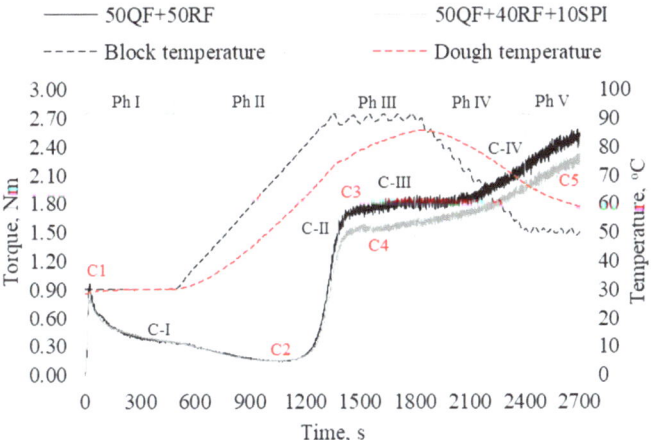

Figure 2. Mixolab curve of dough prepared mixtures of quinoa flour (QF), rice flour (RF) and soy protein isolate (SPI).

Table 2. The influence of gluten-free flour supplementation with soy proteins or peptides on the dough consistency measured at the end of each Mixolab phase.

Sample/Phase	Phase I 0–480 s 30 °C C-I, Nm	Phase II 480–1380 s 30–90 °C C-II, Nm	Phase III 1380–1800 s 90 °C C-III, Nm	Phase IV 1800–2400 s 90–50 °C C-IV, Nm	Phase V 2400–2700 s 50 °C C-V, Nm
50QF + 50RF	0.34 ± 0.01 [d]	1.38 ± 0.02 [d]	1.86 ± 0.02 [b]	2.18 ± 0.02 [a]	2.51 ± 0.02 [b]
50QF + 40RF + 10SPI	0.34 ± 0.01 [d]	1.30 ± 0.02 [e]	1.58 ± 0.02 [e]	1.98 ± 0.02 [c]	2.27 ± 0.01 [d]
50QF + 40RF + 10sSPI-B	0.04 ± 0.01 [f,g]	0.32 ± 0.01 [i]	1.23 ± 0.02 [h]	1.51 ± 0.02 [e]	1.73 ± 0.01 [h]
50QF + 40RF + 10sSPI-N	0.06 ± 0.01 [f]	0.27 ± 0.01 [j]	1.23 ± 0.02 [h]	1.45 ± 0.02 [f]	1.62 ± 0.02 [i]
50QF + 40RF + 10sSPI-T	0.02 ± 0.01 [g]	0.26 ± 0.01 [j]	1.29 ± 0.02 [g]	1.53 ± 0.02 [e]	1.79 ± 0.02 [g]
50QF + 40RF + 10rSPI-B	0.75 ± 0.02 [b]	1.48 ± 0.02 [c]	1.73 ± 0.01 [c]	2.00 ± 0.02 [c]	2.40 ± 0.02 [c]
50QF + 40RF + 10rSPI-N	0.59 ± 0.01 [c]	1.62 ± 0.02 [b]	1.74 ± 0.01 [c]	1.90 ± 0.02 [d]	2.53 ± 0.02 [b]
50QF + 40RF + 10rSPI-T	0.99 ± 0.02 [a]	1.78 ± 0.02 [a]	1.92 ± 0.02 [a]	1.53 ± 0.01 [e]	2.74 ± 0.01 [a]
50QF + 40RF + 10hSPI-B	0.24 ± 0.01 [e]	0.90 ± 0.02 [g]	1.53 ± 0.01 [f]	1.89 ± 0.02 [d]	2.16 ± 0.01 [e]
50QF + 40RF + 10hSPI-N	0.23 ± 0.01 [e]	0.82 ± 0.02 [h]	1.51 ± 0.01 [f]	2.16 ± 0.01 [a]	2.05 ± 0.02 [f]
50QF + 40RF + 10hSPI-T	0.35 ± 0.01 [d]	1.07 ± 0.02 [f]	1.68 ± 0.02 [d]	2.06 ± 0.02 [b]	2.39 ± 0.02 [c]

QF—quinoa flour; RF—rice flour; SPI—soy protein isolate; sSPI, rSPI and hSPI—soluble, insoluble and total soy protein hydrolysate fraction, respectively, obtained with bromelain (B), Neutrase (N) and trypsin (T). Different superscript letters associated with values in same column indicate significant differences among results, as resulted from ANOVA and Tukey post hoc test at $p < 0.05$.

SPI addition to the composite flour resulted in no changes in the Ph I and Ph II of the Mixolab curve (Figure 2). The dough consistency increased rapidly in the first 30 s of kneading and decreased at a faster rate in the phase run at 30 °C, from 0.80 Nm to 0.34 Nm in 480 s, and more slowly in the next 600 s, from 0.34 (C-I) to 0.15 Nm (C2), in Ph II when the temperature increased constantly (Tables 2 and 3).

The main changes caused by the replacement of 10% RF by SPI can be observed in Ph III, IV and V of the Mixolab curves (Figure 2). The addition of SPI lowered the values of the dough consistency at the beginning and end of each of the phases III–V. This behavior can be attributed to starch dilution in the dough system, caused by SPI incorporation. Under these conditions, the water released by the dough system as a result of protein weakening at increasing temperature is higher, therefore causing the dough consistency decrease [28].

Table 3. The influence of gluten-free flour supplementation with soy proteins or peptides on the Mixolab parameters of dough.

Sample/Parameter	C1, Nm	C2, Nm	C3, Nm	C4, Nm	C5—C-III, Nm
50QF + 50RF	0.88 ± 0.02 [d]	0.15 ± 0.01 [e]	1.74 ± 0.02 [b]	1.83 ± 0.02 [a]	0.65 ± 0.002 [e]
50QF + 40RF + 10SPI	0.80 ± 0.02 [e]	0.15 ± 0.01 [e]	1.58 ± 0.02 [c]	1.60 ± 0.02 [c]	0.69 ± 0.002 [c,d]
50QF + 40RF + 10sSPI-B	0.16 ± 0.01 [j]	0.01 ± 0.01 [f,h]	1.15 ± 0.01 [e]	1.18 ± 0.02 [e]	0.50 ± 0.008 [h]
50QF + 40RF + 10sSPI-N	0.23 ± 0.01 [i]	0.03 ± 0.01 [f]	1.16 ± 0.01 [e]	1.18 ± 0.02 [e]	0.39 ± 0.002 [i]
50QF + 40RF + 10sSPI-T	0.18 ± 0.01 [j]	0.00 ± 0.01 [h]	1.17 ± 0.01 [e]	1.20 ± 0.02 [e]	0.50 ± 0.001 [h]
50QF + 40RF + 10rSPI-B	1.96 ± 0.02 [b]	0.30 ± 0.01 [b]	1.71 ± 0.02 [b]	1.66 ± 0.01 [b]	0.67 ± 0.013 [d]
50QF + 40RF + 10rSPI-N	1.56 ± 0.02 [c]	0.27 ± 0.01 [c]	1.70 ± 0.02 [b]	1.64 ± 0.01 [b]	0.79 ± 0.003 [b]
50QF + 40RF + 10rSPI-T	2.17 ± 0.02 [a]	0.42 ± 0.02 [a]	1.91 ± 0.02 [a]	1.83 ± 0.02 [a]	0.82 ± 0.002 [a]
50QF + 40RF + 10hSPI-B	0.46 ± 0.01 [h]	0.11 ± 0.01 [f]	1.45 ± 0.02 [d]	1.44 ± 0.02 [d]	0.63 ± 0.002 [f]
50QF + 40RF + 10hSPI-N	0.71 ± 0.01 [f]	0.10 ± 0.01 [f]	1.44 ± 0.02 [d]	1.44 ± 0.02 [d]	0.55 ± 0.010 [g]
50QF + 40RF + 10hSPI-T	0.50 ± 0.01 [g]	0.18 ± 0.01 [d]	1.55 ± 0.02 [c]	1.57 ± 0.02 [c]	0.71 ± 0.006 [c]

QF—quinoa flour; RF—rice flour; SPI—soy protein isolate; sSPI, rSPI and hSPI—soluble, insoluble and total soy protein hydrolysate fraction, respectively, obtained with bromelain (B), Neutrase (N) and trypsin (T). Different superscript letters associated with values in same column indicate significant differences among results, as resulted from ANOVA and Tukey post hoc test at $p < 0.05$.

Regardless of SPI addition, dough consistency increased over Ph III: the increase rate was lower for the sample with SPI addition (from 1.30 to 1.66 Nm) compared to the control dough (from 1.38 to 1.88 Nm). On the other hand, during Ph IV when the temperature dropped from 90 to 70 °C (from 1800 to 2100 s), the consistency of the sample with SPI increased by 0.13 Nm (from 1.58 to 1.71 Nm), whereas the control registered no important variation in the consistency (from 1.86 to 1.87 Nm) (Table 2). This behavior could be due to the aggregation of soy proteins during heating. The main soybean proteins, glycinin and β-conglycinin, dissociate into monomers with increasing temperature, exposing a number of hydrophobic patches which might lead to the formation of aggregates and ordered gel structures [29].

In the last phase, the dough consistency increase in the two samples was rather similar: 0.29 Nm (from 1.98 to 2.27 Nm) for the sample with SPI and 0.33 Nm (from 2.18 to 2.51 Nm) for the control sample (Table 2).

Marco and Rosell [29] mentioned that dough consistency during heating and cooling depends on the amount of soy protein isolate substituting rice flour, and Bonet et al. [30] emphasized the importance of soy protein processing conditions on the properties of SPI. When adding 13% SPI to rice flour, Marco and Rosell [29] obtained a decrease in C3 and C4 values and an increase in C5 and (C5–C4) values. On the other hand, Patrascu et al. [28] reported a decrease in C3, C4 and C5 values when incorporating 15% SPI into rice flour; a similar trend was observed by Nogueira et al. [31] when SPI was added to the wheat flour.

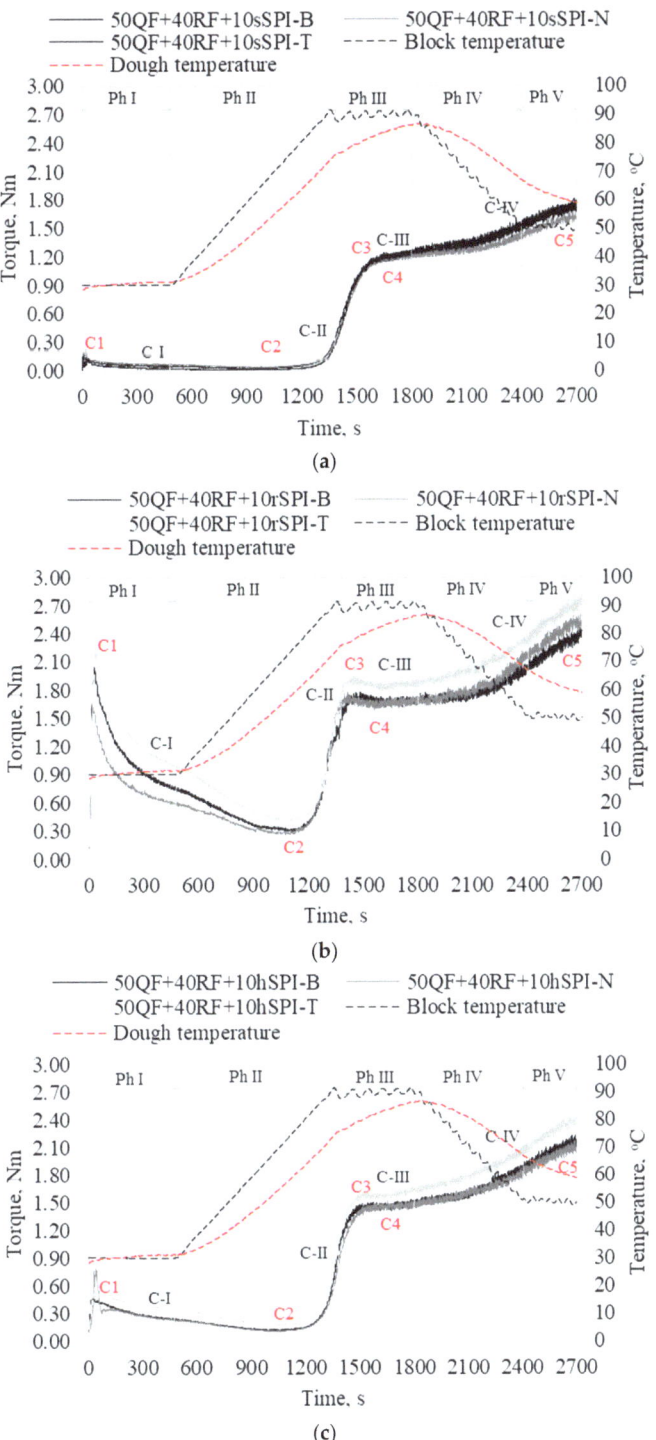

Figure 3. Mixolab curve of dough prepared mixtures of quinoa flour (QF), rice flour (RF) and (**a**) soluble fraction of soy protein hydrolysates (sSPIs), (**b**) insoluble fraction of soy protein hydrolysates (rSPIs) and (**c**) total soy protein hydrolysates (hSPIs) obtained with bromelain (B), Neutrase (N) and trypsin (T).

3.3. The Influence of Soluble Soy-Derived Peptides' Addition on the Thermo-Mechanical Properties of the Gluten-Free Mixtures

Regardless of the exogenous enzyme used for SPI hydrolysis, the addition of sSPI to the composite flour notably reduced the consistency of the dough during Ph I and Ph II (Figure 3). It is known that the protein from quinoa contains mainly globulins and albumins, stabilized by disulfide bonds, the cysteine residues from these fractions playing, therefore, an important role in protein contribution to the overall behavior of the dough matrix [32]. In the composite flour, this behavior is balanced by the presence of high-molecular-mass glutelins, which are the proteins prevailing in the rice flour [33]. The addition of sSPI, containing low-molecular-mass peptides (Figure 1), disturbed the protein network of the dough obtained from the composite flour. Guo et al. [34] reported a deterioration of the protein network in the dough prepared from wheat flour supplemented with soy protein hydrolysates. The authors speculated that the interactions established between the soy peptides and the low-molecular-weight glutenin fractions and gliadin involved disulfide bonds, therefore interfering with the formation of the typical protein network within the dough. Although the number of hydrogen and disulfide bonds was reported to increase, the addition of soy protein hydrolysate weakened the gluten network. Schmiele et al. [35] noted that soy protein hydrolysates could modify ionic and hydrophobic interactions and covalent and hydrogen bonding, preventing the complete hydration of proteins during dough formation.

Another possible explanation for the decrease in dough consistency in Ph I and Ph II could be related to the fact that peptides with low molecular weight have a low water binding capacity. Lamsal et al. [36] reported the decrease in the apparent viscosity of the soy protein hydrolysate obtained with bromelain as a result of the high solubility of the low-molecular-weight peptides formed. Similar observations were collected by Tsumura et al. [37] for the soy protein hydrolysate obtained with papain. Taking into account that all Mixolab measurements were carried out at the same WA of 65%, the excess of the water in the system significantly reduced the consistency of the soy peptide-rich dough samples.

The minimum consistency of the dough samples enriched in sSPI ranged between 0 and 0.03 Nm and was registered after 1080 s at temperature of 55 °C. When compared with the control and the dough supplemented with SPI, one can observe that sSPI addition delayed the start of starch gelatinization (Figure 3). Thus, after 1380 s of mixing and reaching the maximum heating temperature specific to the Chopin+ protocol, the C-II of the soy peptide-enriched doughs was 0.32, 0.27 and 0.26 Nm, for samples supplemented with sSPI-B, sSPI-N and sSPI-T, respectively (Table 2). The gelatinization maximum was reached only after 1620 s, at a temperature of 83 °C, the C4 of the samples ranging between 1.15 and 1.17 Nm (Table 3). At the end of Ph III, the dough reached a temperature of 87 °C, and a C-III of 1.23 Nm was recorded for the samples with sSPI-B and sSPI-N and of 1.29 Nm for the sample with sSPI-T (Table 2). Li et al. [38] investigated the possibility of improving the quality of steamed bread by supplementing flour with soy protein hydrolysate obtained with pepsin. They observed that the addition of soy protein hydrolysate interfered with the orderly structure of the starch granules, therefore affecting the swelling properties and causing the increase in the gelatinization temperature and the decrease in the gelatinization enthalpy.

The dough consistency continued to increase until the end of Ph IV, but also over the entire Ph V (Figure 3). The strength of the gel was higher in the case of dough with sSPI-T, followed by sSPI-B and sSPI-N. It should be noted that trypsin ensured the lowest hydrolysis degree among all enzymes used for soy protein hydrolysis [10]. Moreover, as indicated by the SDS-PAGE analysis, sSPI-T included higher-molecular-weight peptides compared to sSPI-B and sSPI-N (Figure 1). Ashaolu [39] mentioned the importance of the size of the peptide molecules within protein hydrolysates and appreciated that lowering the molecular weight can affect the strength of the gels in which they are incorporated. Anyway, the degree of hydrolysis and the nature of the enzyme used for hydrolysis are critically important. In this respect, Hou and Zhao [40] observed that soy protein hydrolysates have better gel-forming properties than the original protein isolate but also the fact that gel properties depend on the enzyme used for hydrolysis and the degree of hydrolysis. The authors reported that the protein hydrolysate obtained with Neutrase allowed for gels with higher strength compared to those obtained with trypsin. In addition, it is mentioned that at lower degrees of hydrolysis, the strength of the gel is higher. Huang et al. [41] observed that the peptide fraction with low molecular weight (0.5–10 kDa), separated from the hydrolysate obtained from soy protein isolate, allowed for obtaining gels at temperatures above 65 °C, and the obtained gel had high strength compared to the control.

3.4. The Influence of the Insoluble Soy-Derived Peptides' Addition on the Thermo-Mechanical Properties of the Gluten-Free Mixtures

The addition of rSPI considerably increased the consistency of the dough in Ph I and Ph II, compared to the control and dough sample supplemented with SPI (Figure 3). The abrupt drop in the dough consistency in the first 8 minutes while kneaded at 30 °C can be attributed to the slower formation of the dough network. The consistency at the end of Ph I (C-I) was 0.99, 0.75 and 0.59 Nm, while the minimum C2 consistency reached in Ph II was 0.42, 0.30 and 0.27 Nm for the dough samples supplemented with rSPI-T, rSPI-B and sSPI-N, respectively (Table 3). The dough sample prepared with the residue of the hydrolysate obtained with trypsin, which also had the lowest degree of hydrolysis, exhibited the highest consistency after 480 s of kneading at 30 °C (C-I), the steepest consistency drop (C1–C2 of −1.75 Nm) and the highest consistency value (C2) measured at 55 °C (Table 3). The dough samples supplemented with rSPI-B and rSPI-N recorded consistency decreases of −1.66 Nm and −1.29 Nm, respectively. All dough samples supplemented with rSPI had higher C2 values compared to the control and the sample with SPI. The peptide profile of the three hydrolysates used to supplement the gluten-free dough and their water related properties can explain the different consistency variations in the dough [36]. We can assume that the peptides from rSPI have a higher water binding capacity [39], compared to the sample with and without the addition of SPI, but the development of the dough network is more difficult and the stability lower. However, the higher C2 values registered for the samples with rSPI may indicate a greater stability of the protein network upon heating, as a result of molecular interactions through hydrogen bonding and hydrophobic contacts, involving the proteins of rice and quinoa and soy peptides, which improved the resistance of the dough to kneading and heating.

The addition of rSPI to the gluten-free composite flour did not delay starch gelatinization, as observed in case of the addition of soluble soy peptides, but increased the consistency of the gel compared to the control and the sample with SPI. The highest maximum dough consistency of 1.91 Nm, associated with starch gelatinization at 76 °C, was measured in the case of the sample supplemented with rSPI-T derived from the hydrolysate with the lowest degree of hydrolysis, whereas the addition of rSPI-N and rSPI-B to the gluten-free composite flour resulted in the maximum gelatinization values of 1.70–1.71 Nm, obtained at 78 °C (Table 3). As the temperature of the dough increased to 81 °C for the sample with rSPI-T, and to 83 °C for samples with rSPI-N or rSPI-B, the consistency decreased by 0.08 Nm and 0.05–0.06 Nm, respectively. This downward trend was noticed only when supplementing the gluten-free flour with the rSPI fraction of the hydrolysates. When studying the gelling properties of the soy protein hydrolysates obtained with bromelain, Lamsal et al. [36] noted that, although they present gelling capacity, the gel resistance is reduced. The authors assigned the reduced resistance of the gels to the limited hydrophobic interactions and sulfhydryl exchange reactions established between peptides during gelation.

At the maximum dough temperature of 85–87 °C, the consistency reached a higher value in the case of rSPI-T (C-III of 1.92 Nm), compared to rSPI-B and rSPI-N (C-III of 1.74–1.73 Nm) (Table 2). Further on, at the end of Ph V, the consistency of the dough sample supplemented with rSPI-T reached 2.74 Nm, and for the samples with rSPI-N and rSPI-B, with a higher degree of hydrolysis, the consistency reached 2.53 and 2.40 Nm, respectively (Table 2). The difference between the maximum consistency of the dough, recorded at the end of Ph III, and the consistency at the end of Ph V, which can be considered an indicator for starch degradation, had the lowest value of 0.67 Nm, when supplementing the gluten-free composite flour with the residue fraction of the soy protein hydrolysate with the highest degree of hydrolysis (rSPI-B). In the case of the dough with rSPI-N and rSPI-T, the C5-C-III difference was 0.79 and 0.82 Nm, respectively (Table 3).

3.5. The Influence of the Total Soy Protein Hydrolysate Addition on the Thermo-Mechanical Properties of the Gluten-Free Mixtures

The addition of hSPI produced the smallest changes in the dough behavior during mixing at 30 °C and during heating up to 55 °C, compared to the control and the sample supplemented with SPI. The curves obtained for the samples supplemented with hSPI-B and hSPI-N showed almost identical consistency values in the mentioned temperature range but lower compared to the control and the sample supplemented with SPI (Figure 2). Thus, the C-I consistency at the end of Ph I was 0.23–0.24 Nm, and the minimum C2 consistency of 0.10–0.11 Nm was recorded at 55 °C. In contrast, the dough sample with hSPI-T presented a C-I of 0.35 Nm and a minimum C2 consistency of 0.18 Nm, values extremely close to the control and the sample supplemented with SPI (Table 2). Practically, among all studied peptide-enriched dough samples, the dough including soy protein hydrolysate with the lowest degree of hydrolysis (hSPI-T) altered to the lowest extent the interactions between the main macromolecules of the rice and quinoa flours, which are essential for the dough system.

Zhang et al. [42] investigated the effect of wheat flour supplementation with soy protein hydrolysates obtained with Neutrase and reported a decrease in the dough stability, an increase in dough softening and the faringographic number and the obtained dough being more resistant and less extensible and elastic. The authors also noted the decrease in the content of high-molecular-weight protein fractions with the addition of protein hydrolysate, suggesting that the peptides from the hydrolysate formed disulfide bonds with the wheat protein fractions, interfering with glutenin polymerization during dough formation. Li et al. [38] reported a decrease in intermolecular and anti-parallel β-sheets and an increase in β-turns upon the addition of soy protein hydrolysate, stating that the intermolecular β-sheets are an indicator of protein polymerization. The same authors noted that during fermentation, the content of α-helices and β-turns decreased, but the content of anti-parallel β-sheets increased.

When compared with the sample supplemented with SPI, the addition of sSPI delayed starch gelatinization, causing an important reduction in the gelatinization maximum and the decrease in starch retrogradation; rSPI facilitated the faster achievement of the gelatinization maximum by approximately 120 s while increasing starch retrogradation, whereas hSPI addition resulted in delaying the maximum gelatinization by about 60–120 s. The maximum dough consistency and starch retrogradation varied with the nature of the enzyme used for soy protein hydrolysis and consequently with the degree of hydrolysis. Thus, flour supplementation with hSPI-T, which presented the lowest degree of hydrolysis among all investigated samples, resulted in a gelatinization maximum (C3 of 1.55 Nm after 1500 s of Chopin+ protocol) comparable to the dough with SPI (C3 of 1.58 Nm after 1560 s) (Table 3). The samples prepared with hSPI-B and hSPI-N exhibited significantly lower gelatinization maxima, a C3 of 1.44–1.45 Nm being recorded at 1620 s (Table 3). The slight delay in starch gelatinization could be attributed to the effect of soluble peptides present in the hydrolysate, which produced important changes in the behavior of the dough in which SPIH was incorporated.

The dough consistency continued to increase throughout Ph III, IV and V (Figure 3, Table 2). The starch retrogradation of the dough samples with hSPI-B and hSPI-N was significantly lower compared to the control ($p < 0.05$), whereas the starch retrogradation behavior of the dough supplemented with hSPI-T was similar to the sample with SPI (Table 3). Taking into account the overall behavior of the dough supplemented with soy protein hydrolysates, over the five phases of the Mixolab curves (Figure 3), it can be concluded that hSPI-B and hSPI-N ensured the most important improvements to the dough, compared to the sample with SPI. In agreement with our results, Schmiele et al. [35] reported a decrease in C3, C4 and C5 upon the addition of soy protein hydrolysate to wheat flour. However, the authors noted that the presence of protein hydrolysate accelerated starch gelatinization, compared to the control sample and reduced (C5–C4). As in the case of the dough samples prepared with hSPI-B and hSPI-N in the present study, Schmiele et al. [35] mentioned the anti-retrogradation effect exerted by protein hydrolysates on starch, a very important aspect for preventing the staling of bakery products.

3.6. The Influence of Soy Protein Hydrolysate Addition on the Quality of Muffins

The gluten-free flour mixtures supplemented with SPI and hSPI were further used for preparing muffins (Figure 4).

Figure 4. The appearance of the upper surface and crumb of the muffin samples (M—control muffin; M-SPI—muffins enriched with soy protein isolate; M-hSPI-B—muffins enriched with soy protein hydrolysate produced with bromelain; M-hSPI-N—muffins enriched with soy protein hydrolysate produced with Neutrase; M-hSPI-T—muffins enriched with soy protein hydrolysate produced with trypsin). Images were taken with the Canon PowerShot G16 digital camera (Canon Inc., Tokyo, Japan).

Among all investigated muffin formulations, the samples supplemented with M-hSPI-B or M-hSPI-N exhibited the lowest values of weight loss during baking, suggesting the good water holding capacity of the products and finally the advantageous production process [15]. Analyzing the results presented in Table 4, one can observe that flour supplementation with SPI resulted in a significant improvement in the specific volume of the muffins. On the other hand, the specific volume of the hSPI-supplemented muffins decreased significantly compared to the control ($p < 0.05$), as a result of the reduced ability of the network to retain the CO_2 released by the baking powder during baking. The total height of the muffins was significantly influenced by the soy protein or hydrolysate addition ($p < 0.05$). The largest height values were registered in the case of the samples supplemented with SPI, hSPI-B or hSPI-T (Table 4). Regardless of soy protein isolate or hydrolysate addition, no important differences were noted in terms of the firmness of the final product.

Table 4. The influence of gluten-free flour supplementation with soy proteins/peptides on the quality characteristics of the muffins (M—control muffin; M-SPI—muffins enriched with soy protein isolate; M-hSPI-B—muffins enriched with soy protein hydrolysate produced with bromelain; M-hSPI-N—muffins enriched with soy protein hydrolysate produced with Neutrase; M-hSPI-T—muffins enriched with soy protein hydrolysate produced with trypsin).

Muffin Sample	Weight Loss during Baking, g/100 g	Moisture, g/100 g	Height, mm	Specific Volume, cm^3/g	Firmness, N
M	9.90 ± 0.14 [a,b]	20.70 ± 0.05 [a]	3.00 ± 0.00 [c]	159.48 ± 1.43 [b]	6.49 ± 0.89 [a]
M-SPI	10.50 ± 0.14 [a]	20.10 ± 0.02 [c]	3.40 ± 0.00 [a]	181.73 ± 0.59 [a]	6.63 ± 0.66 [a]
M-hSPI-B	9.80 ± 0.00 [b]	20.66 ± 0.05 [a]	3.25 ± 0.07 [a,b]	152.84 ± 0.20 [c]	5.60 ± 0.77 [a]
M-hSPI-N	9.70 ± 0.14 [b]	20.36 ± 0.05 [b]	3.10 ± 0.00 [b,c]	151.93 ± 2.14 [c]	5.74 ± 0.20 [a]
M-hSPI-T	10.20 ± 0.28 [a,b]	20.28 ± 0.02 [b]	3.25 ± 0.07 [a,b]	136.99 ± 0.31 [d]	7.00 ± 0.21 [a]

Different superscript letters associated with values in same column indicate significant differences among results, as resulted from ANOVA and Tukey post hoc test at $p < 0.05$.

Taking into account that color is a sensory attribute decisive for the consumers' choice, the color characteristics of the muffin samples were measured both on the crust and crumb (Table 5). Taking into account the muffins are baked products, the color properties are due both to the individual contribution of the ingredient used for preparation and to their interactions, especially during baking [13,43]. No important differences were observed between samples in terms of crust lightness, except for the muffin supplemented with M-hSPI-T that registered a significant increase in L* (Table 5). The crust of all muffins presented positive a* values, the red color being more intense in the case of the samples with SPI and hSPI-B ($p < 0.05$). No important differences were noted between samples in terms of yellowness, C* and H angle (Table 5). A different trend was noticed in the color parameters of the crumb when supplementing the gluten-free flour formulations with soy protein isolate or hydrolysates (Table 5). The hSPI-enriched samples presented a significantly lower L* and more intense yellow shades compared to the control and muffins with SPI. Moreover, hSPI addition resulted in significantly higher C* values ($p < 0.05$), as a measure of color purity in the CIELAB space [44].

Table 5. The influence of gluten-free flour supplementation with soy proteins/peptides on the color characteristics of the crust and crumb of muffin samples (M—control muffin; M-SPI—muffins enriched with soy protein isolate; M-hSPI-B—muffins enriched with soy protein hydrolysate produced with bromelain; M-hSPI-N—muffins enriched with soy protein hydrolysate produced with Neutrase; M-hSPI-T—muffins enriched with soy protein hydrolysate produced with trypsin).

Muffin Sample	L*	a*	b*	C*	H, °
Crust					
M	46.21 ± 1.94 [b]	10.48 ± 1.01 [c]	26.03 ± 0.68 [a]	28.07 ± 0.30 [a]	68.06 ± 2.96 [a]
M-SPI	45.10 ± 0.20 [b]	13.09 ± 0.41 [a]	26.31 ± 0.56 [a]	29.39 ± 0.39 [a]	63.54 ± 1.42 [a]
M-hSPI-B	44.85 ± 0.92 [b]	12.50 ± 0.07 [a,b]	26.01 ± 0.67 [a]	28.85 ± 0.73 [a]	64.32 ± 0.76 [a]
M-hSPI-N	46.83 ± 0.46 [b]	11.42 ± 0.05 [b,c]	26.62 ± 0.38 [a]	28.97 ± 0.41 [a]	66.78 ± 0.25 [a]
M-hSPI-T	49.87 ± 0.48 [a]	10.88 ± 0.24 [c]	26.56 ± 0.23 [a]	28.70 ± 0.00 [a]	67.72 ± 0.46 [a]
Crumb					
M	54.86 ± 0.17 [a,b]	2.42 ± 0.17 [a]	18.32 ± 0.30 [c]	18.48 ± 0.39 [b]	82.48 ± 0.48 [a]
M-SPI	55.46 ± 0.23 [a]	2.45 ± 0.09 [a]	19.00 ± 0.09 [b]	19.16 ± 0.12 [a,b]	82.66 ± 0.22 [a]
M-hSPI-B	54.24 ± 0.26 [b,c]	2.70 ± 0.27 [a]	19.85 ± 0.38 [a]	20.03 ± 0.49 [a]	82.28 ± 0.75 [a]
M-hSPI-N	53.77 ± 0.57 [c]	2.66 ± 0.09 [a]	19.73 ± 0.26 [a]	19.91 ± 0.31 [a]	82.32 ± 0.20 [a]
M-hSPI-T	54.84 ± 0.73 [a,b]	2.73 ± 0.01 [a]	19.85 ± 0.10 [a]	20.03 ± 0.04 [a]	82.17 ± 0.04 [a]

Different superscript letters associated with values in same column indicate significant differences among results regarding crust or crumb, as resulted from ANOVA and Tukey post hoc test at $p < 0.05$.

A diet rich in antioxidant compounds might help alleviate the prevalence of lifestyle and degenerative diseases [45]. The antioxidant activity of the muffin samples was measured using the DPPH and ABTS radical scavenging activity methods, and the results are presented in Figure 5. Both methods indicated that gluten-free composite flour supplementation with SPI resulted in a significant increase in the antioxidant activity ($p < 0.05$). Meanwhile, the increase was even higher when SPI was replaced by hSPI ($p < 0.05$). The DPPH-based method indicated no differences between muffins supplemented with hSPI, whereas the ABTS-based method suggested that samples with hSPI-B and hSPI-N exhibited the highest antioxidant activity ($p < 0.05$). Previous studies highlighted that possibility of using exogenous enzymes for enhancing the physiological roles of proteins, by releasing the encrypted bioactive peptides responsible for antioxidant properties, immunomodulatory activity, antimicrobial properties, hypocholesterolemic and antihypertensive effects, etc. [11]. The antioxidant activity was mainly assigned to the 2–20 amino acid long peptides with a molecular weight below 6 kDa [11]. Moreover, the exogenous enzyme used for protein hydrolysis, the propensity of hydrophobic amino acids and their location in the peptides highly influence the antioxidant activity. Brumă et al. [10] observed that the hydrolysis of soy protein isolate with bromelain, Neutrase and trypsin resulted in a significant increase in the antioxidant activity. Both DPPH-RSA and ABTS-RSA methods indicated that Neutrase released the highest number of peptides with antioxidant activity. They reported no important differences between samples treated with trypsin and bromelain in terms of ABTS-RSA. In agreement with our observations (Figure 5), Brumă et al. [10] measured higher radical scavenging activity when using the ABTS- compared to DPPH-based method.

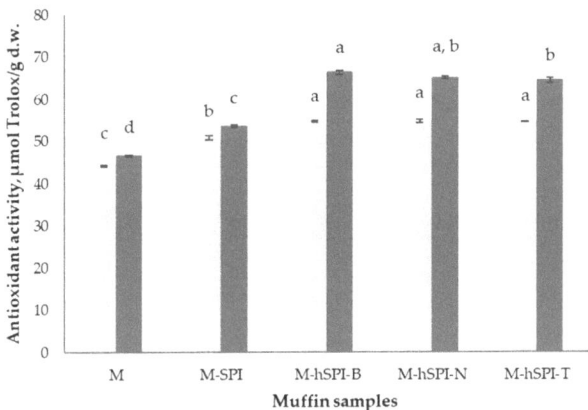

Figure 5. The antioxidant activity of the muffin samples (M—control muffin; M-SPI—muffins enriched with soy protein isolate; M-hSPI-B—muffins enriched with soy protein hydrolysate produced with bromelain; M-hSPI-N—muffins enriched with soy protein hydrolysate produced with Neutrase; M-hSPI-T—muffins enriched with soy protein hydrolysate produced with trypsin). The results of the DPPH- and ABTS-based assays are represented with light and dark gray, respectively. Different superscript letters associated with mean antioxidant activity values determined with the same method indicate significant differences at $p < 0.05$.

4. Conclusions

The gluten-free composite flour consisting of a mixture of rice and quinoa flour was used as the basis for supplementation with a 10% soy protein isolate of hydrolysate obtained with bromelain, Neutrase or trypsin. Empirical rheological measurements were employed to determine the thermo-mechanical behavior of the gluten-free doughs. The flour mixtures enriched with soluble soy peptide fractions, with a molecular weight lower than 20 kDa, exhibited delayed starch gelatinization. The addition of the insoluble soy peptide fraction resulted on the other hand in a dough consistency increase. Based on the thermo-mechanical behavior, it was concluded that, among all tested formulations, the most important improvements in the dough behavior were observed in the case of supplementing the gluten-free flour with 10% soy protein hydrolysates obtained with bromelain and trypsin. The gluten-free muffins' supplementation with soy protein hydrolysate resulted in a more vivid color of the crumb and no important changes in terms of texture, but the specific volume was significantly lower compared to the sample with soy protein isolate. The antioxidant activity of the gluten-free muffins enriched with soy protein hydrolysate was significantly higher compared to the control. In conclusion, soy protein hydrolysate can be successfully used to obtain baked gluten-free products with an improved nutritional and bioactive profile.

Author Contributions: Conceptualization, I.A.; methodology, I.B., I.V. and L.G.-G.; software, I.A., L.D. and I.B.; validation, M.B., L.G.-G., L.D. and I.V.; formal analysis, I.A.; investigation, M.B., L.G.-G., I.V., I.B., L.D. and I.A.; resources, I.A. and L.G.-G.; data curation, L.G.-G., I.V. and I.A; writing—original draft preparation, I.B., L.G.-G., I.V., M.B. and I.A.; writing—review and editing, L.D. and I.A.; visualization, I.A. and L.G.-G.; supervision, I.A.; project administration, I.A.; funding acquisition, I.A. All authors have read and agreed to the published version of the manuscript.

Funding: This research received no external funding.

Institutional Review Board Statement: Not applicable.

Informed Consent Statement: Not applicable.

Data Availability Statement: The original contributions presented in the study are included in the article, further inquiries can be directed to the corresponding author.

Acknowledgments: Ioannis Karpathakis from Merck Romania is acknowledged for technical support regarding the protein electrophoresis analysis using the mPage Lux system. The Integrated Center for Research, Expertise and Technological Transfer in Food Industry is acknowledged for providing technical support.

Conflicts of Interest: The authors declare no conflicts of interest.

References

1. El Khoury, D.; Balfour-Ducharme, S.; Joye, I.J. A review on the gluten-free diet: Technological and nutritional challenges. *Nutrients* **2018**, *10*, 1410. [CrossRef]
2. do Nascimento, A.B.; Fiates, G.M.R.; Dos Anjos, A.; Teixeira, E. Analysis of ingredient lists of commercially available gluten-free and gluten-containing food products using the text mining technique. *Int. J. Food Sci. Nutr.* **2013**, *64*, 217–222. [CrossRef] [PubMed]
3. Skendi, A.; Papageorgiou, M.; Varzakas, T. High protein substitutes for gluten in gluten-free bread. *Foods* **2021**, *10*, 1997. [CrossRef]
4. Phongthai, S.; D'Amico, S.; Schoenlechner, R.; Homthawornchoo, W.; Rawdkuen, S. Effects of protein enrichment on the properties of rice flour based gluten-free pasta. *LWT* **2017**, *80*, 378–385. [CrossRef]
5. Navruz-Varli, S.; Sanlier, N. Nutritional and health benefits of quinoa (Chenopodium quinoa Willd.). *J. Cereal. Sci.* **2016**, *69*, 371–376. [CrossRef]
6. Xu, X.; Luo, Z.; Yang, Q.; Xiao, Z.; Lu, X. Effect of quinoa flour on baking performance, antioxidant properties and digestibility of wheat bread. *Food Chem.* **2019**, *294*, 87–95. [CrossRef] [PubMed]
7. Repo-Carrasco, R.; Espinoza, C.; Jacobsen, S.E. Nutritional value and use of the Andean crops quinoa (*Chenopodium quinoa*) and kañiwa (*Chenopodium pallidicaule*). *Food Rev. Int.* **2003**, *19*, 179–189. [CrossRef]
8. Thamnarathip, P.; Jangchud, K.; Nitisinprasert, S.; Vardhanabhuti, B. Identification of peptide molecular weight from rice bran protein hydrolysate with high antioxidant activity. *J. Cereal. Sci.* **2016**, *69*, 329–335. [CrossRef]
9. Coscueta, E.R.; Campos, D.A.; Osório, H.; Nerli, B.B.; Pintado, M. Enzymatic soy protein hydrolysis: A tool for biofunctional food ingredient production. *Food Chem. X* **2019**, *1*, 100006. [CrossRef]
10. Brumă, M.; Dumitrașcu, L.; Vasilean, I.; Banu, I.; Patrașcu, L.; Aprodu, I. Effect of enzymolysis on the antioxidant activity and functional properties of the soluble soy proteins. *Ann. Univ. Dunarea Jos Galati. Fascicle VI-Food Technol.* **2023**, *47*, 44–63. [CrossRef]
11. Sarmadi, B.H.; Ismail, A. Antioxidative peptides from food proteins: A review. *Peptides* **2010**, *31*, 1949–1956. [CrossRef] [PubMed]
12. SR ISO 712:2005; SR 91:2007. Romanian Standards Catalog for Cereal and Milling Products Analysis. ASRO: Bucharest, Romania, 2008.
13. Singh, J.P.; Kaur, A.; Singh, N. Development of eggless gluten-free rice muffins utilizing black carrot dietary fibre concentrate and xanthan gum. *J. Food Sci. Technol.* **2016**, *53*, 1269–1278. [CrossRef] [PubMed]
14. Banu, I.; Patrașcu, L.; Vasilean, I.; Dumitrașcu, L.; Aprodu, I. Influence of the Protein-Based Emulsions on the Rheological, Thermo-Mechanical and Baking Performance of Muffin Formulations. *Appl. Sci.* **2023**, *13*, 3316. [CrossRef]

15. Banu, I.; Aprodu, I. Investigation on functional, thermo-mechanical and bread-making properties of some white and black rice flours. *Appl. Sci.* **2022**, *12*, 4544. [CrossRef]
16. Matos, M.E.; Sanz, T.; Rosell, C.M. Establishing the function of proteins on the rheological and quality properties of rice based gluten free muffins. *Food Hydrocolloid* **2014**, *35*, 150–158. [CrossRef]
17. Banu, I.; Vasilean, I.; Aprodu, I. Effect of lactic fermentation on antioxidant capacity of rye sourdough and bread. *Food Sci. Technol. Res.* **2010**, *16*, 571–576. [CrossRef]
18. Sui, X.; Zhang, T.; Jiang, L. Soy protein: Molecular structure revisited and recent advances in processing technologies. *Annu. Rev. Food Sci. Technol.* **2021**, *12*, 119–147. [CrossRef] [PubMed]
19. Wilson, S.; Blaschek, K.; De Mejia, E.G. Allergenic proteins in soybean: Processing and reduction of P34 allergenicity. *Nutr. Rev.* **2005**, *63*, 47–58. [CrossRef] [PubMed]
20. Mulalapele, L.T.; Xi, J. Detection and inactivation of allergens in soybeans: A brief review of recent research advances. *Grain Oil Sci. Technol.* **2021**, *4*, 191–200. [CrossRef]
21. Yang, H.; Qu, Y.; Li, J.; Liu, X.; Wu, R.; Wu, J. Improvement of the protein quality and degradation of allergens in soybean meal by combination fermentation and enzymatic hydrolysis. *LWT* **2020**, *128*, 109442. [CrossRef]
22. Wen, L.; Bi, H.; Zhou, X.; Zhu, H.; Jiang, Y.; Ramadan, N.S.; Zheng, R.; Wang, Y.; Yang, B. Structure and activity of bioactive peptides produced from soybean proteins by enzymatic hydrolysis. *Food Chem. Adv.* **2022**, *1*, 100089. [CrossRef]
23. Hanafusa, K.; Murakami, H.; Ueda, T.; Yano, E.; Zaima, N.; Moriyama, T. Worm wounding increases levels of pollen-related food allergens in soybean (*Glycine max*). *Biosci. Biotech Biochem.* **2018**, *82*, 1207–1215. [CrossRef]
24. Mo, X.; Wang, D.; Sun, X.S. Physicochemical properties of β and α′ α subunits isolated from soybean β-conglycinin. *J. Agric. Food Chem.* **2011**, *59*, 1217–1222. [CrossRef] [PubMed]
25. de Souza, C.C.T.; Rosário Filho, N.A.; Camargo, J.F.D.; Godoi, R.H.M. Levels of airborne soybean allergen (Gly m 1) in a Brazilian soybean production city: A pilot study. *Int. J. Environ. Res. Pub. Health* **2020**, *17*, 5381. [CrossRef] [PubMed]
26. Pi, X.; Sun, Y.; Fu, G.; Wu, Z.; Cheng, J. Effect of processing on soybean allergens and their allergenicity. *Trends Food Sci. Technol.* **2021**, *118*, 316–327. [CrossRef]
27. Biscola, V.; de Olmos, A.R.; Choiset, Y.; Rabesona, H.; Garro, M.S.; Mozzi, F.; Chobert, J.-M.; Drouet, M.; Haertie, T.; Franco, B.D.G.D.M. Soymilk fermentation by Enterococcus faecalis VB43 leads to reduction in the immunoreactivity of allergenic proteins β-conglycinin (7S) and glycinin (11S). *Benef. Microbes* **2017**, *8*, 635–643. [CrossRef]
28. Pătrașcu, L.; Banu, I.; Vasilean, I.; Aprodu, I. Effect of gluten, egg and soy proteins on the rheological and thermo-mechanical properties of wholegrain rice flour. *Food Sci. Technol. Int.* **2017**, *23*, 142–155. [CrossRef]
29. Marco, C.; Rosell, C.M. Breadmaking performance of protein enriched, gluten-free breads. *Eur. Food Res. Technol.* **2008**, *227*, 1205–1213. [CrossRef]
30. Bonet, A.; Blaszczak, W.; Rosell, C.M. Formation of homopolymers and heteropolymers between wheat flour and several protein sources by transglutaminase-catalyzed cross-linking. *Cereal Chem.* **2006**, *83*, 655–662. [CrossRef]
31. Nogueira, A.D.C.; Aguiar, E.V.D.; Capriles, V.D.; Steel, C.J. Correlations among SRC, Mixolab®, process, and technological parameters of protein-enriched biscuits. *Cereal Chem.* **2021**, *98*, 716–728. [CrossRef]
32. Dakhili, S.; Abdolalizadeh, L.; Hosseini, S.M.; Shojaee-Aliabadi, S.; Mirmoghtadaie, L. Quinoa protein: Composition, structure and functional properties. *Food Chem.* **2019**, *299*, 125161. [CrossRef] [PubMed]
33. Cao, X.; Wen, H.; Li, C.; Gu, Z. Differences in functional properties and biochemical characteristics of congenetic rice proteins. *J. Cereal Sci.* **2009**, *50*, 184–189. [CrossRef]
34. Guo, X.; Sun, X.; Zhang, Y.; Wang, R.; Yan, X. Interactions between soy protein hydrolyzates and wheat proteins in noodle making dough. *Food Chem.* **2018**, *245*, 500–507. [CrossRef] [PubMed]
35. Schmiele, M.; Felisberto, M.H.F.; Clerici, M.T.P.S.; Chang, Y.K. Mixolab™ for rheological evaluation of wheat flour partially replaced by soy protein hydrolysate and fructooligosaccharides for bread production. *LWT-Food Sci. Technol.* **2017**, *76*, 259–269. [CrossRef]
36. Lamsal, B.P.; Jung, S.; Johnson, L.A. Rheological properties of soy protein hydrolysates obtained from limited enzymatic hydrolysis. *LWT-Food Sci. Technol.* **2017**, *40*, 1215–1223. [CrossRef]
37. Tsumura, K.; Saito, T.; Tsuge, K.; Ashida, H.; Kugimiya, W.; Inouye, K. Functional properties of soy protein hydrolysates obtained by selective proteolysis. *LWT-Food Sci. Technol.* **2017**, *38*, 255–261. [CrossRef]
38. Li, W.; Cao, W.; Wang, P.; Li, J.; Zhang, Q.; Yan, Y. Selectively hydrolyzed soy protein as an efficient quality improver for steamed bread and its influence on dough components. *Food Chem.* **2021**, *359*, 129926. [CrossRef]
39. Ashaolu, T.J. Applications of soy protein hydrolysates in the emerging functional foods: A review. *Int. J. Food Sci. Technol.* **2020**, *55*, 421–428. [CrossRef]
40. Hou, Y.; Zhao, X.H. Limited hydrolysis of two soybean protein products with trypsin or neutrase and the impacts on their solubility, gelation and fat absorption capacity. *Biotechnology* **2011**, *10*, 190–196. [CrossRef]
41. Huang, X.; Li, C.; Yang, F.; Xie, L.; Xu, X.; Zhou, Y.; Pan, S. Interactions and gel strength of mixed myofibrillar with soy protein, 7S globulin and enzyme-hydrolyzed soy proteins. *Eur. Food Res. Technol.* **2010**, *231*, 751–762. [CrossRef]
42. Zhang, Y.; Guo, X.; Shi, C.; Ren, C. Effect of soy proteins on characteristics of dough and gluten. *Food Chem.* **2020**, *318*, 126494. [CrossRef] [PubMed]

43. Walker, R.; Tseng, A.; Cavender, G.; Ross, A.; Zhao, Y. Physicochemical, nutritional, and sensory qualities of wine grape pomace fortified baked goods. *J. Food Sci.* **2014**, *79*, S1811–S1822. [CrossRef] [PubMed]
44. Kane, A.M.; Lyon, B.G.; Swanson, R.B.; Savage, E.M. Comparison of two sensory and two instrumental methods to evaluate cookie color. *J. Food Sci.* **2003**, *68*, 1831–1837. [CrossRef]
45. Nath, P.; Kale, S.J.; Kaur, C.; Chauhan, O.P. Phytonutrient composition, antioxidant activity and acceptability of muffins incorporated with red capsicum pomace powder. *J. Food Sci. Technol.* **2018**, *55*, 2208–2219. [CrossRef]

Disclaimer/Publisher's Note: The statements, opinions and data contained in all publications are solely those of the individual author(s) and contributor(s) and not of MDPI and/or the editor(s). MDPI and/or the editor(s) disclaim responsibility for any injury to people or property resulting from any ideas, methods, instructions or products referred to in the content.

Article

Gluten-Free Cookies Enriched with Baobab Flour (*Adansonia digitata* L.) and Buckwheat Flour (*Fagopyrum esculentum*)

Sylvestre Dossa [1], Christine Dragomir [1], Loredana Plustea [1], Cosmin Dinulescu [1], Ileana Cocan [1], Monica Negrea [1], Adina Berbecea [2], Ersilia Alexa [1,*] and Adrian Rivis [1]

[1] Faculty of Food Engineering, University of Life Sciences "King Mihai I" from Timisoara, Aradului Street No. 119, 300645 Timisoara, Romania; sylvestredossa04@gmail.com (S.D.); christine.dragomir98@gmail.com (C.D.); loredanapaven@yahoo.com (L.P.); cosmin.dinulescu@yahoo.com (C.D.); ileanacocan@usvt.ro (I.C.); monicanegrea@usvt.ro (M.N.); adrianrivis@usvt.ro (A.R.)

[2] Faculty of Agriculture, University of Life Sciences "King Mihai I" from Timisoara, Aradului Street No. 119, 300645 Timisoara, Romania; adina_berbecea@usvt.ro

* Correspondence: ersiliaalexa@usvt.ro

Featured Application: The paper has application potential in the bakery and related industries, considering that the proposed technological solutions can be implemented in the profile units in order to diversify the assortment range of flour products.

Abstract: To provide people with celiac disease with nutrient-rich gluten-free foods, this study aimed to produce cookies based on buckwheat and baobab flours, which were then subjected to nutritional, phytochemical, and sensory analyses. Results demonstrate that baobab flour (BF) and buckwheat flour (BWF) work together to enhance the nutritional properties of the cookies, in that nutrients that BWF is deficient in, BF provides sufficiently, and vice versa. BF is rich in minerals and carbohydrates, while BWF contains comparatively higher fat and protein levels. As for macro- and micro-elements, potassium (K) is the predominant macro-element in BF and BWF, with $13{,}276.47 \pm 174$ mg/kg and 1255.35 ± 58.92 mg/kg, respectively. The polyphenol content is higher in BF than BWF, at 629.7 ± 0.35 mg/100 g as opposed to 283.87 ± 0.06 mg/100 g. Similarly, the total flavonoid content and antioxidant activity of BF was greater than that of BWF, while BF exhibited 213.13 ± 0.08 mg/100 g and $86.62 \pm 0.04\%$, in contrast to BWF, which had 125.36 ± 1.12 mg/100 g and $79.72 \pm 0.01\%$, respectively. BF significantly enhanced the phytochemical composition of the cookies, with the richest sample being BBC3 containing 30% baobab. Buckwheat and baobab have the most abundant phenolic compounds of rutin and epicatechin, respectively. About the analysis of sensory attributes of the cookies, the partial substitution of BWF by BF of up to 20% (BWF3) significantly increased the scores for all attributes. Indeed, the appearance (physical aspect of the cookie: whether it is firm or not) and color (influence of baobab addition on cookie coloration) of the cookies were significantly improved with the addition of BF of up to 20%, but above 20% they were less appreciated. Similarly, up to 20% BF, the texture, flavor, and overall acceptability of the cookies were significantly improved. Taste, on the other hand, was not significantly improved, maybe due to the acidic taste provided by the baobab.

Keywords: gluten-free cookies; *Adansonia digitata*; *Fagopyrum esculentum*; celiac disease; individual polyphenols; antioxidant activity; flavonoids

Citation: Dossa, S.; Dragomir, C.; Plustea, L.; Dinulescu, C.; Cocan, I.; Negrea, M.; Berbecea, A.; Alexa, E.; Rivis, A. Gluten-Free Cookies Enriched with Baobab Flour (*Adansonia digitata* L.) and Buckwheat Flour (*Fagopyrum esculentum*). *Appl. Sci.* **2023**, *13*, 12908. https://doi.org/10.3390/app132312908

Academic Editor: Marco Iammarino

Received: 31 October 2023
Revised: 29 November 2023
Accepted: 29 November 2023
Published: 1 December 2023

Copyright: © 2023 by the authors. Licensee MDPI, Basel, Switzerland. This article is an open access article distributed under the terms and conditions of the Creative Commons Attribution (CC BY) license (https://creativecommons.org/licenses/by/4.0/).

1. Introduction

Cookies are a beloved baked food around the globe, appreciated for their unique flavor and texture [1]. While wheat flour has been the primary ingredient in commercial cookies, there has been a growing interest in using gluten-free cereals due to the high prevalence of celiac disease [2]. Celiac disease is an auto-immune disorder affecting the small intestine. It is a disease that causes permanent intolerance to gluten ingestion in genetically predisposed individuals, in that gluten ingestion leads to the destruction of enterocyte villi in affected patients [3]. About 1% of the global population is affected by celiac disease, and they must adhere to a strict gluten-free diet, which may lack certain essential nutrients [4]. Gluten is a mixture of proteins that ensures the elasticity of the dough and the maintenance of fermentation gases during the technological process of producing bread. The absence of gluten leads to a flattened product with low volume. For people with celiac disease, gluten ingestion causes an abnormal immune response in the small intestine. This reaction not only destroys gluten, as if it is dangerous to the body, but also attacks the lining of the small intestinal mucosa. Inflammatory substances end up destroying the intestinal villi, which allow the nutrient absorption [5]. Thus, the development of gluten-free products for celiacs has become more challenging, requiring the formulation of products that are not only gluten-free but also nourishing.

Originally hailing from the high-altitude regions of southern China, buckwheat (*Fagopyrum esculentum*) now thrives in Asia, Europe, and America. Despite its name, buckwheat is not a true cereal but rather a pseudocereal. Buckwheat, an ancient crop, is abundant in phytochemicals that are salutary to health. High levels of flavonoids and polyphenols are found in buckwheat, with rutin and quercetin being the primary polyphenol group with antioxidant properties [6,7]. Buckwheat has piqued interest in recent years due to its nutritional and medicinal makeup, providing complex carbohydrates, protein, fiber, vitamins, and minerals. Buckwheat proteins are gluten-free and boast a balanced amino acid composition, which is advantageous to those with celiac disease [8,9]. This behavior of buckwheat proteins was previously demonstrated, also emphasizing that the effect of buckwheat flour consumption by celiac patients did not present any toxic prolamin or toxicity to the patients examined in the study [10]. Wheat proteins possess a deficiency of certain amino acids, such as lysine, while buckwheat flour has an excellent protein quality, including specific amino acids like leucine, lysine, histidine, and valine [11]. Generally speaking, every 100 g of buckwheat flour contains 16.66 g of protein, 3.42 g of fat, 72.19 g of carbohydrates, 0.58 g of fiber, and 1.68 g of ash [12]. When compared to wheat, buckwheat contains higher levels of essential minerals such as zinc (Zn), iron (Fe), magnesium (Mg), and calcium (Ca) [6,7].

Researchers seem to be increasingly interested in including buckwheat in the production of healthier foods such as bread, muffins, pasta, cakes, and many other foods [5,13–17], as its composition can have a positive impact on the health of consumers [13]. Taken together, these studies indicate that buckwheat makes an excellent contribution to improving the nutritional and technical quality of gluten-free baked foods due to its content of proteins, lipids, fibers, and minerals as well as bioactive compounds. Buckwheat is an ideal food ingredient for making a variety of foods in general, and bakery products in particular, due to the gluten-free aspect and its abundance of nutrients and health-promoting phenolic compounds [18].

The baobab (*Adansonia digitata* L.) is a big tree native to arid and semi-arid regions of West Africa. Much of its content is important in improving the livelihoods of people in several African countries [19,20]. The most useful part of the baobab tree is the fruit, which is a source of food for a large rural population. The powder obtained from the pulp can be used as a spice in traditional dishes or dissolved in water or milk to make a drink, known as Gunguliz in Sudan [19,21,22]. Baobab fruit is an excellent source of amino acids as it contains all eight essential amino acids and is also rich in vitamins and minerals [23,24]. According to several studies, its pulp contains calcium, phosphorus, potassium, carbohydrates, fiber, protein, lipids, and vitamin C [23,25–28]. It can provide 54–100% of the recommended dietary intake of vitamin C or H. The value is ten times that of oranges. Dried baobab pulp provides between 3 and 499 mg/100 g of vitamin C. It thus contributes to the European Union's (EU) recommended daily requirement for vitamin C (80 mg/day) [23,29]. Baobab pulp is also rich in phenolic compounds, flavonoids, and organic acids [30–33]. These advantages make baobab pulp an ideal carrier for functional food formulations.

Given the functional properties and health benefits of buckwheat and baobab, this study aimed to prepare cookies based on buckwheat flour and baobab pulp in different proportions and to assess the improvement in nutritional and phytochemical values, as well as the sensory properties of the finished products.

2. Materials and Methods

2.1. Preparation of Composite Flours

The baobab (BF) and buckwheat flours were procured, respectively, from Beninese producers and from the SELGROS supermarket in Romania. Composite buckwheat/baobab flours (3) were produced according to [25]. There were BBF1 (10% baobab flour (BF) and 90% buckwheat flour (BWF)), BBF2 (20% BF and 80% BWF), and BBF3 (30% BF and 70% BWF).

2.2. Cookie Preparation

The cookies were prepared according to Farzana et al., 2022, and Mounjouenpou et al., 2018 [18,34], with some modifications. All ingredients (honey, salt, butter, and egg) used for the formulation of the different cookies, except baobab and buckwheat flour, were acquired from the Profile supermarket, in Timisoara, Romania. Four (4) types of cookies (CC, BBC1, BBC2, and BBC3) were formulated with different levels of substitution of BWF by BF (CC—control cookie with 100% BWF; BBC1—biscuit with 10% BF and 90% BWF; BBC2—biscuit with 20% BF and 80% BWF; and BBC3—biscuit with 30% BF and 70% BWF). The dough was obtained by mixing honey, salt, butter, egg, and flour (Figure 1). It was then rolled up in a biodegradable plastic bag and placed in the fridge. After 6 h in the fridge (at 5 °C), the biodegradable plastic bag was removed, the dough cut into rounds, and placed in the oven at 180 °C for 10 min.

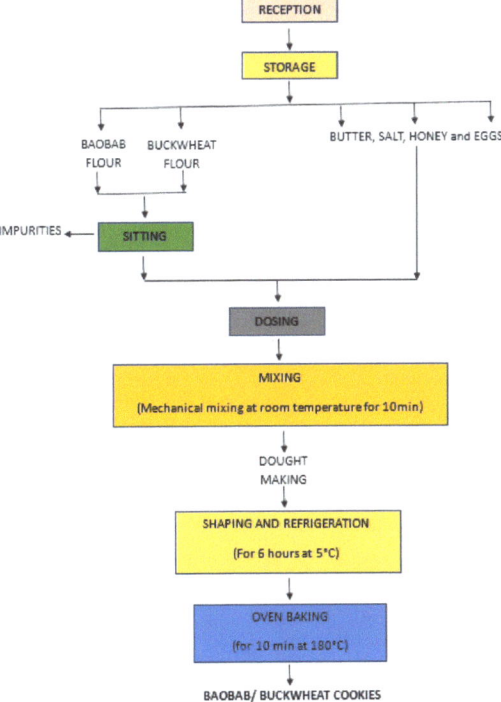

Figure 1. The technological scheme for obtaining cookies.

Table 1 shows the composition of each cookie.

Table 1. Recipe for cookies with composite baobab/buckwheat flour.

Samples	Ingredients					
	Baobab Flour (g)	Buckwheat Flour (g)	Butter (g)	Salt (g)	Honey (g)	Eggs (pcs)
CC	-	340	200	0.5	200	2
BBC1	34	306	200	0.5	200	2
BBC2	68	272	200	0.5	200	2
BBC3	102	238	200	0.5	200	2

Figure 1 shows the technological scheme for obtaining different cookies.
The different types of cookies that resulted from this study are shown in Figure 2.

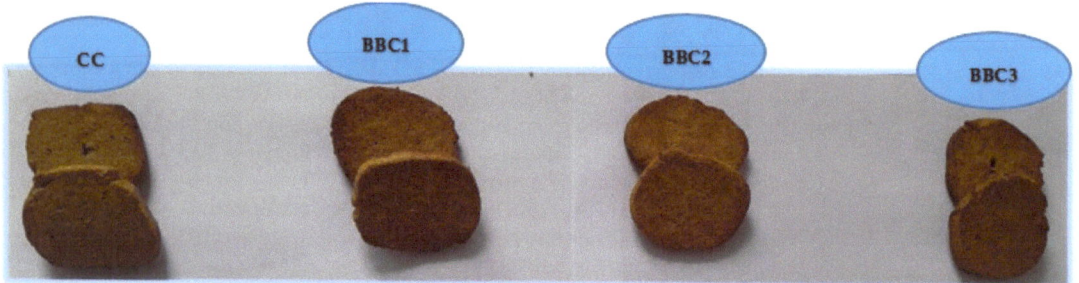

Figure 2. Control cookies and cookies with different proportions of buckwheat/baobab composite flour (CC—control cookie with 100% buckwheat flour; BBC1—cookie with 10% baobab flour and 90% buckwheat flour; BBC2—cookie with 20% baobab flour and 80% buckwheat flour; and BBC3—cookie with 30% baobab flour and 70% buckwheat flour).

2.3. Determination of Proximate Composition

The proximate composition of baobab, buckwheat, and composite baobab/buckwheat flours was determined as part of this study. The same approach was carried out for the various cookies obtained. The methods used to achieve this are presented in Table 2.

Table 2. Methods used to determine the proximate composition of the various samples.

Parameters	Methods	Unit
Ash	ISO 2171/2007 [35]	%
Moisture	Standard Methods of the International Association for Cereal Science and Technology (2003) [36]	%
Protein	Standard Methods of the International Association for Cereal Science and Technology (2003) [36]	%
Fat	Association of Official Analytical Chemists [37]	
Carbohydrate	Carbohydrate content was calculated as the difference between 100 and the sum of moisture, ash, protein, and fat content [38]	g/100 g
Energy value	The energy value was obtained by summing 4 times the protein content, 4 times the carbohydrate content, and 9 times the fat content [38]	kcal/100 g

2.4. Macro- and Micro-Elements

The content of macro- and micro-elements in the various samples was obtained in this study using the method applied in the studies of Plustea et al., 2022 [38]. Results were reported in mg/kg.

2.5. Phytochemical Profile

2.5.1. Preparation of Alcoholic Extracts

Extraction was carried out according to the procedure described by [25]. Alcoholic extracts were prepared by dissolving 1 g of each sample in 10 mL of ethanol (70%) in a hermetically sealed plastic container, then filtering after 30 min of stirring.

2.5.2. Evaluation of Total Phenolic Content (TPC)

The TFC (mg gallic acid equivalents (GAE)/100 g) of different flours (baobab, buckwheat, and baobab/buckwheat) and different cookies (control cookies and cookies with composite baobab/buckwheat flour) was determined using the previously prepared extracts. For this purpose, the Folin–Ciocâlteu method described by Danciu et al., 2019 [39], and Obistioiu et al., 2021 [40], was used as a reference. All determinations were performed three times.

2.5.3. Determination of Total Flavonoid Content (TFC)

The TFC (in mg QE/100 g) of the different flours and cookies in this study was determined using quercetin (QE) as the standard solution and according to the method of [41].

2.5.4. Antioxidant Activity (AA)

The AA of the various cookies and flours was determined using a spectrophotometer at an absorbance of 518 min. The method described by [42] with a few modifications was used. AA was obtained in % according to the following formula:

$$AA\,(\%) = (ControlAbsorbance - SampleAbsorbance/ControlAbsorbance) \times 100$$

where ControlAbsorbance refers to the absorbance values of the control and SampleAbsorbance to the absorbance values of the sample.

Note that ethyl alcohol for the control absorbance was used.

2.5.5. Determination of Individual Polyphenols via LC-MS

To determine individual levels of polyphenols, the method described in [39] was utilized with slight adaptations. A Shimadzu chromatograph equipped with SPD-10A UV and LC-MS 2010 detectors was used with two chromatographic columns. Column I was Adsorbosphere UHS C18, 5 µm, lot 007250, while column II was EC 150/2 NU-CLE-ODUR C18 Gravity SB 150 × 2 mm × 5 µm, ref:760618.20, SN E 15110907, lot 38775055.

The following chromatographic conditions were used:

The mobile phase A was composed of water that was acidified with formic acid to a pH of 3. To create the mobile phase B, a combination of acetonitrile that had been acidified with formic acid to a pH of 3 was utilized.

The grading program was as follows: 0.01 to 20 min at 5% B; 20.01 to 50 min at 5 to 40% B; 50 to 55 min at 40 to 95% B; 55 to 60 min at 95% B.

The solvent flow rate was 0.3 mL/min at 20 °C and the wavelength used for the control was 280 to 320 nm. Calibration curves were performed in the range of 20–50 µg/mL. All measurements were performed in triplicate and results were expressed as mg/kg. All measurements were done in triplicate, and the results were expressed in mg/kg.

2.6. Sensory Analysis

The sensory evaluation of the different cookies was carried out in accordance with ISO 6658:2017 [43]. Twenty-one assessors participated in the sensory analysis. Panel members were trained and aged between 19 and 46. The panel comprised 12 females and 9 males.

2.7. Statistical Analysis

Three replicates were performed for each parameter. All results are expressed as mean ± SD. Microsoft Excel 365 was used to analyze differences between mean values using a two-sample *t*-test assuming equal variances. Differences were considered significant if $p < 0.05$.

3. Results and Discussion

3.1. Proximate Composition of Composite Flours and Cookies

The characteristics of the different flours and formulated cookies are presented in Table 3.

Table 3. Proximal composition of buckwheat flour, baobab flour, buckwheat/baobab composite flours and cookies.

Samples	Nutritional Characteristics					
	Moisture	Mineral Content	Proteins	Lipids	Carbohydrates	Energy Values
	(%)	(%)	(%)	(%)	(g/100 g)	(kcal/100 g)
			Composite flours			
BWF	10.13 ± 1.15 [a]	1.66 ± 0.01 [a]	12.63 ± 0.02 [a]	3.19 ± 0.07 [a]	72.38 ± 1.09 [a]	368.79 ± 4.87
BF	13.79 ± 0.01 [b]	4.00 ± 0.02 [b]	4.31 ± 0.05 [b]	1.56 ± 0.02 [b]	76.34 ± 0.06 [b]	336.62 ± 0.16
BBF1	9.07 ± 0.06 [c]	1.99 ± 0.02 [c]	11.72 ± 0.03 [c]	2.46 ± 0.02 [c]	74.76 ± 0.09 [c]	368.03 ± 0.20
BBF2	9.23 ± 0.04 [c,d]	2.22 ± 0.19 [d]	11.13 ± 0.02 [d]	2.29 ± 0.04 [d]	75.13 ± 0.24 [c,d]	365.65 ± 0.96
BBF3	9.38 ± 0.03 [d]	2.67 ± 0.01 [e]	9.98 ± 0.03 [e]	2.20 ± 0.02 [d]	75.78 ± 0.01 [d]	362.80 ± 0.12
			Cookies			
CC	10.58 ± 0.28 [a]	1.10 ± 0.20 [a]	8.27 ± 0.03 [a]	22.96 ± 0.04 [a]	57.09 ± 0.13 [a]	468.05 ± 0.67
BBC1	11.58 ± 0.01 [b]	1.31 ± 0.04 [b]	7.67 ± 0.03 [b]	21.53 ± 0.08 [b]	57.91 ± 0.04 [a,b]	456.12 ± 0.56
BBC2	12.05 ± 0.08 [b,c]	1.42 ± 0.04 [c]	7.00 ± 0.03 [c]	20.78 ± 0.08 [c]	58.75 ± 0.10 [b,c]	450.05 ± 0.65
BBC3	12.25 ± 0.12 [c]	1.51 ± 0.03 [d]	6.87 ± 0.10 [c]	20.13 ± 0.07 [d]	59.24 ± 0.05 [c]	445.63 ± 0.55

Table values represent the mean ± standard deviation (SD) of three determinations, and different letters (a–e) in the same column for each sample category represent statistically significant differences ($p < 0.05$) detected using the *t* test. BF—baobab flour; BWF—buckwheat flour; BBF1—10% baobab flour and 90% buckwheat flour; BBF2—20% baobab flour and 80% buckwheat flour; and BBF3—30% baobab flour and 70% buckwheat flour; CC—control cookie with 100% buckwheat flour; BBC1—cookie with 10% baobab flour and 90% buckwheat flour; BBC2—cookie with 20% baobab flour and 80% buckwheat flour; and BBC3—cookie with 30% baobab flour and 70% buckwheat flour.

From the analysis of the results in Table 3, it can be seen that BF is richer in minerals, carbohydrates and water than BWF, while BWF is rich in protein and fat. In addition, BWF provides more energy than BF. The baobab and buckwheat flours complement each other, so blending them would give a complete composite flour in terms of nutritional composition. Moisture content was 10.13 ± 1.15% for BWF and 13.79 ± 0.01% for BF. These results are in line with those of [12,44–48]. Composite flour and cookie samples with high proportions of baobab are abundant in terms of moisture. This finding is explained by the high content observed in BF and is similar to those observed by Dossa et al. (2023) [25] and Barakat et al. (2021) [26]. Indeed, in the study by Dossa et al., 2023, from 11.39 ± 0.24% for the flour composed of 10% baobab, moisture content rose to 11.80 ± 0.03% for that with 30% baobab. Similarly, in the study by Barakat et al., 2021, from 11.80 ± 0.35% in the sample with 5% baobab, moisture rose to 11.97 ± 0.42% in the sample with 15% baobab. BF is more than two times richer in mineral substances than BWF (4 ± 0.02% vs. 1.66 ± 0.01%). These results are almost identical to those of Mohajan et al., 2019 [12], and Chopra et al., 2014 [48], who obtained 1.68 ± 0.01% and 1.38 ± 0.01%, respectively, for mineral substances in buckwheat flour. Similar results were also found by Bhinder et al., 2020, where the mineral content for several buckwheat varieties ranged from 1.76 ± 0.26% to 2.80 ± 0.06%. On the other hand, the mineral content obtained in this study for baobab is slightly lower than that obtained by [46], which ranged from 4.9 to 6.4%. This would be due to the variation in Baobab's nutritional values from one region to another [49]. However, other factors such as the species or the method of processing the pulp into flour can also influence nutritional values. For both flours and cookies, the higher the quantity of baobab, the higher the mineral content. From 1.99 ± 0.02% for BBF1, the mineral content rose to 2.67 ± 0.01% for BBF3. In the case of cookies, mineral content increased by 0.41% between CC and BBC3. This increase in mineral content can be explained by the fact that baobab is richer in minerals than buckwheat.

Unlike mineral substances, the protein content of BWF is around four times higher than that of BF. As a result, the protein content of composite flours and cookies decreases as the quantity of BF increases. The same applies to lipid content, which is higher in BWF than in BF. These results are in line with those of [12,25,26,44–47].

Carbohydrate content at BWF was 72.38 ± 1.09 g/100 g. This value is in line with that obtained by [12], which was 72.19 ± 0.09%. It is higher than the BF level (76.34 ± 0.06 g/100 g). This superiority of BF over BWF in carbohydrate content led to an increase in carbohydrate content in samples (flours and cookies) with more BF. From 74.76 ± 0.09 g/100 g for BBF1, it had risen to 75.78 ± 0.01 g/100 g for BBF3, and from 57.09 ± 0.13 g/100 g for CC, it had risen to 59.24 ± 0.05 g/100 g for BBC3.

In terms of energy value, baobab flour and the composite flours provided less than BWF. In addition, the energy value provided by the baobab cookies was lower than that provided by the control cookie. These results suggest that BWF provides a superior source of energy than BF. In the article by [25], between the control sample and the sample containing the highest quantity of baobab (30%), there was a significant decrease in the energy value provided. Similarly, the results of studies by [26,34] show the same observation. This suggests that BF provides less energy in both composite flours and cookies.

A comparison of the nutritional composition of BBC3 with cookies made from 100% wheat flour obtained in the studies by [18,48] shows that BBC3 is richer than cookies made from 100% wheat flour in terms of minerals, lipids, and carbohydrates. It is concluded that BF and BWF contribute to improving the nutritional values of cookies.

3.2. Macro- and Micro-Element Composition of Composite Flours and Cookies

The macro-element and the micro-element contents of buckwheat flour, baobab flour, buckwheat/baobab composite flours, and the resulting cookies are shown in Table 4.

Table 4. Macro- and micro-element content of various samples.

Samples	Macro- and Micro-Element Contents (mg/kg)							
	Fe	Zn	Ni	Cu	K	Mg	Ca	Mn
Composite flours								
BF	155.14 ± 2.95 [a]	14.90 ± 0.01 [a]	0.598 ± 0.002 [a]	8.04 ± 0.05 [a]	13,276.47 ± 174.00 [a]	1066.73 ± 9.97 [a]	1570.67 ± 29.67 [a]	4.84 ± 0.05 [a]
BWF	57.66 ± 0.16 [b]	17.30 ± 0.11 [b]	0.391 ± 0.01 [b]	4.23 ± 0.02 [b]	1255.35 ± 58.92 [b]	287.82 ± 2.01 [b]	181.55 ± 3.24 [b]	10.65 ± 0.04 [b]
BBF1	71.65 ± 0.12 [c,e]	14.12 ± 0.15 [a,c]	0.412 ± 0.01 [b]	4.46 ± 0.14 [b,c]	1389.57 ± 36.28 [a]	288.83 ± 0.48 [b,c]	586.12 ± 2.62 [c]	10.02 ± 0.12 [b]
BBF2	81.96 ± 0.07 [d]	13.60 ± 0.34 [c]	0.467 ± 0.001 [c]	4.57 ± 0.12 [c]	1640.07 ± 135.16 [c]	322.78 ± 23.95 [c,d]	628.74 ± 10.40 [d]	9.10 ± 0.08 [c]
BBF3	89.03 ± 0.1 [e]	12.30 ± 0.13 [d]	0.516 ± 0.02 [d]	4.88 ± 0.06 [d]	2488.81 ± 435.58 [d]	378.57 ± 9.07 [d]	865.82 ± 9.28 [e]	8.26 ± 0.05 [c]
Composite cookies								
CC	29.99 ± 0.11 [a]	7.09 ± 0.10 [a]	0.175 ± 0.03 [a]	3.18 ± 0.17 [a]	951.30 ± 64.88 [a]	272.42 ± 1.00 [a]	144.13 ± 12.19 [a]	6.51 ± 0.15 [a]
BBC1	33.02 ± 0.11 [a,b]	6.91 ± 0.09 [a]	0.243 ± 0.14 [a,b]	3.34 ± 0.14 [a,b]	1192.13 ± 127.20 [b]	275.96 ± 4.55 [a]	442.54 ± 8.15 [b]	6.03 ± 0.22 [a]
BBC2	36.22 ± 0.18 [b,c]	5.72 ± 0.12 [b]	0.347 ± 0.02 [c]	3.55 ± 0.05 [b,c]	1386.30 ± 104.14 [c]	293.20 ± 7.94 [b]	572.73 ± 15.01 [c]	5.16 ± 0.15 [b]
BBC3	39.99 ± 0.03 [c]	5.37 ± 0.08 [b]	0.382 ± 0.01 [c]	3.88 ± 0.04 [c]	2093.07 ± 38.86 [d]	298.72 ± 15.19 [b]	672.97 ± 38.99 [d]	4.67 ± 0.06 [c]

Table values represent the mean ± standard deviation (SD) of three determinations, and different letters (a–e) in the same column for each sample category represent statistically significant differences ($p < 0.05$) detected using the t test. BF—baobab flour; BWF—buckwheat flour; BBF1—10% baobab flour and 90% buckwheat flour; BBF2—20% baobab flour and 80% buckwheat flour; and BBF3—30% baobab flour and 70% buckwheat flour; CC—control cookie with 100% buckwheat flour; BBC1—cookie with 10% baobab flour and 90% buckwheat flour; BBC2—cookie with 20% baobab flour and 80% buckwheat flour; and BBC3—cookie with 30% baobab flour and 70% buckwheat flour.

Minerals are important for maintaining the body's overall physical and mental health and contribute to the maintenance and development of muscles, nerve cells, teeth, bones, tissues, and blood. They also play a crucial role in maintaining the immune system. Conversely, deficiencies in these minerals can lead to poor growth, bone loss, reduced appetite, hypogonadism, and cognitive impairment [50–53]. These results are similar to those of [6,7], who estimated that buckwheat contains higher amounts of essential minerals compared to wheat. These results are also in line with those of [26] for baobab.

The results also reveal that, apart from zinc (Zn) and manganese (Mn), BF is richer than BWF in all the other minerals studied. Moreover, potassium (K) is the dominant macro-element in both baobab and buckwheat, with 13,276.47 ± 174 mg/kg and 1255.35 ± 58.92 mg/kg, respectively. Several authors, [26,27,47,49] on the one hand and [12,44] on the other, have revealed in their respective studies that potassium (K) is the dominant macro-element in BF and BWF. Furthermore, the value found for potassium is in the range of those obtained by [26,27,47,49], i.e., ranging from 9875 to 2390 mg/kg for BF. Similarly, the results for BWF in this study were in the same range as those obtained by [12,44] (between 1087.8 and 6487 mg/kg). K is also the major macro-element in buckwheat/baobab composite flours and cookies according to Table 4. Also, samples (flours and cookies) with high quantities of baobab flour were the richest in K. This can be explained by the abundance of K in BF compared to BWF (Table 4). It can therefore be deduced that BF would contribute more K to buckwheat gluten-free fortified cookies. All these results were also reported by [25,26,34] in their various studies. A total of 100 g BBC3 would cover up to 6% of an adult's daily K requirement recommended by the World Health Organization (3510 mg/day) [54].

Calcium (Ca) is the most prevalent mineral in the body. Beyond its function in keeping bones strong and stable, calcium takes part in a plethora of metabolic processes that include cell adhesion, blood clotting, cell differentiation and growth, the release of hormones and neurotransmitters, muscle constriction, and glycogen metabolism. It serves as a crucial constituent of teeth and bones [53]. In this study, Ca was higher in BF compared to BWF (1570.67 ± 29.67 mg/kg vs. 181.55 ± 3.24 mg/kg). The Ca content obtained for baobab flour in this study does not fall within the range (between 237.03 and 370 mg/100 g) obtained by other researchers [26,27,45–47,49]. Also, the Ca value obtained in BWF is lower than the range found by [44] (30 and 272 mg/100 g). These differences could be attributed to variations in nutritional values of Baobab and buckwheat from one region to another and from one species to another [44,49]. Aside from differences in regional baobab, cultivars state other reasons might influence the varied mineral content, such as cultivation technologies, harvest period, storage, or conditioning conditions. Calcium content increased with the quantity of BF in the formulation of composite flours and cookies. It increased by 376.90% between BWF and BBF3 and by 366.92% between CC and BBC3. These results show that BF improves Ca availability in composite flours and cookies. The EFSA Scientific Panel [55] suggests a maximum daily calcium consumption of 2500 mg for adults, comprising pregnant and breastfeeding women. So, around 3% (67.29 mg Ca) of daily Ca requirements should be covered by 100 g of BBC3.

According to the results of the present study, like Ca and K, BF is richer in magnesium (Mg) and iron (Fe) than BWF. For Mg, BF and BWF were, respectively, 1066.73 ± 9.97 mg/kg and 287.82 ± 2.01 mg/kg. Fe levels were 155.14 ± 2.95 mg/kg and 57.66 ± 0.16 mg/kg for BF and BWF, respectively. Furthermore, as the proportion of baobab in flour and cookie composition increases, the Mg and Fe content becomes more abundant. These results not only agree with those of [12,25–27,44,47,49] but also allow us to deduce that BWF partial substitution by BF results in composite flours and cookies with significantly higher magnesium (Mg) and iron (Fe) contents. Iron's main function is to transport oxygen from the lungs to the tissues. It is also an important constituent of various enzyme systems such as cytochromes, which are involved in oxidative metabolism [53]. Concerning magnesium, the adequate daily intake was estimated [56] at 350, 300, and 170 to 300 mg, respectively, for men, women, and children. This being said, 100 g of the 30% BF cookie from the present study would provide 29.87 mg of Mg, representing, respectively, over 8.53%, 9.97%, and between 17.57 and 9.97% of the adequate Mg intake for men, women, and children.

About the micro-elements zinc (Zn), copper (Cu), manganese (Mn), and nickel (Ni), BWF is more abundant in Zn and Mn than BF. On the other hand, BF is richer in Cu and Ni than BWF. BF had 8.04 ± 0.05, 0.598 ± 0.002, 14.90 ± 0.01, and 4.84 ± 0.05 mg/kg, respectively, for Cu, Ni, Zn, and Mn. These values are close to those of [26,27,47,49]. BWF values for Cu, Ni, Zn, and Mn were 4.23 ± 0.02, 0.391 ± 0.01, 17.3 ± 0.11, and 10.65 ± 0.04 mg/kg, respectively. These values are also close to those of [12,44]. After the exploitation of these results, it is important to conclude that BF and BWF, each in their case, have contributed in one way or another to improving the nutritional composition of the cookies obtained in this study.

3.3. Phytochemical Profile of Composite Flours and Cookies

The results of phytochemical analyses of the various samples are presented in Table 5.

Table 5. Phytochemical composition of various samples.

Samples	Total Polyphenol Content (mg/100 g)	Total Flavonoid Content (mg/100 g)	Antioxidant Activity, DPPH (%)
Flours			
BWF	283.87 ± 0.06 [a]	125.36 ± 1.12 [a]	79.72 ± 0.01 [a]
BF	629.7 ± 0.35 [b]	213.13 ± 0.08 [b]	86.62 ± 0.04 [a,b]
BBF1	292.35 ± 0.35 [a,c]	181.03 ± 0.12 [c]	81.56 ± 0.19 [b]
BBF2	311.62 ± 0.78 [c,d]	194.94 ± 1.78 [d]	83.11 ± 0.02 [b]
BBF3	320.12 ± 2.07 [d]	209.28 ± 0.85 [b]	84.78 ± 0.01 [b]
Cookies			
CC	226.34 ± 0.75 [a]	102.96 ± 4.07 [a]	76.33 ± 0.07 [a]
BBC1	250.06 ± 1.17 [b]	135.74 ± 3.10 [b]	78.74 ± 0.03 [a,b]
BBC2	252.74 ± 0.15 [b]	140.44 ± 0.81 [b,c]	80.32 ± 0.03 [b,c]
BBC3	285 ± 32.82 [c]	147.69 ± 2 [c]	82.52 ± 0.07 [c,d]

Table values represent the mean ± standard deviation (SD) of three determinations, and different letters (a–d) in the same column for each sample category represent statistically significant differences ($p < 0.05$) detected using the t test. BF—baobab flour; BWF—buckwheat flour; BBF1—10% baobab flour and 90% buckwheat flour; BBF2—20% baobab flour and 80% buckwheat flour; and BBF3—30% baobab flour and 70% buckwheat flour; CC—control cookie with 100% buckwheat flour; BBC1—cookie with 10% baobab flour and 90% buckwheat flour; BBC2—cookie with 20% baobab flour and 80% buckwheat flour; and BBC3—cookie with 30% baobab flour and 70% buckwheat flour.

Polyphenols are secondary metabolites produced by plants through the pentose phosphate, phenylpropanoid, and shikimate pathways. They possess various physiological properties, such as antioxidant, antitumoral, antibacterial, and other activities [57,58]. In this study, the total polyphenol content of BF is more than 2 times higher than that of BWF, i.e., 629.7 ± 0.35 mg/100 g versus 283.87 ± 0.06 mg/100 g. For Baobab, [59] reported a total polyphenol content around 1.25 times lower than in the present study. In contrast, [27] reported a total polyphenol content 1.72 times higher than in the present study. For buckwheat, the results of this study are similar to those of [60], who obtained 2.83 ± 0.73 g/mg. However, they are below those obtained by [17,61], which were 7.25 ± 0.2 mg/g and 33.51 ± 0.52 mg/g, respectively. Analysis of the results also shows that the more BF in the various flours and cookies, the higher the total polyphenol content. Thus, there were 292.35 ± 0.35 mg/100 g, 311.62 ± 0.78 mg/100 g, and 320.12 ± 2.07 mg/100 g for BBF1, BBF2, and BBF3, respectively. For the cookie samples, there were, respectively, 226.34 ± 0 mg/100 g, 250.06 ± 1.17 mg/100 g, 252.74 ± 0.15 mg/100 g, and 285 ± 32.82 mg/100 g for CC, BBC1, BBC2, and BBC3. This observation could be explained by BF's abundance of polyphenols compared to BWF (Table 5). The same observation was made by [26]; i.e., polyphenol content increases with the amount of baobab.

Flavonoids are a collection of polyphenolic compounds located in both flora and human sustenance. They possess impressive antitumor, antioxidant, and microcirculation-enhancing attributes [62]. Buckwheat predominantly comprises flavonoids as the main active components [63]. In this study, the total flavonoid content of baobab flour was around 1.70 times higher than that of buckwheat (213.13 ± 0.08 vs. 125.36 ± 1.12 mg/100 g). Chlopicka et al., 2012 [61], found a lower flavonoid value than in the present study

(153 ± 12 µg/g). In this study, flavonoid content was significantly different between BF and BWF. Furthermore, there is a significant increase in flavonoids as the proportion of BF in the different flours and cookies is higher. An increase of 28.25 mg/100 g was observed between BBF1 and BBF3 and of 44.73 between CC and BBC3. Therefore, BF would be responsible for the abundance of flavonoids in the composite flours and cookies obtained compared to the control samples, thanks to its richness in flavonoids compared to buckwheat (Table 5). Similar results have been obtained by [59].

The antioxidant activity (AA) behaved in the same way as the total flavonoid and polyphenol content. In other words, the AA of BF is higher than that of BWF. BF and BWF obtained 86.62 ± 0.04% and 79.72 ± 0.01%, respectively. Similar results were obtained by [64]. In their study of the antioxidant activity of buckwheat extracts, Sun et al. [64] obtained an AA of 78.6 ± 6.2% for buckwheat flour. On the other hand, Antoniewska et al. [17] obtained a result lower (32.06 ± 0.17%) than the present study and [64]. In both buckwheat/baobab composite flours and cookies, AA increases with the proportion of BF in them. From 81.56 ± 0.19% for BBF1, AA rose to 84.78 ± 0.01% for BBF3, while from 76.33 ± 0.07% for CC, it rose to 82.52 ± 0.07% for BBC3. This was also observed by [26]. They highlighted the increase in antioxidant activity and total polyphenol content with the incorporation of a high percentage of baobab. Studies by Bolang et al., 2023 [65], on the combination of several plant ingredients (porang tubers, moringa leaves, and tempe made from black soybeans) in the formulation of functional cookies also reported improved antioxidant activity. It is interesting to note that consuming antioxidant-rich foods plays an important role in preventing degenerative and non-communicable diseases [66].

Following on from all the data concerning bioactive compounds, it should be emphasized that the substitution of BWF by BF significantly increased the phytochemical properties of buckwheat/baobab composite flours and cookies. Baobab could therefore be an alternative for the formulation of flour products with high antioxidant activity and high polyphenol and flavonoid content, without recourse to chemical additives.

3.4. Individual Polyphenols of Composite Flours and Cookies Determined via LC-MS

The quantification of some phenolic compounds in flours and cookies is shown in Table 6.

Table 6. Quantification of individual polyphenols.

Samples	Epicatechin (mg/kg)	Cafeic Acid (mg/kg)	Rutin (mg/kg)	Rosmarinic Acid (mg/kg)	Resveratrol (mg/kg)	Quercitin (mg/kg)
			Composite flours			
BWF	90.03 ± 0.21 [a]	17.57 ± 0.93 [a]	246.93 ± 0.75 [a]	68.13 ± 0.60 [a]	126.30 ± 1.67 [a]	15.80 ± 0.44 [a]
BF	158.6 ± 0.46 [b]	17.67 ± 1.01 [a]	nd *	67.93 ± 0.57 [a]	141.73 ± 0.57 [b]	16.10 ± 0.61 [a]
BBF1	106.63 ± 0.21 [c]	17.63 ± 0.15 [a]	145.80 ± 0.56 [b,c]	68.00 ± 0.50 [a]	128.50 ± 1.14 [a]	15.90 ± 0.10 [a]
BBF2	109.23 ± 0.04 [c]	17.64 ± 0.15 [a]	137.60 ± 0.61 [c,d]	68.01 ± 0.61 [a]	129.90 ± 0.10 [a,c]	15.91 ± 0.08 [a]
BBF3	121.18 ± 0.73 [d]	17.64 ± 0.14 [a]	122.20 ± 0.02 [d]	68.00 ± 0.05 [a]	132.20 ± 0.02 [c]	16.05 ± 0.13 [a]
			Cookies			
CC	nd	17.80 ± 0.1 [a]	259.57 ± 0.40 [a]	68.0 ± 0.53 [a]	116.53 ± 0.55 [a]	15.90 ± 0.10 [a]
BBC1	92.55 ± 0.05 [a]	17.75 ± 0.04 [a]	151.40 ± 0.18 [b]	67.97 ± 0.15 [a]	121.31 ± 0.45 [b]	15.94 ± 0.19 [a]
BBC2	99.79 ± 0.52 [b]	17.77 ± 0.04 [a]	140.98 ± 0.28 [b]	67.83 ± 0.06 [a]	124.12 ± 0.64 [c]	16.20 ± 0.13 [b]
BBC3	108.59 ± 0.62 [c]	17.77 ± 0.03 [a]	130.47 ± 0.51 [b]	67.83 ± 0.12 [a]	126.98 ± 0.25 [c]	16.26 ± 0.09 [b]

Table values represent the mean ± standard deviation (SD) of three determinations, and different letters (a–d) in the same column for each sample category represent statistically significant differences ($p < 0.05$) detected using the t test. BF—baobab flour; BWF—buckwheat flour; BBF1—10% baobab flour and 90% buckwheat flour; BBF2—20% baobab flour and 80% buckwheat flour; and BBF3—30% baobab flour and 70% buckwheat flour; CC—control cookie with 100% buckwheat flour; BBC1—cookie with 10% baobab flour and 90% buckwheat flour; BBC2—cookie with 20% baobab flour and 80% buckwheat flour; and BBC3—cookie with 30% baobab flour and 70% buckwheat flour. * nd: not detected.

Phenolic compounds are vital bioactive substances possessing antioxidant, anti-inflammatory, and antimicrobial properties [67]. Buckwheat and baobab are both commendable sources of these compounds [7,59,67]. The various phenolic compounds identified in flours and cookies are presented in Table 6. This table shows that the most abundant phenolic compounds studied (Table 6) in buckwheat and baobab are rutin and epicatechin, respectively. Several previous studies have identified rutin as the most abundant phenolic compound in buckwheat [7,13,68]. Buckwheat is a significant source of rutin in the human diet [68]. Similarly, Balarabe et al. [59] concluded that the most abundant phenolic compound in baobab fruit pulp was epicatechin.

Epicatechin is a type of phenolic compound that is thought to offer potential health benefits in preventing or mitigating cardiovascular disease [69]. In this study, BF was more abundant in epicatechin (158.6 ± 0.46 mg/kg) than BWF (90.03 ± 0.21 mg/kg). As the amount of BF in composite flours and cookies increases, the sample becomes more abundant in epicatechin. This is explained by the fact that BF is more abundant in epicatechin than BWF. This suggests that partial substitution of BWF by BF results in flours and cookies with higher epicatechin content. Apart from epicatechin, BF is also more abundant in resveratrol. BF had a resveratrol composition of 141.73 ± 0.57 mg/kg versus 126.3 ± 1.67 for BWF. Here too, the higher the BF content in the various flour and cookie samples, the more abundant the resveratrol content. From 128.5 ± 1.14 mg/kg for BBF1, resveratrol content rose to 132.2 ± 0.02 mg/kg for BBF3. Also, from 116.53 ± 0.55 mg/kg for CC, its content rose to 124.12 ± 0.64 mg/kg for BBC3. Unlike resveratrol and epicatechin, BF and BWF have virtually identical values for cafeic, rosmarinic, and quercitin. In the case of these compounds, the values did not practically change from one sample to the next. There is therefore no significant difference between the values obtained for these compounds from one sample to another.

Rutin, also known as vitamin P, is a flavonoid glycoside found in citrus fruits. This biologically active molecule can impact a wide range of both non-reproductive and reproductive processes, offering potential therapeutic benefits for various disorders. Among natural antioxidants, rutin is considered to be one of the most potent within its known class [70,71]. In the present study, BWF contained 246.93 ± 0.75 mg/kg rutin. However, BF did not obtain a value. This may be due to the analysis conditions and, above all, the solvent used. As an example, in the work carried out by [59] where they identified and quantified phenolic acids and flavonoids in baobab fruit powder extracts via HPLC using different solvents (μg/g), when they used ethyl acetate as the solvent, they had not been able to determine a value for rutin. On the other hand, when they used acid methanol, they obtained a value of 12.21 ± 0.86 μg/g. In the case of this study, the greater the quantity of BWF in the various flour and cookie samples, the greater the rutin content. This finding is explained by the fact that BWF is a good source of rutin compared to BWF. The same observation was made by [13]. It can therefore be deduced that BWF provides rutin-rich cookies.

There is also an increase in rutin content in cookies compared with flour. Similar results were reported by [13]. They reported a significantly higher rutin content after baking than before. This is explained by the fact that rutin, which was present in bound form before baking, was released during baking [68].

3.5. Sensory Analysis

The acceptability of the different cookies to consumers was determined by a sensory evaluation according to a five-point hedonic scale. The average scores for the various consumer sensory properties: appearance (the physical aspect of the cookie: whether it is firm or not), color (influence of the addition of baobab on the color of the cookie), flavor, texture, aroma, and overall acceptability of the cookies are presented in the figure below.

Partial substitution of BWF by BF of up to 20% significantly increased scores for all attributes (Figure 3). For each of the attributes, BBC2 was the sample with the highest scores across all samples, including the control sample (CC). It scored 4.53 ± 0.51, 4.65 ± 0.49,

4.53 ± 0.51, 4.41 ± 0.71, 4.47 ± 0.72, and 4.41 ± 0.62, respectively, for appearance, color, texture, taste, flavor, and overall acceptability. Having obtained scores between 4.5 and 5 for appearance, color, and texture, BBC2 is very acceptable for these different criteria. On the other hand, it is acceptable for the rest of the criteria, since its scores for these criteria (taste, flavor, and overall acceptability) were between 3.5 and 4.49. Sample BBC3 had the lowest scores among the samples with different proportions of BF (BBC1, BBC2, and BBC3). It should be noted that for some attributes BBC3 scored higher than CC. These were appearance (4.29 ± 0.69 vs. 3.41 ± 1.33), color (4.24 ± 0.66 vs. 4.12 ± 1.36), and texture (3.94 ± 0.75 vs. 3.71 ± 1.31). From all this information, it should be noted that BF improved all the sensory qualities of buckwheat gluten-free cookies by up to 20%. It should also be noted that up to 30% BF continues to improve the appearance, color, and texture of cookies. However, at 30% BF and above, consumers were less and less appreciative of the taste and flavor of the cookies, probably due to the tangy, lemony flavor that baobab brings to the cookies, which was not necessarily to consumers' taste. In conclusion, the limit of acceptability of baobab in cookies is 20%. It was also reported in [34] that replacing 20% BF in the cookie improved sensory qualities. A comparison of the acceptable limit level of BF substitution in bakery products from several studies [25,26,34], including the present study, allows us to say that Baobab is more suitable for cookie production than cake and bread production from a sensory point of view. Indeed, the substitution limit for BFs was 10% in bread [25] and 15% in cakes [26] but can be as high as 20% for cookies (Figure 3; [34]).

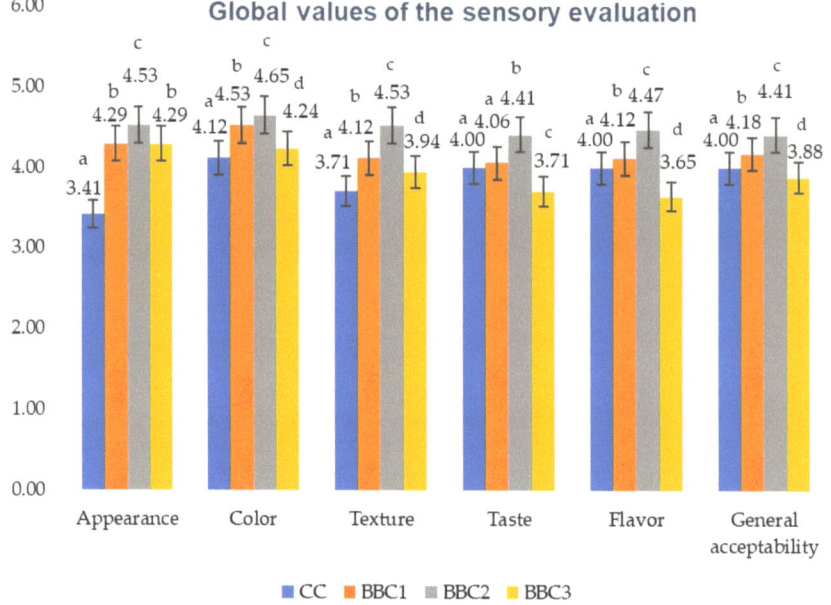

Figure 3. Overall sensory-evaluation values (consumer acceptance). Column values represent the mean of three determinations ± standard deviation (SD). Within each characteristic category, statistically significant differences ($p < 0.05$) are indicated by different letters (a–d) in the columns. CC—control cookie with 100% buckwheat flour; BBC1—cookie with 10% baobab flour and 90% buckwheat flour; BBC2—cookie with 20% baobab flour and 80% buckwheat flour; and BBC3—cookie with 30% baobab flour and 70% buckwheat flour.

4. Conclusions

The first objective of the study was to formulate functional, gluten-free cookies using a composite flour made up of baobab and buckwheat. In addition, the study investigated the nutritional, phytochemical, and sensory attributes of baobab flour in cookie production. The analyses conducted during the study revealed that the combination of baobab and buckwheat can effectively enhance bakery products, particularly cookies. The mineral and carbohydrate composition of cookies was enhanced by baobab, while buckwheat had a significant impact on their protein and lipid composition. In the same vein, the partial substitution of buckwheat flour for baobab flour resulted in composite flours and cookies with a high content of micro- and macro-elements. Rutin and epicatechin were found to be the most abundant phenolic compounds in buckwheat and baobab, respectively. It is worth noting that replacing BWF with BF significantly improved the phytochemical properties of buckwheat/baobab composite flours and cookies. This could make baobab a viable alternative to chemical additives for producing bakery products with high antioxidant activity and high levels of polyphenols and flavonoids. The study also found that substituting baobab improved cookie acceptability by up to 20%. However, when substituted at levels above 30%, although certain nutritional and phytochemical values increased, the resulting cookies were less well liked from an organoleptic perspective. This may be due to baobab's acidic taste, which hurt the cookie's taste. Therefore, the acceptable limit for baobab in cookies is 20%, regardless of appearance, color, taste, texture, flavor, or overall acceptance of the cookies. This formulation results in cookies with an impressive nutrient content and attractive physical-chemical characteristics, which are also highly appreciated by consumers and do not adversely affect their nutritional and technological properties. Therefore, it is concluded that the optimal recipe for producing cookies with higher nutritional quality and suitable sensory characteristics, utilizing baobab/buckwheat composite flours, is 20% baobab and 80% buckwheat. Consequently, the formulation and use of these functional foods can improve the nutritional well-being of consumers.

Author Contributions: All authors contributed to the study conception and design. Conceptualization original draft preparation, S.D., I.C., M.N. and E.A.; methodology: S.D., C.D. (Christine Dragomir), C.D. (Cosmin Dinulescu), L.P., I.C., A.B. and M.N.; formal analysis, S.D., E.A. and A.R.; review and editing and validation, S.D., E.A. and A.R. Supervising, A.R. All authors have read and agreed to the published version of the manuscript.

Funding: This research is supported by the project entitled: "Increasing the impact of excellence research on the capacity for innovation and technology transfer within USAMVB Timisoara", code 6PFE, submitted in the competition Program 1—Development of the national system of research—development, Subprogram 1.2—Institutional performance, Institutional development projects—Development projects of excellence in R.D.I.

Institutional Review Board Statement: Not applicable.

Informed Consent Statement: Not applicable.

Data Availability Statement: The report of the analyses performed for the samples in the paper can be found at the Interdisciplinary Research Platform (PCI) at the University of Life Sciences "King Mihai I", Timisoara.

Acknowledgments: The authors of this paper acknowledge the technical support provided by the Interdisciplinary Research Platform at the University of Life Sciences "King Mihai I", Timisoara, where the analyses were performed.

Conflicts of Interest: The authors declare no conflict of interest.

References

1. Lee, C.J.; Moon, T.W. Structural characteristics of slowly digestible starch and resistant starch isolated from heat–moisture treated waxy potato starch. *Carbohydr. Polym.* **2005**, *125*, 200–205. [CrossRef]
2. Xu, J.; Zhang, Y.; Wang, W.; Li, Y. Advanced properties of gluten-free cookies, cakes, and crackers: A review. *Trends Food Sci. Technol.* **2020**, *103*, 200–213. [CrossRef]

3. Green, P.H.; Cellier, C. Celiac disease. *N. Engl. J. Med.* **2007**, *357*, 1731–1743. [CrossRef]
4. Leonard, M.M.; Sapone, A.; Catassi, C.; Fasano, A. Celiac disease and nonceliac gluten sensitivity: A review. *JAMA* **2017**, *318*, 647–656. [CrossRef]
5. Di Cairano, M.; Galgano, F.; Tolve, R.; Caruso, M.C.; Condelli, N. Focus on gluten-free biscuits: Ingredients and issues. *Trends Food Sci. Technol.* **2018**, *81*, 203–212. [CrossRef]
6. Sakač, M.; Torbica, A.; Sedej, I.; Hadnađev, M. Influence of breadmaking on antioxidant capacity of gluten-free bread based on rice and buckwheat flours. *Food Res. Int.* **2011**, *44*, 2806–2813. [CrossRef]
7. Huda, M.N.; Lu, S.; Jahan, T.; Ding, M.; Jha, R.; Zhang, K.; Zhang, W.; Georgiev, M.I.; Park, S.U.; Zhou, M. Treasure from the garden: Bioactive compounds of buckwheat. *Food Chem.* **2021**, *335*, 127653. [CrossRef] [PubMed]
8. Ikeda, K. Buckwheat composition, chemistry, and processing. *Adv. Food Nutr. Res.* **2002**, *44*, 395–434.
9. Wu, N.N.; Tian, X.H.; Liu, Y.X.; Li, H.H.; Liang, R.P.; Zhang, M.; Liu, M.; Wang, L.P.; Zhai, X.T.; Tan, B. Cooking quality, texture and antioxidant properties of dried noodles enhanced with tartary buckwheat flour. *Food Sci. Technol. Res.* **2017**, *23*, 783–792. [CrossRef]
10. De Francischi, M.L.P.; Salgado, J.; Da Costa, C.P. Immunological analysis of serum for buckwheat fed celiac patients. *Mater. Veg.* **1994**, *46*, 207–211. [CrossRef]
11. Baljeet, S.Y.; Ritika, B.Y.; Roshan, L.Y. Studies on functional properties and incorporation of buckwheat flour for biscuit making. *Int. Food Res. J.* **2010**, *17*, 395–434.
12. Mohajan, S.; Munna, M.M.; Orchy, T.N.; Hoque, M.M.; Farzana, T. Buckwheat flour fortified bread. *Bangladesh J. Sci. Ind. Res.* **2019**, *54*, 347–356. [CrossRef]
13. Brites, L.T.G.F.; Rebellato, A.P.; Meinhart, A.D.; Godoy, H.T.; Pallone, J.A.L.; Steel, C.J. Technological, sensory, nutritional and bioactive potential of pan breads produced with refined and whole grain buckwheat flours. *Food Chem.* **2022**, *13*, 100243. [CrossRef] [PubMed]
14. Bączek, N.; Jarmułowicz, A.; Wronkowska, M.; Haros, C.M. Assessment of the glycaemic index, content of bioactive compounds, and their in vitro bioaccessibility in oat-buckwheat breads. *Food Chem.* **2020**, *330*, 127199. [CrossRef]
15. Choy, A.L.; Morrison, P.D.; Hughes, J.G.; Marriott, P.J.; Small, D.M. Quality and antioxidant properties of instant noodles enhanced with common buckwheat flour. *J. Cereal Sci.* **2013**, *57*, 281–287. [CrossRef]
16. Lin, L.Y.; Liu, H.M.; Yu, Y.W.; Lin, S.D.; Mau, J.L. Quality and antioxidant property of buckwheat enhanced wheat bread. *Food Chem.* **2009**, *112*, 987–991. [CrossRef]
17. Antoniewska, A.; Rutkowska, J.; Pineda, M.M.; Adamska, A. Antioxidative, nutritional and sensory properties of muffins with buckwheat flakes and amaranth flour blend partially substituting for wheat flour. *LWT* **2018**, *89*, 217–223. [CrossRef]
18. Farzana, T.; Hossain, F.B.; Abedin, M.J.; Afrin, S.; Rahman, S.S. Nutritional and sensory attributes of biscuits enriched with buckwheat. *J. Agric. Food Res.* **2022**, *10*, 100394. [CrossRef]
19. De Caluwé, E.; Halamouá, K.; Van Damme, P. *Adansonia digitata* L.—A review of traditional uses, phytochemistry, and pharmacology. *Afr. Focus* **2010**, *23*, 11–51. [CrossRef]
20. Buchmann, C.; Prehsler, S.; Hartl, A.; Vogl, C.R. The importance of baobab (*Adansonia digitata* L.) in rural West African subsistence—Suggestion of a cautionary approach to international market export of baobab fruits. *Ecol. Food Nutr.* **2010**, *49*, 145–172. [CrossRef]
21. Gebauer, J.; Luedeling, E. A note on baobab (*Adansonia digitata* L.) in Kordofan, Sudan. *Genet. Resour. Crop Evol.* **2013**, *60*, 1587–1596. [CrossRef]
22. Ajayi, I.A.; Dawodu, F.A.; Oderinde, R.A.; Egunyomi, A. Fatty acid composition and metal content of *Adansonia digitata* seeds and seed oil. *Riv. Ital. Delle Sostanze Grasse* **2003**, *80*, 41–44.
23. Asogwa, I.S.; Ibrahim, A.N.; Agbaka, J.I. African baobab: Its role in enhancing nutrition, health, and the environment. *Trees For. People* **2021**, *3*, 100043. [CrossRef]
24. Namratha, V.; Sahithi, P. Baobab: A review about "the tree of life". *Int. J. Adv. Herb. Sci. Technol.* **2015**, *1*, 20–26.
25. Dossa, S.; Negrea, M.; Cocan, I.; Berbecea, A.; Obistioiu, D.; Dragomir, C.; Alexa, E.; Rivis, A. Nutritional, Physico-Chemical, Phytochemical, and Rheological Characteristics of Composite Flour Substituted by Baobab Pulp Flour (*Adansonia digitata* L.) for Bread Making. *Foods* **2023**, *12*, 2697. [CrossRef] [PubMed]
26. Barakat, H. Nutritional and Rheological Characteristics of Composite Flour Substituted with Baobab (*Adansonia digitata* L.) Pulp Flour for Cake Manufacturing and Organoleptic Properties of Their Prepared Cakes. *Foods* **2021**, *10*, 716. [CrossRef] [PubMed]
27. Cissé, I.; Montet, D.; Reynes, M.; Danthu, P.; Yao, B.; Boulanger, R. Biochemical and nutritional properties of baobab pulp from endemic species of Madagascar and the African mainland. *Afr. J. Agric. Res.* **2013**, *8*, 6046–6054.
28. Burlando, B.; Verotta, L.; Cornara, L.; Bottini-Massa, E. *Herbal Principles in Cosmetics: Properties and Mechanisms of Action*; CRC Press: Boca Raton, FL, USA, 2010.
29. Brendler, T.; Eloff, J.N.; Gurib-Fakim, A.; Phillips, L.D. (Eds.) *African Herbal Pharmacopoeia, Association for African Medicinal Plants Standards (AAMPS)*; Graphics Press Ltd.: Baie du Tombeau, Mauritius, 2010; Volume 289.
30. Tsetegho Sokeng, A.J.; Sobolev, A.P.; Di Lorenzo, A.; Xiao, J.; Mannina, L.; Capitani, D.; Daglia, M. Metabolite characterization of powdered fruits and leaves from *Adansonia digitata* L. (baobab): A multi-methodological approach. *Food Chem.* **2019**, *272*, 93–108. [CrossRef]

31. Chadare, F.; Linnemann, A.; Hounhouigan, J.; Nout, M.; Van Boekel, M. Baobab food products: A review on their composition and nutritional value. *Crit. Rev. Food Sci. Nutr.* **2009**, *49*, 254–274. [CrossRef]
32. Gebauer, J.; El Siddig, K.; Ebert, G. Baobab (*Adansonia digitata* L.): A review on a multipurpose tree with promising future in the Sudan. *Gartenbauwissenscha* **2002**, *67*, 155–160.
33. Diop, A.; Sakho, M.; Dornier, M.; Cisse, M.; Reynes, M. Le baobab africain (*Adansonia digitata* L.): Principales caractéristiques et utilisations. *Fruits* **2005**, *61*, 55–69. [CrossRef]
34. Mounjouenpou, P.; Ngono Eyenga, S.N.N.; Kamsu, E.J.; Kari, P.B.; Ehabe, E.E.; Ndjouenkeu, R. Effect of fortification with baobab (*Adansonia digitata* L.) pulp flour on sensorial acceptability and nutrient composition of rice cookies. *Sci. Afr.* **2018**, *1*, e00002. [CrossRef]
35. ISO 2171; Determination of Ash Content by Incineration in Cereals, Pulses, and Derived Products, 4th ed. International Standard ISO: London, UK, 2017; p. 11.
36. No. 110/1.ICC; ICC Standard Methods of the International Association for Cereal Science and Technology. ICC: Vienna, Austria, 2003.
37. Association of Official Analytical Chemists (AOAC). *Official Methods of Analysis of AOAC International*, 17th ed.; AOAC International: Gaithersburg, MD, USA, 2000.
38. Plustea, L.; Negrea, M.; Cocan, I.; Radulov, I.; Tulcan, C.; Berbecea, A.; Popescu, I.; Obistioiu, D.; Hotea, I.; Suster, G. Lupin (*Lupinus* sPP.)-Fortified Bread: A Sustainable, Nutritionally, Functionally, and Technologically Valuable Solution for Bakery. *Foods* **2022**, *11*, 2067. [CrossRef]
39. Danciu, C.; Muntean, D.; Alexa, E.; Farcas, C.; Oprean, C.; Zupko, I.; Bor, A.; Minda, D.; Proks, M.; Buda, V.; et al. Phytochemical Characterization and Evaluation of the Antimicrobial, Antiproliferative and Pro-Apoptotic Potential of Ephedra alata Decne. Hydroalcoholic Extract against the MCF-7 Breast Cancer Cell Line. *Molecules* **2019**, *24*, 13. [CrossRef] [PubMed]
40. Obistioiu, D.; Cocan, I.; Tîrziu, E.; Herman, V.; Negrea, M.; Cucerzan, A.; Neacsu, A.G.; Cozma, A.L.; Nichita, I.; Hulea, A. Phytochemical Profile and Microbiological Activity of Some Plants Belonging to the Fabaceae Family. *Antibiotics* **2021**, *10*, 662. [CrossRef]
41. Cocan, I.; Cadariu, A.-I.; Negrea, M.; Alexa, E.; Obistioiu, D.; Radulov, I.; Poiana, M.-A. Investigating the Antioxidant Potential of Bell Pepper Processing By-Products for the Development of Value-Added Sausage Formulations. *Appl. Sci.* **2022**, *12*, 12421. [CrossRef]
42. Ciulca, S.; Roma, G.; Alexa, E.; Radulov, I.; Cocan, I.; Madosa, E.; Ciulca, A. Variation of Polyphenol Content and Antioxidant Activity in Some Bilberry (*Vaccinium myrtillus* L.) Populations from Romania. *Agronomy* **2021**, *11*, 2557. [CrossRef]
43. ISO 6658:2017; Sensory Analysis—Methodology—Overall Guidelines. ISO: Geneva, Switzerland, 2017.
44. Bhinder, S.; Kaur, A.; Singh, B.; Yadav, M.P.; Singh, N. Proximate composition, amino acid profile, pasting and process characteristics of flour from different Tartary buckwheat varieties. *Food Res. Int.* **2020**, *130*, 108946. [CrossRef]
45. Lockett, C.; Calvert, C.; Grivetti, L. Energy and micronutrient composition of dietary and medicinal wild plants consumed during drought. Study of rural Fulani, Northeastern Nigeria. *Int. J. Food Sci. Nutr.* **2000**, *51*, 195–208.
46. Murray, S.; Schoeninger, M.; Bunn, H.; Pickering, T.; MarIen, J. Nutritional composition of some wild plant foods and honey used by Hadza foragers of Tanzania. *Food Compos. Anal.* **2001**, *14*, 3–13. [CrossRef]
47. Osman, M. Chemical and nutrient analysis of baobab (*Adansonia digitata*) fruit and seed protein solubility. *Plant Foods Hum. Nutr.* **2004**, *59*, 29–33. [CrossRef] [PubMed]
48. Chopra, N.; Dhillon, B.; Puri, S. Formulation of buckwheat cookies and their nutritional, physical, sensory and microbiological analysis. *Int. J. Adv. Biotechnol. Res.* **2014**, *5*, 381–387.
49. Stadlmayr, B.; Wanangwe, J.; Waruhiu, C.G.; Jamnadass, R.; Kehlenbeck, K. Nutritional composition of baobab (*Adansonia digitata* L.) fruit pulp sampled at different geographical locations in Kenya. *J. Food Compos. Anal.* **2020**, *94*, 103617. [CrossRef]
50. Wardlaw, G.M.; Kessel, M.W. *Perspectives in Nutrition*, 5th ed.; McGraw-Hill Education: New York, NY, USA, 2002; pp. 162–452.
51. Cozzolino, S.M.F. *Biodisponibilidade de Nutrientes*; Manole: Plovdiv, Bulgaria, 2009; p. 1172.
52. Gupta, U.C.; Gupta, S.C. Sources and deficiency diseases of mineral nutrients in human health and nutrition: A review. *Pedosphere* **2014**, *24*, 13–38. [CrossRef]
53. *WHO Guidelines on Food Fortification with Micronutrients*; WHO: Geneva, Switzerland, 2006. Available online: https://www.who.int/publications/i/item/9241594012 (accessed on 15 August 2023).
54. World Health Organization (WHO). *Guidelines: On Potassium Intake in Adults and Children Summary of Guidance*. WHO/NMH/NHD/13.1; World Health Organization (WHO): Geneva, Switzerland, 2013. Available online: https://apps.who.int/iris/bitstream/handle/10665/85225/WHO_NMH_NHD_13.1_fre.pdf (accessed on 15 August 2023).
55. EFSA Panel on Dietetic Products, Nutrition and Allergies. Scientific Opinion on the Tolerable Upper Intake Level of Calcium. *EFSA J.* **2012**, *10*, 2814.
56. EFSA Panel on Dietetic Products, Nutrition and Allergies. Scientific Opinion on Dietary Reference Values for Magnesium. *EFSA J.* **2015**, *13*, 4186.
57. Reena, R.; Lin, Y.T.; Shetty, K. Phenolics, their antioxidant and antimicrobial activity in dark germinated fenugreek sprouts in response to peptide and phytochemical elicitors. *Asia Pac. J. Clin. Nutr.* **2004**, *13*, 295–307.
58. Balasundram, N.; Sundram, K.; Samman, S. Phenolic compounds in plants and agri-industrial by-products: Antioxidant activity, occurrence, and potential uses. *Food Chem.* **2006**, *99*, 191–203. [CrossRef]

59. Balarabe, B.I.; Yunfeng, P.; Lihua, F.; Munir, A.D.; Mingming, G.; Donghong, L. Characterizing the phenolic constituents of baobab (*Adansonia digitata*) fruit shell by LC-MS/QTOF and their in vitro biological activities. *Sci. Total Environ.* **2019**, *694*, 133387.
60. Şensoy, İ.; Rosen, R.T.; Ho, C.T.; Karwe, M.V. Effect of processing on buckwheat phenolics and antioxidant activity. *Food Chem.* **2006**, *99*, 388–393. [CrossRef]
61. Chlopicka, J.; Pasko, P.; Gorinstein, S.; Jedrysas, A.; Zagrodzki, P. Total phenolic and total flavonoid content, antioxidant activity and sensory evaluation of pseudocereal bread. *LWT Food Sci. Technol.* **2012**, *46*, 548–555. [CrossRef]
62. Cook, N.C.; Samman, S. Flavonoids-chemistry, metabolism, cardioprotective effects, and dietary sources. *J. Nutr. Biochem.* **1996**, *7*, 66–76. [CrossRef]
63. Dorota, D.-S.; Wieslaw, O. Effect of processing on the flavonoid content in buckwheat (*Fagopyrum esculentum* Moench) grain. *J. Agric. Food Chem.* **1999**, *47*, 4383–4387.
64. Sun, T.; Ho, C.T. Antioxidant activities of buckwheat extracts. *Food Chem.* **2005**, *90*, 743–749. [CrossRef]
65. Bolang, A.S.L.; Rizal, M.; Nurkolis, F.; Mayulu, N.; Taslim, N.A.; Radu, S.; Samtiya, M.; Assa, Y.A.; Herlambang, H.A.; Pondagitan, A.S.; et al. Cookies rich in iron (Fe), folic acid, cobalamin (vitamin B12), and antioxidants: A novel functional food potential for adolescent with anemia. *F1000Research* **2023**, *10*, 1075. [CrossRef]
66. Ngadiarti, I.; Nurkolis, F.; Handoko, M.N.; Perdana, F.; Permatasari, H.K.; Taslim, N.A.; Mayulu, N.; Wewengkang, D.S.; Noor, S.L.; Batubara, S.C.; et al. Anti-aging potential of cookies from sea grapes in mice fed on cholesterol- and fat-enriched diet: In vitro with in vivo study. *Heliyon* **2022**, *8*, e09348. [CrossRef] [PubMed]
67. Tembo, D.T.; Holmes, M.J.; Marshall, L.J. Effect of thermal treatment and storage on bioactive compounds, organic acids and antioxidant activity of baobab fruit (*Adansonia digitata*) pulp from Malawi. *J. Food Compos. Anal.* **2017**, *58*, 40–51. [CrossRef]
68. Lee, L.S.; Choi, E.J.; Kim, C.H.; Sung, J.M.; Kim, Y.B.; Seo, D.H.; Choi, H.W.; Choi, Y.S.; Kum, J.S.; Park, J.D. Contribution of flavonoids to the antioxidant properties of common and tartary buckwheat. *J. Cereal Sci.* **2016**, *68*, 181–186. [CrossRef]
69. Schewe, T.; Sies, H. *82—Epicatechin and Its Role in Protection of LDL and Vascular Endothelium*; Preedy, V.R., Ed.; Beer in Health and Disease Prevention; Academic Press: Cambridge, MA, USA, 2009; pp. 803–813.
70. Sirotkin, A.V.; Kolesarova, A. *Chapter 4—Plant Molecules and Their Influence on Health and Female Reproduction*; Sirotkin, A.V., Kolesarova, A., Eds.; Environmental Contaminants and Medicinal Plants Action on Female Reproduction; Academic Press: Cambridge, MA, USA, 2022; pp. 245–399.
71. Patel, K.; Patel, D.K. Chapter 26—The Beneficial Role of Rutin, A Naturally Occurring Flavonoid in Health Promotion and Disease Prevention: A Systematic Review and Update. In *Bioactive Food as Dietary Interventions for Arthritis and Related Inflammatory Diseases*, 2nd ed.; Watson, R.R., Preedy, V.R., Eds.; Academic Press: Cambridge, MA, USA, 2019; pp. 457–479.

Disclaimer/Publisher's Note: The statements, opinions and data contained in all publications are solely those of the individual author(s) and contributor(s) and not of MDPI and/or the editor(s). MDPI and/or the editor(s) disclaim responsibility for any injury to people or property resulting from any ideas, methods, instructions or products referred to in the content.

Article

The Experimental Development of Bread with Enriched Nutritional Properties Using Organic Sea Buckthorn Pomace

Ioana Stanciu [1], Elena Loredana Ungureanu [2,*], Elisabeta Elena Popa [1,*], Mihaela Geicu-Cristea [1], Mihaela Draghici [1], Amalia Carmen Mitelut [1], Gabriel Mustatea [2] and Mona Elena Popa [1]

[1] Faculty of Biotechnology, University of Agronomic Sciences and Veterinary Medicine of Bucharest, 011464 Bucharest, Romania; loryg208@yahoo.com (I.S.); mihaela_geicu@yahoo.com (M.G.-C.); mihaeladraghici38@gmail.com (M.D.); amaliamitelut@yahoo.com (A.C.M.); pandry2002@yahoo.com (M.E.P.)

[2] National Research and Development Institute for Food Bioresources, 6 Dinu Vintila Str., 021102 Bucharest, Romania; gabi.mustatea@bioresurse.ro

* Correspondence: elena_ungureanu93@yahoo.com (E.L.U.); elena.eli.tanase@gmail.com (E.E.P.)

Abstract: In this study, sea buckthorn (*Hippophae rhamnoides* L.) pomace resulting from juice extraction was dried and ground in order to obtain a powder that was further used in bread making. Sea buckthorn pomace, an invaluable by-product of the industry, contains bioactive compounds and dietary fibers that promote health. Dried by-products of sea buckthorn are rich sources of nutritional and bioactive compounds, offering great potential for use as nutraceuticals in animal feed, ingredients in functional food, and the pharmaceutical industry. The utilization of sea buckthorn by-products promotes a circular and sustainable economy by implementing innovative methods and strategic approaches to recover high-value products and minimize waste in multiple ways. For this purpose, three organic sea buckthorn varieties were used, namely Mara (M), Clara (C), and Sorana (S). Further, 6%, 8%, and 10% pomace powder were added to wheat flour to prepare functional bread, and its effects on structural, nutritional, and sensorial characteristics were investigated. The volume, porosity, and elasticity of the obtained bread samples were slightly lower compared to the control sample (white bread). The nutritional characteristics revealed that the developed bread presented higher antioxidant activity, polyphenolic content, and crude fiber compared to the control sample. The acceptability test showed that consumer preferences were directed toward the bread samples containing 8% sea buckthorn powder, regardless of the variety, while the addition of 10% pomace powder led to major sensorial changes. The results of this study showed that sea buckthorn pomace powder can be successfully incorporated into bread in order to obtain a food product with enhanced properties.

Keywords: sea buckthorn pomace; by-product valorization; functional bread

1. Introduction

Recent interest by consumers for fresh, local, and nutritionally enriched food products with a low impact on the environment has led to the development of foods with enhanced properties using functional ingredients obtained from fruit and vegetable by-products. In this context, valorization of food by-products and waste is important both at national and regional levels to help achieve the targets established by the United Nations Sustainable Development Goals in order to fulfill green and circular economy principles and efficient management of resources [1].

The appropriate utilization of food waste and by-products as raw materials or food additives has the potential to yield economic benefits for the industry. Additionally, it can contribute to the mitigation of nutritional issues, generate positive health effects, and reduce the environmental consequences associated with waste mismanagement. Currently, industries are actively seeking innovations to achieve zero waste, wherein generated waste is repurposed as raw materials for new products and applications. Such actions can directly influence the attainment of the Millennium Development Goals, the forthcoming Sustainable Development Goals, the Post-2015 Agenda, and the Zero Hunger Challenge [2]. The nutritional challenges faced by contemporary society emphasize the necessity for alternative sources of nutrients and functional compounds. The significant quantity of waste generated by the food industry not only results in the loss of valuable materials but also presents substantial management difficulties, both in economic and environmental terms. Nevertheless, numerous food waste residues have the potential to be repurposed in alternative production systems. These residues can be utilized directly as powders after the removal of antinutritional factors, or their lipids and antioxidants can be extracted for further use [3].

Bread is a staple food [4,5] and is one of the most consumed foods all over the world [6]. Over the years, studies have been performed for bread enrichment in nutritional composition by adding different flours or ingredients with high amounts of bioactive compounds [7] and health benefits [8].

Sea buckthorn (*Hippophae rhamnoides* L.) belongs to the Elaeagnaceae family. It is a deciduous shrub that is widely found in Northwestern Europe and Asia and is distributed from the coastal regions to the mountains [9]. Its use and processing gained more and more attention recently due to its nutritional properties and health benefits; it is rich in ascorbic acid, carotenoids, tannins, tocopherols, fatty acids, and flavonoids and improves blood circulation, relieving irritability. It is a good immunity enhancer, protecting the cardiovascular system, resisting oxidation, and presenting great antimicrobial properties [9–11]. Sea buckthorn berries can be consumed in their natural state or be processed into juice, with the latter resulting in a significant amount of pomace. Unfortunately, pomace is often regarded as waste [12].

Fruit and vegetable pomace can serve as an excellent source of fiber for enriching bakery products due to its favorable properties, including a balanced soluble/insoluble fiber ratio, good hydration, fermentability, and phytochemical content. The origin and processing of pomace into powder form through various pre-treatment, drying, and size reduction methods can affect its functional properties. The functional properties of fruit and vegetable pomace can enhance the quality of food, and they can be used to enrich a wide range of bakery products such as biscuits, buns, cookies, crackers, cakes, muffins, wheat rolls, and scones. Fruit pomaces are particularly suitable for bakery products as they enhance sensory attributes. In addition, fruit and vegetable pomace contains antioxidants that can improve the storage stability of baked goods. Therefore, fruit and vegetable pomace can be effectively used as a functional ingredient for the development of fiber-rich bakery products [13].

The addition of dried fruit pomace to bakery products can replace flour, sugar, or fat, resulting in reduced energy load while increasing fiber and antioxidant levels. Nonetheless, the high fiber content of fruit pomace can lead to techno-functional interactions that influence physicochemical and sensory characteristics [14].

Sea buckthorn pomace contains dietary fibers that are bioactive and promote health. However, due to the lack of suitable handling or processing facilities, this pomace is often either used as animal feed or discarded [15].

Fortification of bakery-based products with bioactive compounds extracted from sea buckthorn has been reported to inhibit starch digestion, which is beneficial for people with diabetes [16]. Sea buckthorn pomace flour has been shown to increase the porosity of pastries and decrease the wet gluten content, leading to moisture loss. However, the addition of sea buckthorn flour has been found to improve the appearance, color, and consistency of pastries according to organoleptic assessments. Furthermore, samples containing sea buckthorn flour have exhibited microbiological stability due to the chemical composition of sea buckthorn [17].

Therefore, the objective of this study was to enhance the technology of bakery products by incorporating sea buckthorn pomace flour, thereby expanding the range of functional products. The investigation focused on assessing the impact of varying concentrations of sea buckthorn berry flour on the sensory attributes, physicochemical properties, polyphenol content, antioxidant properties, and shelf-life of wheat bread. The aim was to diversify the assortment of bakery products and explore the potential benefits of incorporating sea buckthorn pomace flour. The novelty of this study consists of the studied organic sea buckthorn varieties and the sustainability of the processing method for sea buckthorn pomace powder obtained (from waste resulting after sea buckthorn processing) using the entire quantity of waste (seeds, pulp, and skin).

2. Materials and Methods

2.1. Raw Materials and Chemicals

Wheat flour (type 650) of superior quality (Baneasa, Romania), compressed *Saccharomyces cerevisiae* yeast (Dr. Oetker, Curtea de Argeș, Romania), iodized salt, and water were used in bread making. The ingredients were purchased from local stores. Organic sea buckthorn varieties (Mara, Clara, Sorana) were purchased from Bio Cătina Cooperativa Agricolă—Dâmbovița, Romania. Briefly, sea buckthorn berries were processed using a screw press (Biovita), resulting in the extraction of pomace consisting of skin, pulp, and seeds. This pomace accounted for 20% of the total fruit weight. The obtained pomace was then dried in a hot air dryer, specifically the DEH600D BIOVITA model equipped with 9 stainless trays. The drying process took place for 12 h at a temperature of 50 °C. Subsequently, the dried pomace was ground in a Biovita blender with a power of 1000 W using the maximum speed setting for 2 min and ensuring a thorough blend. To remove any hard seed components, the resulting powder was sifted through a 7 mm sieve. The resulting pomace powder was then used as an ingredient in the recipe for functional bread making. The process of obtaining sea buckthorn pomace powder is presented in Figure 1, illustrating the different stages involved.

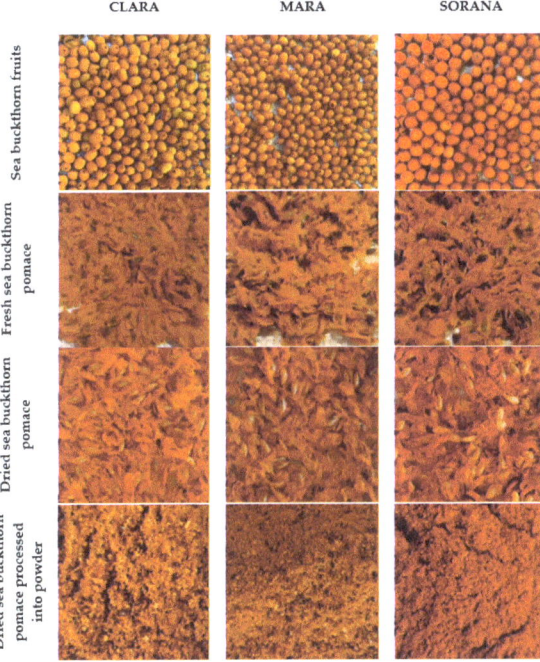

Figure 1. Aspects of various stages of the technological flow for obtaining sea buckthorn pomace powder.

All chemical reagents (DPPH, Folin–Ciocalteu's reagent, quercitin, gallic acid, α—amylase, protease, HCl, NaOH, HNO_3, and solvents) were of analytical grade or chromatographic grade and were purchased from Sigma Aldrich (St. Louis, MO, USA).

2.2. Bread Making

Samples of bread were obtained using 650 type flour and 6%, 8%, and 10% sea buckthorn pomace powder to determine its influence on organoleptic, physical–chemical, and nutritional indicators, as well as the structural parameters of bread. Previously, an initial study was performed using 1% to 5% sea buckthorn pomace powder. The obtained results combined with data in the literature showed that an increase in powder content can be applied, which has a beneficial effect on the nutritional value of the product. Higher amounts of added powder favor the increasing of antioxidants and fiber content within the product. However, concentrations higher than 8–10% lead to the sensory modification of the product, having an aftertaste that is slightly sour, bitter, and not specific to the product. The obtained samples were coded as follows: the control (wheat bread), C6, C8, and C10 (bread with 6%, 8%, and 10% sea buckthorn pomace powder from the Clara variety), M6, M8, and M10 (bread with 6%, 8% and 10% sea buckthorn pomace powder from the Mara variety), and S6, S8, and S10 (bread with 6%, 8% and 10% sea buckthorn pomace powder from the Sorana variety). The amount of sea buckthorn pomace powder was reported per 100 g of flour. The control sample was prepared using only wheat flour. Briefly, each batch of dough was kneaded for 8–9 min in a mixer and fermented at 28 °C for 28–30 min (Figure 2). Afterward, the dough was divided, shaped, and left to rest for the final fermentation at a temperature of 35 °C for 35 min. The bread samples were then baked in a preheated oven at a temperature of 130 °C for 30 min and were then cooled 24 °C for 3 h. Afterward, the bread samples were further analyzed. The technological scheme of bread making is presented in Figure 3 and the aspect of bread samples is presented in Figure 4.

Figure 2. The appearance of the obtained dough samples with the addition of sea buckthorn pomace powder.

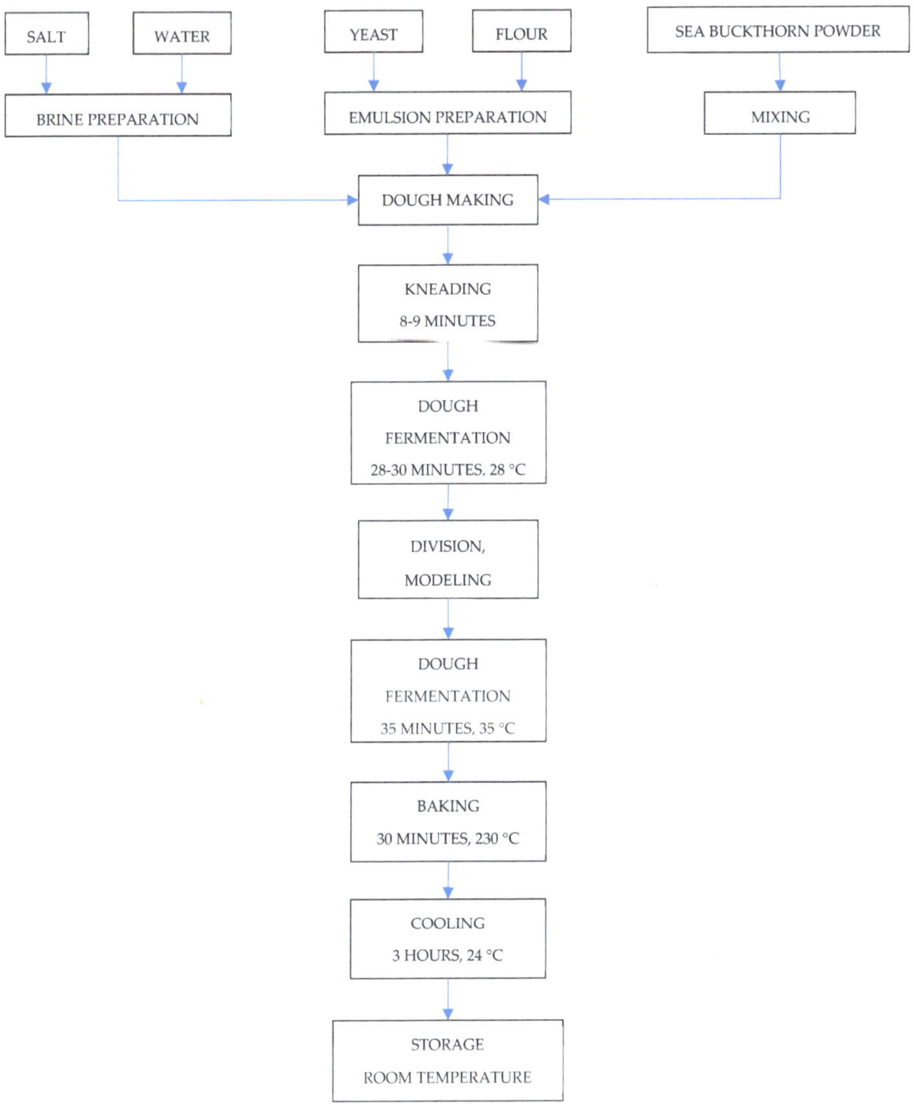

Figure 3. Technological scheme for functional bread preparation.

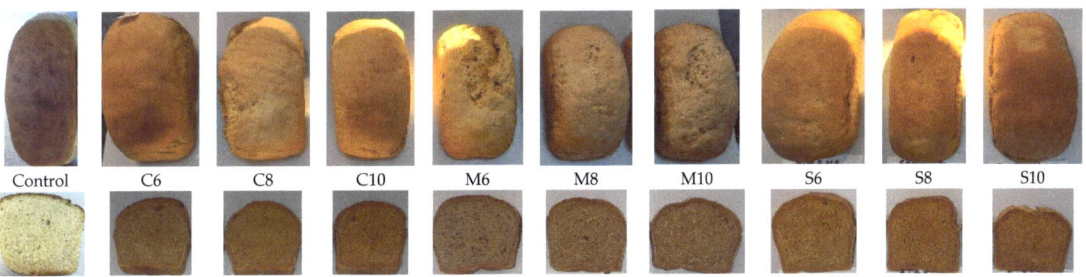

Figure 4. The aspect of the developed bread samples.

2.3. Structural and Physical–Chemical Characteristics Determination

The functional bread weight was established by using an analytical balance with an accuracy of four decimal places to weigh the samples.

The Fornet bread volumeter (Chopin, France) and the rapeseed displacement method [18] were used to measure the bread volume. The volume of the bread was calculated using Equation (1) and expressed as cm^3 per 100 g of product.

$$V = (V_1/m) \times 100 \tag{1}$$

where V_1 is the determined volume of the sample (cm^3) and m is the mass of the sample used (g).

To determine the core elasticity of the bread, a specific-shaped piece of the bread core is pressed, and the rebound to the original position after removing the pressing force is measured [19]. Elasticity is expressed as a percentage ratio between the height after pressing and the rebound to the initial height of the core cylinder.

The porosity of the bread samples was measured according to Romanian standard SR 91:2007 [19] and expressed as a volume percentage. It was calculated using Equation (2).

$$\text{Porosity} = [(V - m/\rho)/V] \times 100 \tag{2}$$

where V is the volume of the core cylinder [cm^3], m is core cylinder mass [g], and ρ is the density of the compact core [g/cm^3]. The compact core density for white flour bakery products has an average value of $\rho = 1.31\ g/cm^3$.

To determine the acidity of the bread samples, the Romanian standard SR 91:2007 [19] was followed. An aqueous extract of the bread was titrated with a 0.1 N sodium hydroxide (NaOH) solution in the presence of phenolphthalein as an indicator. The acidity was then calculated using Equation (3).

$$\text{Acidity} = [V \times (0.1/m)] \times 100 \tag{3}$$

where V is the volume of the 0.1 N NaOH solution used in the titration, m is the mass of the sample taken for analysis, in g, and 0.1 is the normality of the NaOH solution.

To determine the moisture content of the bread samples, the AOAC Official Method 934.06 [20] was used. This involved evaporating water from the sample by heating it in an oven at a constant temperature of $103 \pm 2\ °C$ until a constant mass was achieved. The moisture content was then calculated as a percentage using Equation (4).

$$\text{Moisture content} = [(M_1 - M_0)/(M_1 - M_2)] \times 100 \tag{4}$$

where M_0 is the mass, in grams, of the dish and lid, M_1 is the mass, in grams, of the dish, lid and test portion before drying, and M_2 is the mass, in grams, of the dish, lid, and test portion after drying.

The water activity index was determined using NOVASINA AG (Lachen, Switzerland) equipment. Briefly, the samples were introduced to specific recipients of the equipment and the value of the a_w index was read when stable at $25\ °C$.

2.4. Nutritional Parameters Determination

To measure the protein content, the Kjeldahl method [21] was used along with a conversion factor of nitrogen to protein of 6.25. The protein content was then calculated using Equation (5).

$$N = [(V_1 - V_0) \times T \times 0.014 \times 100]/m \times (100/100 - w) = 140T\,(V_1 - V_0)/m \times (100 - w) \tag{5}$$

where V_0 is the volume of the hydrochloric acid (HCl) used in the control sample (mL), V_1 is the volume of HCl used in the sample (mL), T is the normality of HCl, m is the mass of the sample (g), and w is the moisture of the sample; the protein conversion factor is 6.25.

The fat content was determined using the AOAC 963.15 [22] method by extracting it with a petroleum ether under reflux conditions in Soxhlet equipment. The fat content was then calculated using Equation (6).

$$\text{Fat content} = (m_2 - m_1)/m \times 100 \times 100/(100 - w) \qquad (6)$$

where m_1 is the mass of the empty balloon (g), m_2 is the mass of the balloon with fat (g), m is the mass of the sample taken for analysis (g), and w is the moisture of the sample determined according to the procedure described for moisture (%).

The fiber content was performed using the gravimetric-enzymatic method AOAC 991.43 [23] with a Tris-Mes buffer. In this method, the sample undergoes successive enzymatic digestions using thermally stabilized α-amylase, protease, and amyloglucosidase. Fibertec 1023 equipment (FOSS Analytics, Hillerod, Denmark) was used to carry out the determinations.

The AOAC 923.03 [24] method was used to determine the total ash content, which involved burning the samples in a muffle furnace at 550 °C. The value for this parameter was obtained by applying Equation (7).

$$\text{Total ash content} = (m_2 - m_1) \times (100/m_0) \times 100/(100 - w) \qquad (7)$$

where m_0 is the mass of the working sample (g), m_1 is calcination crucible mass (g), m_2 is the mass of the calcining crucible and the calcined residue (g), and w is the moisture of the sample (%).

The AACC method 10-05.01 [18] was used to express the total carbohydrates as a percentage, which was then calculated using Equation (8). The calorie content was estimated using conversion factors of 9 for fat, 4 for carbohydrates, and 4 for protein, as per the AACC method 10-05.01 [19].

$$\text{Carbohydrate} = 100 - (\% \text{ protein} + \% \text{ fat} + \% \text{ fiber} + \% \text{ ash}) \qquad (8)$$

To determine the heavy metals, microelements, and minerals, the samples were dry ashed in a muffle furnace L1206 (Caloris, Bucharest, Romania) at 550 °C. The heavy metals (Pb, Cd, and Cr) and the microelement (Mn, Cu, Zn, and Fe) content were analyzed using Inductively Coupled Plasma Mass Spectrometry, NexION 300Q (Perkin Elmer, Waltham, MA, USA). The content of minerals, on the other hand, was determined using HR-CS-AAS high-resolution continuum source atomic absorption spectrometry with ContrAA700 (Analytik Jena, Jena, Germany).

To determine the content of total phenols, the Folin–Ciocalteu method was used [25,26]. The concentration of polyphenols reported in gallic acid equivalents per 100 g of product was calculated according to Equation (9).

$$\text{Total polyphenols (GAE) mg}/100\text{ g} = (A_{765} - 0.0082)/0.001 \; (R2 = 0.9995) \qquad (9)$$

where $A765$ is the sample absorbance read at 765 nm, GAE is the concentration in gallic acid equivalents (mg/L), and R2 = 0.9995 is the correlation coefficient.

For antioxidant activity determination, the free radical scavenging effect on the 1,1-diphenyl-2-picrylhydrazyl (DPPH) radical was evaluated [25,26]. The samples were prepared by macerating them in ethanol (75%) for 48 h at room temperature in the dark. The results were expressed in quercetin equivalents per 100 g of product using Equation (10).

$$\text{AAR (QE)} = (\%\Delta A_{515} - 3.4954)/0.0811 \qquad (10)$$

where AAR (QE) is the antiradical activity expressed in quercetin equivalents, $\%\Delta A515 = [(A_{515(t=0)} - A_{515(t=30)})/A_{515(t=0)}] \times 100$.

2.5. Sensory Analysis

A group of ten trained panelists, consisting of 70% females and 30% males, conducted a sensory evaluation of bread to assess its overall acceptability using a five-point hedonic scale. The bread samples were placed on white plates and labeled with random three-digit numbers. The sensory evaluation criteria are outlined in Table 1.

Table 1. Criteria for the sensory evaluation of bread containing sea buckthorn pomace powder.

Sensory Characteristic		Scale	Product Description
Flavor	Of sea buckthorn	0–5	Flavorless—very aromatic
Taste	Of sea buckthorn	0–5	Unnoticed—very strong
	Sour taste	0–5	
	Bitter taste	0–5	
External appearance of the crust	Color	0–5	Very light—very brown
	Hardness	0–5	Very crispy—very soft
Core aspect	Freshness	0–5	Dry core—very soft, fresh core
	Crumbliness	0–5	Compact, non-crumbling—crumbly core
	Color	0–5	Very light—very brown
	Elasticity	0–5	Inelastic—elastic
Aftertaste	Of sea buckthorn	0–5	Flavorless—very aromatic

2.6. Statistical Analysis

The functional bread parameters were determined by averaging the results from three parallel extraction experiments and calculating their standard deviations. To compare the means, one-way ANOVA with Tukey's test and Fisher's test was applied, with a significance level of $p < 0.05$. The statistical calculations were performed using Minitab 15 (Minitab LTD., Coventry, UK) and Microsoft Excel 2010.

3. Results and Discussion

3.1. Structural and Physical–Chemical Characteristics of Functional Bread

The characteristics regarding the structural parameters of the obtained functional bread samples are presented in Table 2.

Table 2. The characteristics of functional bread with sea buckthorn pomace powder (mean ± SD).

Sample	Bread Weight (g)	Volume (cm^3/100 g)	Porosity (%)	Elasticity
Control	738.13 ± 0.40	249.28 ± 0.20	72.9 ± 0.32 [a]	96.7 ± 0.61
C6	732.15 ± 0.55 [a]	226.72 ± 0.08 [a]	73.2 ± 0.53 [a]	85.2 ± 0.33
C8	733.87 ± 0.35 [a]	208.48 ± 0.13 [a]	67.3 ± 0.46 [b]	70.5 ± 0.46 [a]
C10	728.72 ± 0.05	209.96 ± 0.25	69.8 ± 0.77 [b]	71.2 ± 0.48 [a]
M6	724.66 ± 0.07	205.61 ± 0.28	67.2 ± 0.61 [b,c]	73.3 ± 0.67
M8	736.91 ± 0.06	185.23 ± 0.22	62.4 ± 0.55 [d,e]	66.7 ± 0.35 [b]
M10	740.03 ± 0.08	171.61 ± 0.18	60.8 ± 0.41 [e]	64.4 ± 0.52 [c]
S6	726.83 ± 0.09	208.44 ± 0.37	64.8 ± 0.63 [c,d]	67.2 ± 0.28 [b]
S8	735.56 ± 0.06	172.66 ± 0.42	61.6 ± 0.48 [e]	67.8 ± 0.69 [b]
S10	726.05 ± 0.08	166.66 ± 0.24	58.6 ± 0.39	63.3 ± 0.73 [c]

All the results are mean ± standard deviation. Means followed by a common letter are not significantly different ($p > 0.05$). Values followed by different superscript letters in the same column are significantly different ($p < 0.05$). Means that do not share a letter in a column are significantly different ($p < 0.05$).

3.1.1. Weight of Functional Bread

The functional bread samples with sea buckthorn pomace powder showed varying weights in the range of 740.03 g to 724.66 g, as presented in Table 2. The addition of sea buckthorn flour significantly influenced ($p < 0.05$) the weight of the obtained samples. The bread samples with a 10% sea buckthorn powder addition from the Mara variety

(740.03 g) and the control sample (738.13 g) had the highest weights, followed closely by the bread sample with an 8% sea buckthorn powder addition from the Sorana variety (735.56 g), the bread sample with an 8% sea buckthorn powder addition from the Clara variety (733.87 g), and the bread sample with a 6% sea buckthorn powder addition from the Clara variety (732.15 g). On the other hand, the samples of bread with 6% added sea buckthorn powder from the Mara variety (724.66 g), 10% added sea buckthorn powder from the Sorana variety (726.05 g), and 6% added sea buckthorn powder from the Sorana variety (726.83 g) registered the lowest weights.

3.1.2. Specific Volume

The addition of sea buckthorn powder to functional bread resulted in varying volumes from 166.66 to 226.72 cm^3 m^{-1} (Table 2). The control sample had a volume of 249.28 cm^3 m^{-1}, which was significantly different ($p < 0.05$) from the samples with sea buckthorn powder. The C6 sample had the highest volume (226.72 cm^3 m^{-1}), followed by the C10 sample (209.96 cm^3 m^{-1}), the C8 sample (208.48 cm^3 m^{-1}), and the S6 sample (208.44 cm^3 m^{-1}). On the other hand, the S10 (166.66 cm^3 m^{-1}) and M10 (171.61 cm^3 m^{-1}) samples had lower volumes. The rise in specific volume could be attributed to changes in the dough's strength, allowing for more extensive deformation and better correlation to increase dough height. Enhancing the specific volume of bread may lead to better gas retention, thereby allowing the dough to rise to a greater volume [12]. However, Nilova and Malyutenkova's [27] findings revealed that incorporating sea buckthorn seed powder, marc powder, and peel powder in bakery products improved their specific volume. Similarly, in 2019 a study by Guo et al. [12], sea buckthorn pulp was found to increase the specific volume of bread, with the highest value observed for bread with 10% sea buckthorn pulp. In addition to sea buckthorn, Nilova and Malyutenkova, 2021 [28] also reported that incorporating marc powder obtained from blueberry, cloudberry, and rowan increased the specific volume of bakery products. The highest specific volume was observed for products with cloudberry (380.40 ± 8.0%), followed by those with sea buckthorn (372.5 ± 8.6%).

3.1.3. Porosity

In terms of porosity, the analysis of functional bread samples (Table 2) revealed that the S6 sample had a higher porosity (73.2%) compared to the control sample (72.9%). The addition of sea buckthorn pomace powder significantly influences the porosity of bread obtained from the Mara and Sorana varieties ($p < 0.05$). Conversely, all other functional bread samples had lower porosity values, ranging from 67.3% to 58.6%, compared to the control. The bread sample with a 10% addition of Sorana variety sea buckthorn pomace powder (S10) had the lowest porosity (58.6%).

Porosity plays a crucial role in the digestibility of bakery products since higher porosity in the bread core makes it easier for the consumer's body to digest it. The vitamins and simple carbohydrates found in sea buckthorn stimulate the fermentative activity of bakery yeast, which affects the porosity of the bread core [29].

Similar to the results obtained in this research, a study conducted by Sturza et al., 2016 [17], showed that the addition of sea buckthorn flour at 2% and 4% increased the porosity of pastry crumbs from 70% in the control sample to 72.2% and 72.1%, respectively. Additionally, Nilova and Malyutenkova, 2018 [27] found that the porosity of bakery products increased when seed powder, marc powder, and peel powder from sea buckthorn fruits were added. The porosity of the samples ranged from 74.5 ± 1.3% to 72.8 ± 1.8%, which was higher than the control sample. The addition of marc powder obtained from blueberry, cloudberry, rowan, and sea buckthorn also increased the porosity of bakery products, with cloudberry showing the highest porosity (75.4 ± 1.4%) followed by sea buckthorn (73.8 ± 1.8%) [28]. In contrast, Ghendov-Mosanu et al., 2020 [29] found that the porosity of the samples decreased by 5.7% and 17.4% in the samples with 3% and 5% sea buckthorn, respectively, compared to the control. This indicates that as the percentage of added by-products increased, the porosity decreased. This may be due to the low extensibility of the dough, which contributed to the reduction in gas retention capacity during fermentation [29].

3.1.4. Elasticity

In terms of the elasticity of the analyzed bread samples (Table 2), it was noted that the control sample exhibited the highest elasticity (96.7%), followed closely by the C6 sample (85.2%). The elasticity of the functional bread samples ranged from 63.3% to 85.2%. The S10 bread sample (63.3%) and the M10 bread sample (64.4%) had the lowest elasticity. The addition of sea buckthorn pomace powders significantly influences ($p < 0.05$) the elasticity of the obtained bread samples.

Table 3 presents the physical–chemical characteristics of the obtained functional bread with sea buckthorn pomace powder.

Table 3. The physical–chemical characteristics of functional bread with sea buckthorn pomace powder (mean ± SD).

Sample	Moisture (%)	Acidity (deg)	a_w
Control	34.23 ± 0.71 [c]	1.3 ± 0.01	0.949 ± 0.001 [b,d,e]
C6	45.24 ± 0.35 [a,b]	2.2 ± 0.02 [a,b]	0.935 ± 0.001
C8	44.75 ± 0.43 [a,c]	1.5 ± 0.02	0.964 ± 0.001 [a]
C10	44.33 ± 0.28 [c]	2.2 ± 0.03 [a,b]	0.960 ± 0.001 [a,b,c]
M6	44.61 ± 0.34 [a,c]	2.2 ± 0.01 [a,b]	0.945 ± 0.001 [d,e]
M8	44.62 ± 0.43 [a,c]	2.2 ± 0.02 [a,b]	0.939 ± 0.001 [e]
M10	44.48 ± 0.39 [a,c]	1.7 ± 0.01 [b]	0.960 ± 0.001 [a,b,d]
S6	44.58 ± 0.33 [a,c]	2.2 ± 0.02 [a]	0.941 ± 0.001 [e]
S8	44.47 ± 0.42 [b,c]	2.6 ± 0.03	0.942 ± 0.001 [c,d,e]
S10	43.89 ± 0.22 [c]	2.8 ± 0.01	0.937 ± 0.001 [e]

All the results are mean ± standard deviation. Means followed by a common letter are not significantly different ($p > 0.05$). Values followed by different superscript letters in the same column are significantly different ($p < 0.05$). Means that do not share a letter in a column are significantly different ($p < 0.05$).

3.1.5. Moisture Content

Regarding the moisture content of the analyzed bread samples (as shown in Table 3), it was observed that the control sample had the lowest moisture content of 34.23%. However, all functional bread samples with added sea buckthorn pomace powder had significantly ($p < 0.05$) higher moisture content than the control sample, ranging between 45.24% and 43.89%. There were no significant differences in the moisture content observed among the functional bread samples with added sea buckthorn pomace powder.

Guo et al., 2019 [12] also found that the moisture content in bread samples increased as the percentage of sea buckthorn pulp added increased, with values ranging from 32.84 ± 0.56% for the control sample to 36.23 ± 0.32% for bread with 10% sea buckthorn pulp added. This suggests that an increase in the percentage of sea buckthorn leads to higher moisture content. In biscuits with sea buckthorn by-products, Muresan et al., 2019 [30] found that the moisture content was 95.23 ± 0.57% for those obtained at 50 °C for 12 h and 96.55 ± 0.57% for those with thermal treatment at 80 °C for 5 h. In contrast, Nilova and Malyutenkova, 2021 [28] found that the moisture content of bakery products obtained by adding marc powder obtained from blueberry, cloudberry, rowan, and sea buckthorn was similar to or slightly lower than the control sample. Furthermore, Sturza et al., 2016 [17] noted that as the proportion of sea buckthorn flour increased in pastry products; the moisture content decreased from 22% in the control sample to 20% for gingerbread with 4% sea buckthorn, while for sponge cakes, the moisture content decreased from 36.6% in the control sample to 36.12% in the sample with 4% sea buckthorn flour. The authors explained that the addition of sea buckthorn flour led to a decrease in the amount of wet gluten in the product, which resulted in a reduction in moisture in the final product.

3.1.6. Acidity

In terms of acidity, the bread samples analyzed showed that the control sample had the lowest acidity value of 1.30, while the functional bread samples had significantly higher ($p < 0.05$) acidity values ranging from 1.5 to 2.8 degrees. Based on the data presented in Table 3, it can be observed that the bread samples with the lowest acidity values were those with an 8% addition of sea buckthorn powder from the Clara variety (C8) (1.50 deg. acidity) and those with a 10% addition of sea buckthorn powder from the Mara variety (M10) (1.70 deg. acidity). On the other hand, the bread samples with the highest acidity values were the S8 and S10 bread samples with values of 2.60 acidity and 2.80 acidity, respectively. According to Ghendov-Mosanu et al., 2020 [29], there is a positive correlation between the acidity of bakery products and the concentration of added sea buckthorn berry flour. This can be attributed to the presence of organic acids and sugars in the berry flour, which speeds up the fermentation of the dough.

Furthermore, Nilova and Malyutenkova, 2018 [27], Ghendov-Mosanu et al., 2020 [29], and Nilova and Malyutenkova, 2021 [28] demonstrate that the addition of sea buckthorn in bakery products increases their acidity. The study by Nilova and Malyutenkova, 2018 [27] showed an increase in acidity from 1.7 ± 0.1 to 3.2 ± 0.1, 3.4 ± 0.1, and 3.6 ± 0.1 for bakery products made with sea buckthorn seeds powder, marc powder, and peel powder, respectively; values that are close to the ones obtained in this study. Similarly, the research conducted by Ghendov-Mosanu et al., 2020 [29] reported a significant increase in acidity by 100–291.7% compared to the control samples. Nilova and Malyutenkova, 2021 [28] found that the addition of marc powder obtained from blueberry, cloudberry, rowan, and sea buckthorn in bakery products resulted in higher acidity than the control sample, with the highest acidity of 3.4 ± 0.1 recorded for the bakery products with sea buckthorn marc powder.

3.1.7. Water Activity

The a_w index (Table 3) did not significantly ($p > 0.05$) vary between the developed bread samples. In general, a higher value of this index was obtained for the samples prepared using sea buckthorn pomace powder. Rajeswari et al., 2018 [31] observed that the water activity of bread samples enriched with banana, aonla, and sapota powders varied between 0.77 (for 5% aonla powder) and 0.82 (for 20% banana and 20% sapota powder). In contrast, Nikolau et al., 2022 [32] found that the water activity of bakery products with bioactive compounds from black Corianthian grape was lower than that of the respective dough, due to direct heat contact during baking. The authors noted that water activity was highly dependent on the enrichment type and leavening agent. Conversely, Costa et al., 2018 [33] did not observe any significant difference in water activity for pan bread containing pumpkin seed flour. Lim et al., 2011 [34] reported a significant decrease in the water activity of bread as the turmeric powder level in the recipe increased.

3.2. Nutritional Characteristics of Functional Bread

3.2.1. Protein Content

Changes were observed in the protein content, which ranged between 13.07% and 13.82%, obtaining higher values for the bread samples containing sea buckthorn pomace powder than those of the control sample (Table 4). However, with the exception of the S8 sample, it was observed that the protein content increased as the amount of sea buckthorn powder added to the samples increased.

Similar results were obtained by Guo et al., 2019 [12], who found that adding 5% sea buckthorn pulp to bread resulted in a protein content similar to the control sample, but adding 10% sea buckthorn pulp decreased the protein content to 11.11 ± 0.32 compared to 11.23 ± 0.08 in the control sample. On the other hand, Odunlade et al., 2017 [35] observed an increase in protein content from 9.5% to 13.09% as the substitution level of leafy vegetable powder increased.

Table 4. The content of crude fiber, total ash, protein, fat, and the energy value of the developed bread samples.

Sample	Crude Fiber (%)	Total Ash (%)	Protein (%)	Fat (%)	Energy Value (Kcal/100 g)
Control	0.12 ± 0.01	1.37 ± 0.01	13.07 ± 0.10 [d]	0.17 ± 0.01	260 ± 0.15
C6	1.33 ± 0.01	1.26 ± 0.01	13.29 ± 0.06 [d]	0.80 ± 0.01	217 ± 0.12 [b]
C8	1.68 ± 0.01 [a]	1.54 ± 0.01	13.50 ± 0.04 [b,c,d]	1.14 ± 0.01 [a]	219 ± 0.19 [b]
C10	2.04 ± 0.01	1.63 ± 0.01 [a]	13.75 ± 0.07 [a,b]	1.56 ± 0.01	221 ± 0.22 [a,b]
M6	1.46 ± 0.01	1.24 ± 0.01 [b]	13.58 ± 0.09 [a,b,c]	0.68 ± 0.01	219 ± 0.10 [b]
M8	1.51 ± 0.01	1.58 ± 0.01	13.72 ± 0.05 [a,b]	0.93 ± 0.01	219 ± 0.14 [b]
M10	1.59 ± 0.01	1.65 ± 0.01 [a]	13.82 ± 0.06 [a]	1.10 ± 0.01 [a]	220 ± 0.12 [b]
S6	1.07 ± 0.01	1.18 ± 0.01 [b]	13.39 ± 0.05 [c,d]	0.97 ± 0.01	221 ± 0.15 [a,b]
S8	1.41 ± 0.01	1.25 ± 0.01 [b]	13.36 ± 0.07 [c,d]	1.30 ± 0.01 [a]	221 ± 0.13 [a,b]
S10	1.68 ± 0.01 [a]	1.33 ± 0.01	13.46 ± 0.06 [c,d]	1.60 ± 0.01	224 ± 0.18 [a]

All the results are mean ± standard deviation. Means followed by a common letter are not significantly different ($p > 0.05$). Values followed by different superscript letters in the same column are significantly different ($p < 0.05$). Means that do not share a letter in a column are significantly different ($p < 0.05$).

3.2.2. Fat Content

The functional bread samples showed a significant increase ($p < 0.05$) in fat content when compared to the control sample (Table 4). The control sample had a fat content of 0.17%, while the functional bread samples had values ranging from 0.68% to 1.6%. The sample with the lowest lipid content was the one with a 6% addition of the Mara variety, while the sample with the highest lipid content was the sample with a 10% addition of Sorana sea buckthorn powder. It can be observed that the functional bread with the addition of sea buckthorn powder from the Sorana variety was richer in lipids compared to the samples that had sea buckthorn powder from the Mara and Clara varieties added. The fat content obtained in the present study was lower than the findings of Odunlade et al., 2017 [35], who reported an increase in fat content from 1.3% to 2.0% in bread enriched with leafy vegetable powder.

Furthermore, Janotkova et al., 2021 [36] found that the total fat content in biscuits with a 10% addition of sea buckthorn biomass was 16.7% compared to the control. They attributed these differences to the higher thermal treatment of the biscuits. The seeds and fruits of sea buckthorn are rich in fatty substances and acids, which make the biscuits containing this by-product have a higher fat content than other types of biscuits. Muresan et al., 2019 [30] reported that the fat content of biscuits made with sea buckthorn by-products was 30.11 ± 0.65 when baked at 50 °C for 12 h, and 30.48 ± 0.06 when baked at 80 °C for 5 h. This is due to the high fat content of sea buckthorn fruits and seeds. Biscuits containing dried by-product powder had higher fat content compared to other types of biscuits, and samples with powders dried at 80 °C contained more fat than those dried at 50 °C.

3.2.3. Total Ash Content

There were no significant differences ($p > 0.05$) in the ash content among the bread samples analyzed (Table 4), which ranged from 1.18% to 1.63%. The sample with a 6% addition of Sorana sea buckthorn pomace powder had the lowest ash content, while the sample with a 10% addition of Clara sea buckthorn pomace powder had the highest ash content. The ash content values obtained in the present study were higher than those reported by Muresan et al., 2019 [30] for biscuits containing sea buckthorn by-products, with ash content ranging from 0.23 ± 0.02 to 0.25 ± 0.03, depending on the thermal treatment conditions. On the other hand, Odunlade et al., 2017 [35] reported a significant increase ($p < 0.05$) in ash content from 1.1 to 2.4% when adding leafy vegetable powders to bread samples, attributing this increase to the content of green leafy vegetables.

3.2.4. Crude Fiber

Concerning the crude fiber, the results indicate a significant increase ($p < 0.05$) in crude fiber content in the functional bread samples compared to the control sample. Specifically, the control sample had a raw fiber content of 0.12%, whereas the functional bread samples showed values ranging from 1.07 to 2.04%. The results are in accordance with the findings of Guo et al., 2019 [12], who demonstrated that bread samples containing 10% sea buckthorn pulp had twice the amount of total dietary fiber content compared to the control sample. Meanwhile, Odunlade et al., 2017 [35] observed that the crude fiber content of their bread samples ranged from 1.8% to 4.0%. They also reported a significant ($p < 0.05$) increase in fiber content with the addition of leafy vegetable powder.

The functional characteristics of food are linked to the presence of non-digestible carbohydrates in the form of dietary fiber, which serves as a source of nourishment for symbiotic bacteria in the large intestine. The role of dietary fiber in the human body is influenced by its quantity in the diet, as well as its fractional composition, which can vary based on factors such as plant species, developmental stage, anatomical part of the plant, and the specific technological processes employed [37]. For example, adding apple pomace to bakery products is believed to enhance their dietary fiber content and health benefits. However, in most cases, the addition of pomace leads to a decrease in the quality and sensory properties of the baked products. The incorporation of apple pomace has been shown to increase the total dietary fiber (TDF), total phenolics content (TPC), antioxidant activity, total flavonoid content (TFC), and proanthocyanidins content (PAC) [38].

Sea buckthorn berries are rich in dietary fiber, with a substantial amount of 6.55 g per 100 g. This content exceeds 25% of the recommended daily intake of dietary fiber [39]. The amount of dietary fiber and the proportion of soluble and insoluble fiber in pomace are influenced by the origin of the pomace and the specific processing conditions employed during fiber extraction [40]. Furthermore, the fiber content of sea buckthorn depends on weather conditions and the maturity stage of the berries. The crude fiber level in sea buckthorn typically ranges from 62 g/kg to 100 g/kg of dry weight [41]. According to a study by Jaroszewska et al., 2018 [42], sea buckthorn berries are rich in dietary fibers. The distribution of dietary fiber fractions in sea buckthorn berries is as follows: 160–200 g/kg dry weight of neutral detergent fiber, 120–145 g/kg dry weight of acid detergent fiber, 50–70 g/kg dry weight of acid detergent lignin, 45–55 g/kg dry weight of hemicellulose, and 60–75 g/kg dry weight of cellulose. The raw fiber content in sea buckthorn seeds was measured to be 130 g/kg dry weight, while the peel and the pulp contained 66 g/kg dry weight and 47 g/kg dry weight of raw fiber [43]. The dried sea buckthorn pomace exhibited a significant crude fiber content, with values of 19.86% [44].

Considering the high fiber content in sea buckthorn berries and pomace, their incorporation in bread significantly enhances the crude fiber content of the final product.

3.2.5. Energetic Value

The energetic value of the obtained bread samples ranged between 217 ± 0.12 and 260 ± 0.15 Kcal/100 g (Table 4). It was observed that all samples containing sea buckthorn pomace powder presented lower energetic values compared to the control sample. Between these samples, the highest energetic value was obtained for the S10 bread sample (224 ± 0.18) and the lowest value was obtained for sample C6 (217 ± 0.12). However, no significant differences were observed between the bread samples containing sea buckthorn pomace powder.

3.2.6. Contaminants

Table 5 shows that samples C8 and M10 had higher levels of lead, with 23.08 ± 0.10 µg/kg and 25.59 ± 0.17 µg/kg, respectively, compared to the control sample. All samples had higher levels of cadmium than the control sample, with sample S6 having the highest level of cadmium at 18.88 ± 0.18 µg/kg. Regarding Cr levels, C10 and S8 had higher Cr content than the control sample, with levels of 301.66 ± 2.36 µg/kg and 343.08 ± 1.02 µg/kg, respectively. The Clara variety pomace powder incorporated in bread samples had lower lead content compared to the Mara and Sorana varieties. The samples added with Mara variety pomace powder were the least contaminated with cadmium. The addition of sea buckthorn powder increased the cadmium levels in the bread samples, which may have been due to the contamination of sea buckthorn with cadmium. Therefore, it is necessary rigorously control the powders and flours obtained from different vegetable and fruit residues used in the development of new food products.

Table 5. The content of heavy metals in functional bread samples (mean ± SD).

Sample	Pb (µg/kg) ± SD	Cd (µg/kg) ± SD	Cr (µg/kg) ± SD
Control	19.20 ± 0.19 [a]	4.70 ± 0.01	173.48 ± 1.91
C6	18.87 ± 0.01 [a]	12.89 ± 0.07	87.07 ± 3.98
C8	23.08 ± 0.10	10.70 ± 0.13 [b]	142.09 ± 2.51
C10	18.08 ± 0.19	13.78 ± 0.03 [a]	301.66 ± 2.36
M6	13.79 ± 0.01	10.29 ± 0.10 [b]	224.36 ± 2.55
M8	14.89 ± 0.12 [b]	8.39 ± 0.08	6.39 ± 0.90
M10	25.59 ± 0.17	13.50 ± 0.10 [a]	115.16 ± 2.04 [a]
S6	10.29 ± 0.03	18.88 ± 0.18	116.35 ± 2.38 [a]
S8	15.09 ± 0.17 [b]	6.95 ± 0.03	343.08 ± 1.02
S10	8.43 ± 0.10	13.79 ± 0.09 [a]	154.28 ± 2.30

All the results are mean ± standard deviation. Means followed by a common letter are not significantly different ($p > 0.05$). Values followed by different superscript letters in the same column are significantly different ($p < 0.05$). Means that do not share a letter in a column are significantly different ($p < 0.05$).

According to studies conducted by Petrescu-Mag et al., 2021 [45] and Dudarev et al., 2019 [46], it was found that the levels of Pb, Cd, and Cr were higher in cultivated sea buckthorn compared to wild berries. Additionally, Petrescu-Mag et al., 2021 [45] reported that the heavy metal levels in sea buckthorn samples collected from non-mining areas were lower than those from mining areas, indicating contamination of wild berries due to soil contamination. Furthermore, Vaitkeviciene et al., 2019 [47] showed that lead and cadmium content was higher in cultivated sea buckthorn samples compared to wild sea buckthorn berries.

Concerning the microelements content, Table 6 presents the results for Cu, Zn, Mn, and Fe. All samples showed higher levels of Cu compared to the control samples. The bread samples with sea buckthorn pomace powder from the Mara variety had the highest copper content (9.25 mg Cu/kg), zinc content (11.55 mg Zn/kg), and manganese content (21.75 mg Mn/kg). On the other hand, the bread samples with sea buckthorn pomace powder from the Clara variety had the highest iron content (54.89%).

Table 6. The content of microelements in functional bread samples (mean ± SD).

Sample	Cu (mg/kg) ± SD	Zn (mg/kg) ± SD	Mn (mg/kg) ± SD	Fe (mg/kg) ± SD
Control	2.64 ± 0.01 [g]	12.31 ± 0.10 [a,b]	11.88 ± 0.26 [b,e]	1.11 ± 0.02
C6	3.24 ± 0.01 [d,f]	9.16 ± 0.80 [c,d]	12.97 ± 0.32 [a,b,c,d]	3.49 ± 0.08 [a]
C8	4.67 ± 0.24 [a]	10.77 ± 0.43 [c]	14.03 ± 0.32 [a]	3.92 ± 0.32 [a,b]
C10	4.50 ± 0.12 [a]	9.75 ± 0.30 [c,d]	11.86 ± 0.06 [e]	3.35 ± 0.10 [a,c]
M6	3.39 ± 0.10 [d,e]	8.35 ± 0.10 [b,d]	11.48 ± 0.75 [c,e]	1.67 ± 0.12 [b,e]
M8	3.32 ± 0.18 [c,d,f]	9.39 ± 0.69 [c,d]	12.71 ± 0.53 [a,e]	3.43 ± 0.11 [a]
M10	4.02 ± 0.04 [b]	9.14 ± 0.03 [d]	11.91 ± 0.10 [e]	2.56 ± 0.25 [c,d]
S6	3.89 ± 0.11 [b,c]	8.98 ± 0.22 [d]	11.80 ± 0.06 [e]	1.90 ± 0.18 [d,e]
S8	2.93 ± 0.11 [f,g]	10.54 ± 0.12 [c]	11.30 ± 0.07 [e]	1.62 ± 0.12 [e]
S10	2.79 ± 0.21 [e,f,g]	10.38 ± 0.74 [a,c,d]	10.96 ± 0.87 [d,e]	7.33 ± 0.13

All the results are mean ± standard deviation. Means followed by a common letter are not significantly different ($p > 0.05$). Values followed by different superscript letters in the same column are significantly different ($p < 0.05$). Means that do not share a letter in a column are significantly different ($p < 0.05$).

The findings revealed that all three sea buckthorn varieties used in the functional bread samples exhibited an increase in copper and iron content, while a decrease was observed in zinc content compared to the control sample. With regard to manganese content, a decline was noted in the functional bread samples, except for those containing 6% and 8% sea buckthorn pomace powder from the Clara variety, as well as 8% and 10% sea buckthorn pomace powder from the Mara variety.

The amount of Zn detected in the current study is higher than the levels found in bread samples fortified with beetroot powder (2–15%), as reported by Ani et al., 2022 [48]. On the other hand, the levels of Fe in the present study are lower than those reported by Ani et al., 2022 [48]. The study demonstrated that the Fe and Zn content in bread enriched with beetroot powder was lower than those in the control sample. In contrast, the results of the current study indicate that sea buckthorn pomace powder is a viable option for enhancing bread with Fe and Cu. However, the values of Zn and Mn in the sea buckthorn-enriched bread are similar or lower compared to those in the control samples.

The iron content observed in the current study is lower than the iron content found in bread fortified with fluted pumpkin, amaranth, or African eggplant leafy powders, as reported by Odunlade et al., 2017 [35]. However, the levels of zinc obtained in this study are similar to those reported by Odunlade et al., 2017 [35]. The authors showed that the iron and zinc content increased as the percentage of leafy powder added increased.

According to Luntraru et al., 2022 [49], the microelement content of sea buckthorn by-products is 4.51 ± 0.090 mg/100 g for Zn, 1.23 ± 0.073 mg/100 g for Cu, 3.51 ± 0.098 mg/100 g for Mn, and 3.6 ± 0.079 mg/100 g for Fe. Vaitkeviciene et al., 2019 [47] reported that the microelement content of sea buckthorn varies depending on the source of the raw material. They demonstrated that the tested cultivated sea buckthorn had higher Fe and Mn content compared to wild sea buckthorn. On the contrary, wild berries were found to have higher Cu, Zn, and Ni content.

In terms of sodium and potassium content (Table 7), it is noticeable that the bread samples added with sea buckthorn pomace powder from the Sorana variety contain the highest amount (227.1 mg Na/100 g and 360.3 mg K/100 g, respectively). As for calcium content, the highest amount is found in the bread samples containing sea buckthorn pomace powder from the Clara variety (23.87 mg Ca/100 g), while the bread samples containing sea buckthorn pomace powder from the Mara variety has the highest magnesium content (77.94 mg Mg/100 g).

Table 7. The content of minerals in functional bread samples (mean ± SD).

Sample	Na (mg/100 g) ± SD	K (mg/100 g) ± SD	Ca (mg/100 g) ± SD	Mg (mg/100 g) ± SD
Control	63.78 ± 0.19 [b,c]	52.99 ± 0.21 [b]	5.58 ± 0.08 [c]	10.60 ± 0.11 [c]
C6	63.75 ± 0.06 [b,c]	54.74 ± 0.22 [a]	6.94 ± 0.07 [a]	10.89 ± 0.15 [a,b]
C8	63.65 ± 0.06 [c]	84.15 ± 0.34	6.43 ± 0.23 [a,b]	10.92 ± 0.11 [a,b]
C10	64.41 ± 0.06 [a]	54.01 ± 0.22 [a,b]	6.95 ± 0.08 [a]	11.19 ± 0.13 [a]
M6	63.97 ± 0.13 [b,c]	55.11 ± 0.33 [a]	5.75 ± 0.12 [b,c]	10.98 ± 0.05 [a,b]
M8	64.32 ± 0.19 [a,b]	55.0 ± 0.11 [a]	7.09 ± 0.10 [a]	10.94 ± 0.11 [a,b]
M10	64.09 ± 0.06 [b]	54.92 ± 0.33 [a]	6.24 ± 0.29 [a,b,c]	11.02 ± 0.07 [a]
S6	63.72 ± 0.10 [b,c]	54.12 ± 0.11 [a,b]	5.77 ± 0.05 [b,c]	10.81 ± 0.17
S8	62.13 ± 0.06	54.16 ± 0.32 [a,b]	5.56 ± 0.14 [c]	10.95 ± 0.16 [a,b]
S10	61.06 ± 0.18	53.63 ± 0.11 [b]	5.66 ± 0.10 [c]	10.82 ± 0.03 [a,b]

All the results are mean ± standard deviation. Means followed by a common letter are not significantly different ($p > 0.05$). Values followed by different superscript letters in the same column are significantly different ($p < 0.05$). Means that do not share a letter in a column are significantly different ($p < 0.05$).

The analyzed bread samples exhibited noteworthy distinctions in sodium content when compared to the control sample, ranging from 61.06 to 64.41 mg Na/100 g. Similar observations were noted for magnesium content, which varied between 10.81 and 11.19 mg Mg/100 g. Conversely, the potassium content of the functional bread was greater than that of the control sample. Specifically, the control sample contained 52.99 mg K/100 g, whereas the functional bread demonstrated an elevated content ranging from 53.63 to 84.15 mg K/100 g. With respect to calcium content, all functional bread samples exhibited higher values than the control sample, except for the bread containing 8% sea buckthorn pomace powder from the Sorana variety. The control sample had a value of 5.58 mg Ca/100 g, while the functional bread samples recorded higher values ranging from 5.66 to 7.09 mg Ca/100 g.

The sodium content found in this study was higher than the sodium content reported by Odunlade et al., 2017 [35], who added leafy powder to bread samples. On the other hand, the levels of calcium and magnesium obtained in the present study were significantly lower than those obtained by Odunlade et al., 2017 [35]. Their results showed that an increase in the percentage of leafy powder led to an increase in the content of sodium, magnesium, and calcium.

According to Vaitkeviciene et al., 2019 [47], cultivated sea buckthorn contains 10.30 ± 0.42 g/kg of potassium (K), 0.844 ± 0.07 g/kg of calcium (Ca), and 0.701 ± 0.03 g/kg of magnesium (Mg). Therefore, sea buckthorn can be considered a valuable source of minerals for food products with high mineral content.

3.2.7. Total Polyphenolic Content

The polyphenolic content of the bread samples increased as the amount of sea buckthorn powder added to them increased, as shown in Figure 5. Additionally, the bread samples containing sea buckthorn powder had a higher polyphenolic content compared to the control sample.

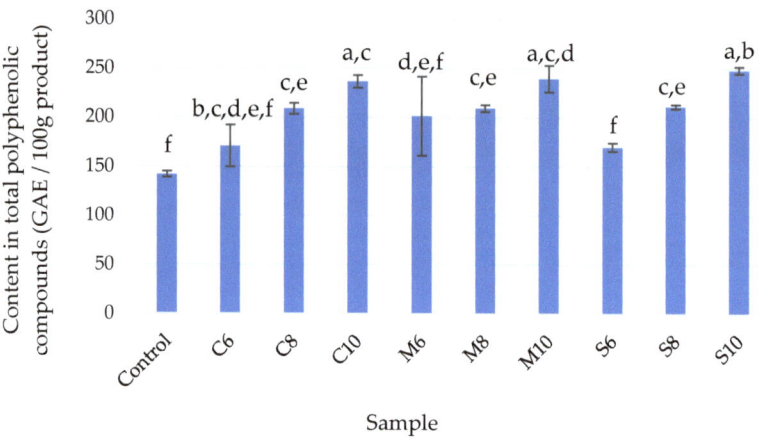

Figure 5. The content of polyphenols for the samples of functional bread with the addition of sea buckthorn pomace powder. Means followed by a common letter are not significantly different ($p > 0.05$). Values followed by different superscript letters in the same column are significantly different ($p < 0.05$). Means that do not share a letter in a column are significantly different ($p < 0.05$).

Similar to the results obtained in this study, Nilova and Malyutenkova, 2018 [27] discovered that the addition of sea buckthorn to bread samples increased the phenolic content in both the crust and crumb. Additionally, they found that decreasing the baking temperature resulted in a higher number of phenolic compounds and flavonoids in both the crust and crumb. At a temperature of 220 °C, the loss of phenolic compounds was between 23.8 and 25.6%, while at 200 °C, the loss was between 7 and 13%, which was significantly lower. They also demonstrated that a high baking temperature completely destroyed the ascorbic acid in all enriched samples. In a separate study by Akbas and

Kilmanoglu, 2022 [50], they observed a significant increase ($p < 0.05$) in the phenolic content of all bread that had fruit and vegetables added when compared to the control group.

3.2.8. Antioxidant Activity

After assessing the antioxidant activity of the bread samples (as depicted in Figure 6), it was observed that all samples incorporated with sea buckthorn pomace powder exhibited elevated values for this parameter (over 200 QE/100 g product) in comparison to the control sample. This augmentation is attributed to the presence of polyphenols, such as phenolic acid and flavonoids, which are acknowledged as the most active components responsible for the antioxidant characteristics of various sea buckthorn varieties. In addition to polyphenols, the antioxidant ability of sea buckthorn is linked to the combined influence of ascorbic acid and carotenoids [36]. Based on the obtained results, it can be concluded that the addition of sea buckthorn pomace powder led to a significant ($p < 0.05$) escalation in the antioxidant activity of the developed products. The C10 bread sample had the highest value, whereas the M10 sample had the lowest value. Similar results were obtained by Janotkova et al., 2021 [36]; namely, when the percentage of sea buckthorn was increased to 20% in cereal biscuits, a noticeable decrease in antioxidant activity was observed. This decline may be due to some antioxidants that can act as prooxidants at higher concentrations [36]. The antioxidant analysis of the bread samples confirmed that the products exhibit substantial antioxidant potential and constitute a rich source of antioxidant compounds, which may impart many beneficial effects on the human body.

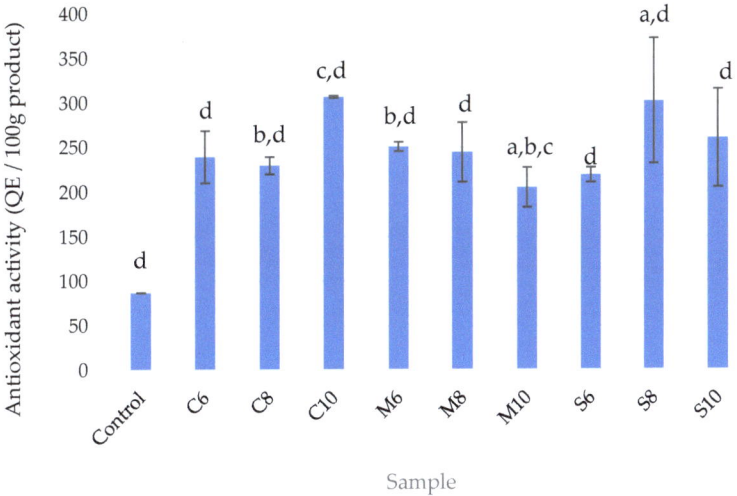

Figure 6. Antioxidant activity of functional bread samples with the addition of sea buckthorn pomace powder. Means followed by a common letter are not significantly different ($p > 0.05$). Values followed by different superscript letters in the same column are significantly different ($p < 0.05$). Means that do not share a letter in a column are significantly different ($p < 0.05$).

Sturza et al., 2016 [17] reported that the presence of antioxidants in bread samples with sea buckthorn reduces the degree of lipid oxidation due to the fat content of sea buckthorn. Additionally, Janotkova et al., 2021 [36] found that cereal biscuits with 15% sea buckthorn biomass had increased antioxidant activity.

According to the study conducted by Sturza et al., 2016 [17], the antioxidant activity (expressed as the percentage of free inhibited DPPH radicals) of pastry products increased from $-18.08 \pm 1.21\%$ to $73.52 \pm 0.63\%$ for 4% sea buckthorn flour gingerbread, and from $-12.98 \pm 0.91\%$ to $17.86 \pm 0.61\%$ for 4% sea buckthorn flour sponge cake. On the other hand,

Ghendov-Mosanu et al., 2020 [29] demonstrated that the addition of 1% sea buckthorn berry flour slightly increased the antioxidant activity of their samples, but it remained negative (−8.65 ± 0.62%). When 5% sea buckthorn biomass flour was added, the antioxidant activity of the samples increased to 20.05 ± 0.51%, which was significantly lower than the antioxidant activity observed in the study conducted by Sturza et al., 2016 [17].

Muresan et al., 2019 [30] showed that the antioxidant activity of sea buckthorn powder used in biscuit preparation was affected by the drying temperature. Specifically, biscuits made with sea buckthorn powder dried at 50 °C for 12 h exhibited an antioxidant activity of 88.97 ± 0.82, while those made with powder dried at 80 °C for 5 h had an activity of 89.14 ± 0.27.

In a separate study by Nilova and Malyutenkova, 2018 [27], it was observed that the type and amount of sea buckthorn powder used in bakery products influenced their antioxidant activity. Bakery products made with sea buckthorn peel powder had the highest antioxidant activity, followed by those made with marc powder and seed powder. This was due to the peel powder's high content of ascorbic acid, total phenolic content, and total flavonoid content, followed by the marc powder and seed powder.

3.3. Sensory Analysis

Based on the data presented in Figure 7 for bread samples added with Clara sea buckthorn pomace powder, it can be observed that the sample with 10% sea buckthorn had a strong aroma and taste of sea buckthorn, with a slightly sour taste and a distinct aftertaste. The internal appearance of the bread with 10% sea buckthorn was judged to be the darkest, but it showed good elasticity and a compact, non-crumbling core. In contrast, the sample with 8% sea buckthorn had a pleasant appearance, with a crispy crust and the lightest crust color among the samples with powder from the Clara variety.

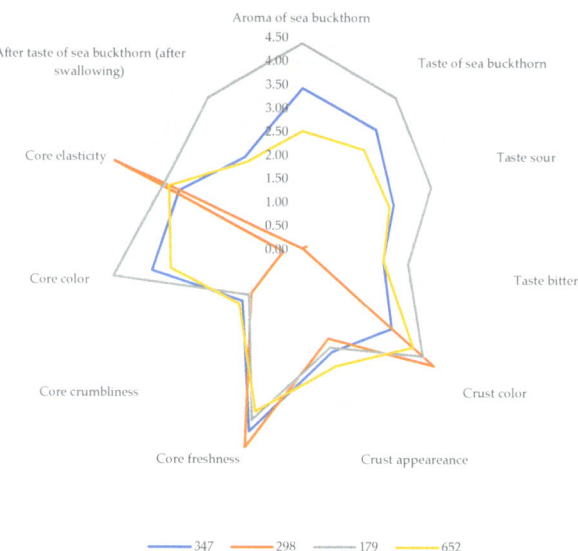

Figure 7. Graphic representation of the sensory characteristics of the functional bread samples obtained by adding 6%, 8%, and 10% sea buckthorn powder from the Clara variety (legend: 347—C8 sample, 298—control sample, 179—C10 sample, 652—C6 sample).

Based on the data presented in Figure 8 for bread added with Mara sea buckthorn pomace powder, it was observed that the sample containing 10% sea buckthorn pomace powder had a robust sea buckthorn aroma and a taste that was slightly sour and bitter, with a potent aftertaste. On the other hand, the sample with 8% sea buckthorn powder had a light-colored crust that was crisp and fresh, while the core was compact and non-crumbly.

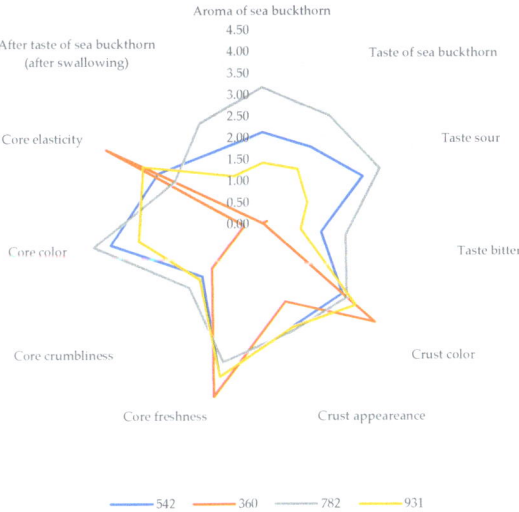

Figure 8. Graphic representation of the sensory characteristics of the functional bread samples obtained by adding 6%, 8%, and 10% sea buckthorn powder from the Mara variety (legend: 542—M8 sample, 360—control sample, 782—M10 sample, 931—M6 sample).

Based on the presented data in Figure 9 for bread added with Sorana sea buckthorn pomace powder, it is observed that the samples containing sea buckthorn powder had a mild sea buckthorn aroma and a distinctive sea buckthorn flavor, with a slightly sour and bitter undertone. In particular, the sample with 8% sea buckthorn pomace powder had a pale crust and a dark, dense, and elastic core that did not crumble easily.

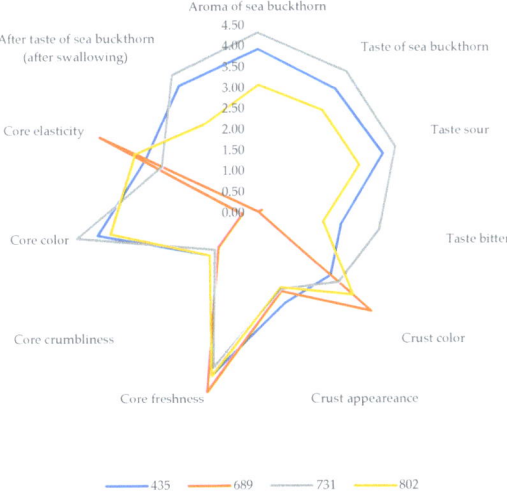

Figure 9. Graphic representation of the sensory characteristics of the functional bread samples obtained by adding 6%, 8%, and 10% sea buckthorn powder from the Sorana variety (legend: 435—S8 sample, 689—control sample, 731—S10 sample, 802—S6 sample).

According to the organoleptic analysis conducted by the panelists, a specific smell of yeast was noted in the samples, and when a larger amount of sea buckthorn powder was added, a slight smell of fermentation was also detected. However, the sea buckthorn flavor was found to be generally pleasant.

Regarding the assessment of the overall acceptability of functional bread samples produced by adding 6%, 8%, and 10% sea buckthorn pomace powder, the findings are illustrated in Figure 10. The samples with a 6% addition of sea buckthorn pomace powder received a unanimous acceptance with a maximum score of 100%. The samples containing 8% sea buckthorn powder showed a high level of acceptability, with the sample containing sea buckthorn powder from the Mara variety being the most preferred, scoring 93.9%. The sample with 8% sea buckthorn pomace powder from the Clara variety was also highly appreciated, with a score of 92.6%. The sample with 8% sea buckthorn powder from the Sorana variety received a slightly lower score of 86.58%. However, consumer preferences were not met by the samples containing 10% sea buckthorn pomace powder, as their scores were below 42%, with the sample containing 10% sea buckthorn pomace powder from the Mara variety obtaining a score of 41.46%, the sample with 10% sea buckthorn pomace powder from the Sorana variety obtaining a score of 34.14%, and the sample with 10% sea buckthorn pomace powder from the Clara variety obtaining a score of 29.2%.

Furthermore, the addition of wine grape pomace (WGP) to food products can enhance the dietary fiber (DF) and polyphenol content in the diet of consumers. However, the addition of WGP may have negative impacts on the sensory characteristics of food, such as flavor, color, and texture, as well as water and fat absorption. To mitigate these potential negative effects, strategies such as increasing hydration, using other ingredients to mask WGP flavors, reducing the particle size of WGP powders, and optimizing the formulation can be employed [51].

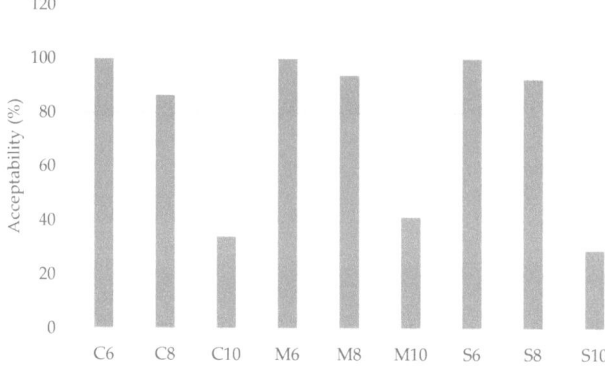

Figure 10. Acceptability test results of functional bread samples obtained by adding 6%, 8%, and 10% sea buckthorn powder.

According to Sturza et al., 2016 [17], the addition of 2% sea buckthorn flour to gingerbread improved its external appearance, color, and texture in comparison to the control. Guo et al., 2019 [12] demonstrated that consumers favored the soft texture of bread with a larger volume obtained with sea buckthorn. The bread sample with 5% sea buckthorn pulp was rated the highest in terms of taste (5.6), indicating it was the most palatable among the samples. However, bread with 8% sea buckthorn pulp received a lower score (4.47) compared to the control (4.67), indicating that flavor might not be favored by consumers when higher concentrations of residues are added. In terms of appearance, sea buckthorn pulp addition had a negative effect on the bread's appearance, which could be attributed to the high crude fiber content of the residues.

Ghendov-Mosanu et al., 2020 [29] found that adding 1% sea buckthorn concentration to the sample had a favorable effect on the organoleptic index, resulting a product with a golden crust, an elastic core, well-developed porosity, and pleasant taste and aroma. In contrast, samples with 3% and 5% sea buckthorn berry flour had a dark crust, dry crumb, poorly developed porosity, and a specific sea buckthorn berry flavor and odor. However, the total scores for the organoleptic analysis of the samples containing 3% and 5% sea buckthorn berry flour were within the range of 24.1–30.0, indicating that the products were of very good quality. The values for product shape and volume, crust appearance and color, baking degree, state and appearance of the bread core, bread core porosity and pore structure, aroma, and taste of the bread obtained with 1% sea buckthorn berry flour were identical to those of the control sample, indicating that this percentage did not significantly affect the characteristics of the obtained product. As the percentage of sea buckthorn berry flour added increased, the values of these parameters decreased.

Muresan et al., 2019 [30] observed that the taste and odor of by-products used in bakery products are influenced by the temperature of by-product drying. By-product samples dried at both 50 °C and 80 °C retained a characteristic odor and taste of fresh berries. In terms of sensory analysis, the biscuits made with sea buckthorn biomass powder obtained the highest score of 9, while those made with blueberry powder obtained a score of 8. The sensory analysis revealed that all three fruit pastes (including sea buckthorn and blueberry) had very good acceptability in terms of appearance, flavor, consistency, and overall aspect.

Janotkova et al., 2021 [36] demonstrated that the addition of 20% sea buckthorn to biscuits resulted in the most pleasant acidity and taste. The authors also observed that the addition of sea buckthorn biomass increased the acidity of the biscuits, which can be attributed to the presence of organic acids and sugars. Biscuits without any biomass addition were the hardest and most fragile, while those with 20% sea buckthorn were considered hard and brittle at a minimum level.

According to Stoin et al., 2014 [52], adding 10% sea buckthorn into digestive cookies resulted in a crispy texture, as opposed to the hard and brittle control sample. Furthermore, adding 15% sea buckthorn biomass resulted in the most enjoyable overall experience. The authors concluded that the digestive cookies with 10% sea buckthorn had higher scores (up to 19.71 out of a possible maximum of 20 points) compared to the control sample (17.85 points). The authors reported that adding 10% fresh sea buckthorn, dried sea buckthorn fruit powder, and sea buckthorn syrup to digestive cookies resulted in a brownish color, as opposed to the brighter (golden yellow) color of the control sample.

Based on the presented results, the bread samples containing 8% sea buckthorn pomace powder had the most favorable properties in terms of nutritional parameters and sensorial acceptability and were chosen as having the optimum quantity of sea buckthorn pomace. Therefore, these samples were further subjected to shelf-life analysis consisting of microbiological evaluation and textural analysis in order to determine the aging degree of the enriched bread samples.

3.3.1. Microbiological Evaluation

The microbiological safety of the developed bread samples consisted of the evaluation of the appearance of mold on the surface of the bread samples during storage at 21 °C (room temperature). Therefore, the obtained bread samples were packed individually in polyethylene bags and evaluated daily for signs of mold on the surface of the bread sample. As the sample evaluation revealed (Table 8), the bread samples containing sea buckthorn pomace powder presented lower mold spoilage on the surface compared to the control sample. Molds appeared on the sample surface after 72 h for the control sample, 96 h for the M8 sample, and 120 h for the C8 and S8 samples. This fact could be due to the polyphenolic content and antioxidant activity of the bread samples containing sea buckthorn powder, properties that could lead to a reduction in microbial development. As stated in many studies, polyphenols present inhibitory effects over both food-borne pathogens and food spoilage microorganisms [53–55]. Therefore, the addition of sea buckthorn pomace powder led to an increased shelf-life of 24 h for the M8 sample and 48 h for the C8 and S8 samples. Similar results were obtained by Ghendov-Mosanu et al., 2020 [29], who added sea buckthorn flour to wheat bread and observed an increased shelf-life of 3 (1% sea buckthorn flour), 4 (3% sea buckthorn flour), and 5 days (5% sea buckthorn flour) for the obtained bread samples.

Table 8. Spoilage characteristics of the bread samples containing 8% sea buckthorn pomace powder.

Sample	24 h	48 h	72 h	96 h	120 h	144 h
Control	-	-	+	++	+++	++++
C8	-	-	-	-	+	++
M8	-	-	-	+	++	+++
S8	-	-	-	-	+	++

- absent microbial alteration; + early signs of microbial alteration; ++ moderate signs of microbial alteration; +++ high spoilage on the bread surface; ++++ advanced mold growth on the sample surface.

3.3.2. Texture Analysis

The *texture* characteristics of the crumb were assessed using the Instron Texture Analyzer (Illinois Tool Works Inc., Norwood, MA, USA) as per the procedure outlined by Stamatie et al., 2022 [56]. A compression piston with a 12 mm diameter and a 50 N load cell were used to conduct the test, which involved compressing the middle of a 20 mm slice of bread with a compression speed of 2 mm/min. The test comprised two cycles with 20 mm depth compressions. The analysis was carried out at room temperature on the 3rd and 5th days of storage. The results of the texture analysis of functional bread enriched with 8% sea buckthorn pomace powder are presented in Table 9.

Table 9. Texture properties of functional bread samples with 8% sea buckthorn pomace powder (mean ± SD).

Sample	Hardness (N)		Cohesiveness		Elasticity		Chewiness (N)	
	Day 3	Day 5	Day 3	Day 5	Day 3	Day 5	Day 3	Day 5
Control	3.7 ± 0.09 [b]	4.23 ± 0.19 [a]	0.25 ± 0.07 [a]	0.04 ± 0.09 [a]	1.06 ± 0.11 [c]	1.18 ± 0.28 [a]	0.98 ± 0.31 [b]	0.16 ± 0.4 [a]
C8	7.14 ± 0.13 [a]	7.01 ± 0.09	0.35 ± 0.02 [a]	0.2 ± 0.08 [a]	1.07 ± 0.16 [a,c]	1.08 ± 0.15 [a]	2.69 ± 0.44 [a]	1.47 ± 0.43 [a]
M8	7.31 ± 0.44 [a]	8.55 ± 0.22	0.33 ± 0.05 [a]	0.24 ± 0.06 [a]	0.97 ± 0.03 [b,c]	1.13 ± 0.26 [a]	2.34 ± 0.4 [a,b]	2.33 ± 0.72 [a]
S8	4.77 ± 0.41 [b]	5.63 ± 0.44 [a]	0.28 ± 0.12 [a]	0.1 ± 0.07 [a]	1.08 ± 0.17 [a,b]	1.11 ± 0.16 [a]	1.38 ± 0.47 [a,b]	0.73 ± 0.62 [a]

All the results are mean ± standard deviation. Means followed by a common letter are not significantly different ($p > 0.05$). Values followed by different superscript letters in the same column are significantly different ($p < 0.05$). Means that do not share a letter in a column are significantly different ($p < 0.05$).

During the two days of analysis, the M8 bread sample exhibited the highest firmness (7.31 N day 3 and 8.55 N day 5), followed by the C8 bread sample (7.14 N day 3 and 7.01 N day 5), while the S8 bread sample had the least firmness (4.77 N day 3 and 5.63 N day 5). The control sample showed the lowest firmness, indicating the highest softness (3.7 N day 3 and 4.23 N day 5). The elasticity of the samples was similar among the samples and between the two days of analysis, with values ranging from 0.97 for the M8 sample on day 3 to 1.18 for the control sample on day 5. The C8 bread sample (1.08 on day 3) exhibited the values closest to the control sample (1.06 on day 3). The cohesiveness of the samples decreased significantly ($p < 0.05$) between the two days of analysis, indicating that the samples offered less resistance during the second deformation. The highest cohesiveness was observed in the C8 bread sample on day 3, with a value of 0.35, while the lowest cohesiveness was recorded in the control sample on day 5, with a value of 0.04. The S8 bread sample (0.28 day 3) exhibited worse cohesiveness than the control sample (0.25 day 3). The C8 bread sample exhibited the highest chewiness (2.69 N day 3), while the S8 bread sample (1.38 N day 3) showed values closest to the control sample (0.98 N day 3). The addition of sea buckthorn pomace powder at 8% increased the texture of the samples. On the 3rd day of storage, the M8 bread sample exhibited the highest values for hardness, elasticity, cohesiveness, and chewiness. The S8 sample showed values closest to the control sample.

According to Guo et al., 2019 [12], the bread samples enriched with sea buckthorn pulp showed increased values of springiness, cohesiveness, and resilience, while the values of hardness, gumminess, and chewiness were lower compared to the control sample. The addition of sea buckthorn pulp led to a significant decrease in the hardness of bread, which could be attributed to the denser crumb structure formed during the fermentation process. The cohesiveness of the bread was increased to 38.03% in bread samples containing 5% sea buckthorn pulp compared to the control. The chewiness of the bread obtained with 10% sea buckthorn pulp was similar to that of the control. However, the resilience decreased with the increasing addition of residues, indicating a less elastic gluten structure of bread. The bread sample containing 5% sea buckthorn pulp showed the most significant effect on resilience, with a difference of 65.17% compared to the control. On the contrary, the addition of sea buckthorn biomass to biscuits led to a decrease in their hardness, most likely due to a gradual decrease in gluten content, which hindered the formation of gluten matrices, as well as competition for water between sugar and flour, as mentioned by Janotkova et al., 2021 [36]. The reduction in hardness of bakery products enriched with sea buckthorn may also be due to the presence of organic acids, such as malic acid, fumaric acid, or citric acid, as well as limited activity of amylases in the dough, along with the content of phenolic compounds [36]. According to a study conducted by Sturza et al. (2016) [17], the addition of sea buckthorn flour in sponge cakes resulted in a decreased occurrence of starch and protein modifications, leading to a slower progression of changes in the structural, mechanical, and sensory characteristics of the fortified samples.

Furthermore, according to the research conducted by Muresan et al., 2019 [30], the hardness of biscuits was impacted by the drying temperature of the powder they contained. The biscuits that contained powder dried at 80 °C had higher hardness values but lower total work at the same temperature. This was attributed to the formation of more brittle structures, which led to product fracture.

Janotkova et al., 2021 [36] highlighted that texture analysis parameters, such as hardness and brittleness, should be kept as low as possible in bakery products since they are closely linked to consumers' perception of freshness. These parameters are crucial for evaluating the quality of bakery products.

4. Conclusions

The increasing demand from consumers for healthier food products, capable of preventing nutrition-related diseases and enhancing physical and mental well-being, has determined the rapid growth of the functional foods market. In response, the bakery industry is shifting its focus toward enhancing the health attributes of its products.

This study presented a novel approach to bread baking by adding sea buckthorn pomace powder. The bread was prepared by a direct method by adding 6, 8, and 10% sea buckthorn powder. The obtained products were subjected to physical–chemical, nutritional, organoleptic, and consumer acceptability analyses.

The resulting bread contained high amounts of fibers, proteins, and lipids, with a lower energy value than white bread. The analysis revealed strong antioxidant and antimicrobial properties, which increased as the percentage of sea buckthorn in the bread increased. The phenolic fraction and aroma substances in the raw materials were attributed to the prolonged shelf-life of the product due to the known antimicrobial activity of polyphenols.

Additionally, the results obtained indicate that sea buckthorn by-products are a viable way to enhance the mineral and microelement content of functional food.

Sensory tests showed that bread samples with an 8% addition of sea buckthorn powder were widely accepted, but high concentrations could lead to negative sensory attributes. In summary, this study recommends adding *Hippophae rhamnoides* pomace powder to wheat bread to obtain a product enriched in health-promoting biomolecules, with better sensory properties and a longer shelf-life. Additionally, future research could explore the addition of sea buckthorn pomace powder to other bakery products, including gluten-free options.

There are many future research possibilities that could be explored in the bakery industry in order to obtain fortified products with enhanced properties regarding their nutritional value. One of these applications could be encapsulation technology to obtain better preservation of bioactive compounds. Furthermore, given the potential of microalgae as a promising candidate to be used in food products manufacturing, a future approach involves the development of bakery products incorporating microalgae due to its ability to prevent oxidative deterioration and enhance the nutritional value of food products. Another application involves the development of bakery products using cricket powder as a novel source of innovative ingredients, a powder that has the potential to serve as a protein-enriching ingredient, providing essential amino acids and minerals for improved nutritional value in bakery products.

Author Contributions: Conceptualization, E.E.P., M.E.P. and A.C.M.; methodology, I.S., E.L.U., E.E.P. and M.G.-C.; resources, I.S., E.L.U. and E.E.P.; data curation, M.D., A.C.M. and G.M.; writing—original draft preparation, I.S., E.L.U. and M.G.-C.; writing—review and editing, E.L.U. and E.E.P.; project administration, M.E.P. All authors have read and agreed to the published version of the manuscript.

Funding: This work was supported by contract 186/2020, project acronym MILDSUSFRUIT. The authors acknowledge the financial support for this project provided by transnational funding bodies, partners of the H2020 ERA-NETs SUSFOOD2 and CORE Organic Cofund, under the Joint SUSFOOD2/CORE Organic Call 2019.

Conflicts of Interest: The authors declare no conflict of interest.

References

1. Sharma, M.; Shehzad, H.; Tatsiana, S.; Riina, A.; Rajeev, B. Valorization of sea buckthorn pomace to obtain bioactive carotenoids: An innovative approach of using green extraction techniques (ultrasonic and microwave-assisted extractions) synergized with green solvents (edible oils). *Ind. Crops Prod.* **2023**, *175*, 114257. [CrossRef]
2. Torres-Leon, C.; Ramirez-Guzman, N.; Londono-Hernandez, L.; Martinez-Medina, G.A.; Diaz-Herrera, N.; Navarro-Macias, V.; Alvarez-Perez, O.B.; Picazo, B.; Villarreal-Vazquez, M.; Ascacio-Valdes, J.; et al. Food Waste and Byproducts: An Opportunity to Minimize Malnutrition and Hunger in Developing Countries. *Front. Sustain. Food Syst.* **2018**, *2*, 52. [CrossRef]
3. Torres-Leon, C.; Rojas, R.; Contreras-Esquivel, J.C.; Serna-Cock, L.; Belmares-Cerda, R.E.; Aguilar, C.N. Mango seed: Functional and nutritional properties. *Trends Food Sci. Technol.* **2016**, *55*, 109–117. [CrossRef]
4. Compaore-Sereme, D.; Fatoumata, H.B.; Fidèle, W.B.T.; Heikki, M.; Ndegwa, H.M.; Mamoudou, H.D.; Hagrétou, S.L. Production and sensory evaluation of composite breads based on wheat and whole millet or sorghum in the presence of *Weissella confusa* A16 exopolysaccharides. *Heliyon* **2023**, *9*, e13837. [CrossRef]
5. Mafu, A.; Ketnawa, S.; Phongthai, S.; Schonlechner, R.; Rawdkuen, S. Whole Wheat Bread Enriched with Cricket Powder as an Alternative Protein. *Foods* **2022**, *11*, 2142. [CrossRef]
6. Dymchenko, A.; Milan, G.; Tomáš, G. 2023. Trends in bread waste utilization. In *Trends in Food Science & Technology*, 1st ed.; Toldra, F., Yada, R.Y., Eds.; Elsevier Ltd.: Amsterdam, The Netherlands, 2023; Volume 132, pp. 93–102.
7. Ivanišová, E.; Čech, M.; Hozlár, P.; Zaguła, G.; Gumul, D.; Grygorieva, O.; Makowska, A.; Kowalczewski, P.Ł. Nutritional, Antioxidant and Sensory Characteristics of Bread Enriched with Wholemeal Flour from Slovakian Black Oat Varieties. *Appl. Sci.* **2023**, *13*, 4485. [CrossRef]
8. Pinto, D.; Castro, I.; Vicente, A.; Bourbon, A.I.; Cerqueira, M.A. Functional Bakery Products—An Overview and Future Perspectives. In *Bakery Products Science and Technology*, 2nd ed.; Zhou, W., Hui, H., De Leyn, I., Pagani, M.A., Rosell, C.M., Selman, D., Therdthai, N., Eds.; John Wiley & Sons: Hoboken, NJ, USA, 2014; pp. 431–452.
9. Xie, X.; Yongcheng, S.; Xiufang, B.; Xiaocui, L.; Yage, X.; Zhenming, C. Characterization of sea buckthorn polysaccharides and the analysis of its regulatory effect on the gut microbiota imbalance induced by cefixime in mice. *JFF* **2023**, *104*, 105511. [CrossRef]
10. Kumar, A.; Pankaj, K.; Ajit, S.; Dharam, P.S.; Manisha, T. Scientific insights to existing know-how, breeding, genetics, and biotechnological interventions pave the way for the adoption of high-value underutilized super fruit Sea buckthorn (*Hippophae rhamnoides* L.). *S. Afr. J. Bot.* **2022**, *145*, 348–359. [CrossRef]
11. Ursache, F.M.; Ghinea, I.O.; Turturică, M.; Aprodu, I.; Râpeanu, G.; Stănciuc, N. Phytochemicals content and antioxidant properties of sea buckthorn (*Hippophae rhamnoides* L.) as affected by heat treatment–Quantitative spectroscopic and kinetic approaches. *Food Chem.* **2017**, *233*, 442–449. [CrossRef]
12. Guo, X.; Shi, L.; Yang, S.; Yang, R.; Dai, X.; Zhang, T.; Liu, R.; Chang, M.; Jin, Q.; Wang, X. Effect of sea-buckthorn pulp and flaxseed residues on quality and shelf life of bread. *Food Funct.* **2019**, *10*, 4220. [CrossRef]

13. Prashant, S.; Shere, D.M. Utilization of fruit and vegetable pomace as functional ingredient in bakery products: A review. *Asian J. Diary Food Res.* **2018**, *37*, 202–211.
14. Quiles, A.; Campbell, G.M.; Struck, S.; Rohm, H.; Hernando, I. Fiber from fruit pomace: A review of applications in cereal-based products. *Food Rev. Int.* **2016**, *34*, 162–181. [CrossRef]
15. Hussain, S.; Sharma, M.; Bhat, R. Valorisation of Sea Buckthorn Pomace by Optimization of Ultrasonic-Assisted Extraction of Soluble Dietary Fibre Using Response Surface Methodology. *Foods* **2021**, *10*, 1330. [CrossRef] [PubMed]
16. Gani, A.; Jan, R.; Ashwar, B.A.; Ashraf-ul, Z.; Shah, A.; Gani, A. Encapsulation of saffron and sea buckthorn bioactives: Its utilization for development of low glycemic baked product for growing diabetic population of the world. *LWT Food Sci. Technol.* **2021**, *142*, 111035. [CrossRef]
17. Sturza, R.A.; Ghendov-Mosanu, A.A.; Deseatnicov, O.I.; Suhodol, N.F. Use of sea buckthorn fruits in the pastry manufacturing. *Sci. Study Res. Chem. Chem. Eng. Biotechnol. Food Ind.* **2016**, *17*, 35–43.
18. *AACC Method 10-05.01 Guidelines for Measurement of Volume by Rapeseed Displacement*, 11th ed.; AACC International: Washington, DC, USA, 2010.
19. *SR 91:2007*; Bread and Fresh Pastries. Analysis Methods. ASRO—Romanian Standardization Association: Bucharest, Romania, 2007.
20. *AOAC Official Method 934.06 Moisture in Dried Fruits*; AACC International: Washington, DC, USA, 1935.
21. *AOAC Official Method 979.09 Protein in Grains*; AACC International: Washington, DC, USA, 1994.
22. *AOAC Official Method 963.15 Fat in Cacao Products-Soxhlet Extraction Method*; AACC International: Washington, DC, USA, 1973.
23. *AOAC Official Method 991.43 Total, Soluble, and Insoluble Dietary Fibre in Foods*; AACC International: Washington, DC, USA, 1994.
24. AOAC Method 923.03 Ash of Flour (Direct Method). In *Official Methods of Analysis*, 18th ed.; AOAC International Publisher: Gaithersburg, MD, USA, 2005.
25. Popa, M.E.; Stan, A.; Popa, V.; Tanase, E.E.; Mitelut, A.C.; Badulescu, L. Postharvest quality changes of organic strawberry Regina cultivar during controlled atmosphere storage. *Qual. Assur. Saf. Crops Foods* **2019**, *11*, 631–638. [CrossRef]
26. Stanciu, I.; Dima, R.; Popa, E.E.; Popa, M.E. Nutritional characterization of organic sea buckthorn pomace. *Sci. Pap. Ser. B Hortic.* **2022**, *LXVI*, 913–918.
27. Nilova, L.; Malyutenkova, S. The possibility of using powdered sea-buckthorn in the development of bakery products with antioxidant properties. *Agron. Res.* **2018**, *16*, 1444–1456.
28. Nilova, L.; Malyutenkova, S. The influence of plant ingredients on the composition of antioxidants in bakery products. *JHED* **2021**, *34*, 77–82.
29. Ghendov-Mosanu, A.; Cristea, E.; Patras, A.; Sturza, R.; Padureanu, V.; Deseatnicova, O.; Turculet, N.; Boestean, O.; Niculaua, M. Potential Application of *Hippophae Rhamnoides* in Wheat Bread Production. *Molecules* **2020**, *25*, 1272. [CrossRef]
30. Muresan, E.A.; Chis, S.; Cerbu, C.G.; Man, S.; Pop, A.; Marc, R.; Muresan, V.; Muste, S. Development and characterization of biscuits based on sea buckthorn and blueberries by-products. *JAP&T* **2019**, *25*, 59–63.
31. Rajeswari, H.; Jagadeesh, S.L.; Suresh, G.J. Physicochemical and sensory qualities of bread fortified with banana, aonla and sapota powders. *JNHFE* **2018**, *8*, 487–492.
32. Nikolau, E.N.; Karvela, E.D.; Marini, E.; Panagopoulou, E.A.; Chiou, A.; Karathanos, V.T. Enrichment of bakery products with different formulations of bioactive microconstituents from black Corinthian grape: Impact on physicochemical and rheological properties in dough matrix and final product. *J. Cereal Sci.* **2022**, *108*, 103566. [CrossRef]
33. Costa, L.L.; Tome, P.H.F.; Jardim, F.B.B.; Silva, V.P.; Castilho, V.P.; Damasceno, K.A.; Campagnol, P.C.B. Physicochemical and rheological characterization of pan bread made with pumpkin seed flour. *Int. Food Res. J.* **2018**, *25*, 1489–1496.
34. Lim, H.S.; Park, S.H.; Ghafoor, K.; Hwang, S.Y.; Park, J. Quality and antioxidant properties of bread containing turmeric (*Curcuma longa* L.) cultivated in South Korea. *Food Chem.* **2011**, *124*, 1577–1582. [CrossRef]
35. Odunlade, T.V.; Famuwagun, A.A.; Taiwo, K.A.; Gbadamosi, S.O.; Oyedele, D.J.; Adebooye, O.C. Chemical Composition and Quality Characteristics of Wheat Bread Supplemented with Leafy Vegetable Powders. *J. Food Qual.* **2017**, *2017*, 9536716. [CrossRef]
36. Janotkova, L.; Potocnakova, M.; Kreps, F.; Krepsova, Z.; Acsova, A.; Haz, A.; Jablosky, M. Effect of sea buckthorn biomass on oxidation stability and sensory attractiveness of cereal biscuits. *Bioresources* **2021**, *16*, 5097–5105. [CrossRef]
37. Biel, W.; Jaroszewska, A. The nutritional value of leaves of selected berry species. *Sci. Agric.* **2016**, *74*, 405–410. [CrossRef]
38. Lyu, F.; Luiz, S.F.; Perdomo Azeredo, D.R.; Cruz, A.G.; Ajlouni, S.; Renadheera, C.S. Apple Pomace as a Functional and Healthy Ingredient in Food Products: A Review. *Processes* **2020**, *8*, 319. [CrossRef]
39. El-Sohaimy, S.A.; Shehata, M.G.; Mathur, A.; Darwish, A.G.; El-Aziz, N.M.; Gauba, P.; Upadhyay, P. Nutritional Evaluation of Sea Buckthorn "Hippophae rhamnoides" Berries and the Pharmaceutical Potential of the Fermented Juice. *Fermentation* **2022**, *8*, 391. [CrossRef]
40. Jureviciute, I.; Kersiene, M.; Basinskiene, L.; Leskauskaite, D.; Jasutiene, I. Characterization of Berry Pomace Powders as Dietary Fiber-Rich Food Ingredients with Functional Properties. *Foods* **2022**, *11*, 716. [CrossRef] [PubMed]
41. Gatlan, A.M.; Gutt, G. Sea Buckthorn in Plant Based Diets. An Analytical Approach of Sea Buckthorn Fruits Composition: Nutritional Value, Applications, and Health Benefits. *Int. J. Envion. Res. Public Health* **2021**, *18*, 8986. [CrossRef] [PubMed]
42. Jaroszewska, A.; Biel, W.; Telesiński, A. Effect of mycorrhization and variety on the chemical composition and antioxidant activity of sea buckthorn berries. *J. Elem.* **2018**, *23*, 673–684. [CrossRef]

43. Piłat, B.; Bieniek, A.; Zadernowski, R. Chemical composition of individual morphological parts of the sea buckthorn fruit (*Hippophae rhamnoides*, L.). In *Producing Sea Buckthorn of High Quality, Proceedings of the 3rd European Workshop on Sea Buckthorn, Helsinki, Finland, 10–14 October 2014*; Kauppinen, S., Petruneva, E., Eds.; Natural Resources Institute Finland: Helsinki, Finland, 2014; pp. 79–82.
44. Nour, V.; Panaite, T.D.; Corbu, A.R.; Ropota, M.; Turcu, R.P. Nutritional and Bioactive Compounds in Dried Sea-Buckthorn Pomace. *Erwebsobstbau* **2021**, *63*, 93–98. [CrossRef]
45. Petrescu-Mag, R.M.; Vermeir, I.; Roba, C.; Petrescu, D.C.; Bican-Brisan, N.; Martonos, I.M. Is "Wild" a Food Quality Attribute? Heavy Metal Content in Wild and Cultivated Sea Buckthorn and Consumers' Risk Perception. *Int. J. Environ. Res. Public Health* **2021**, *18*, 9463. [CrossRef]
46. Dudarev, A.A.; Chupakhin, V.S.; Vlasov, S.V.; Yamin-Pasternak, S. Traditional Diet and Environmental Contaminants in Coastal Chukotka III: Metals. *Int. J. Environ. Res. Public Health* **2019**, *16*, 699. [CrossRef]
47. Vaitkeviciene, N.; Jariene, E.; Danilcenko, H.; Kulaitiene, J.; Mazeika, R.; Hallmann, E.; Blinstrubiene, A. Comparison of mineral and fatty acid composition of wild and cultivated sea buckthorn berries from Lithuania. *J. Elem.* **2019**, *24*, 1101–1113. [CrossRef]
48. Ani, I.F.; Adetola, A.B.; Alfa, T.; Ajuzie, N.C.; Ibagere, O.F.; Akinlade, A.R.; Omotoye, A.A. Quality assessment of bread made from whole wheat flour and betroor powder. *JDAN* **2022**, *13*, 11–18. [CrossRef]
49. Luntraru, C.M.; Apostol, L.; Oprea, O.B.; Neagu, M.; Popescu, A.F.; Tomescu, J.A.; Multescu, M.; Susman, I.E.; Gaceu, L. Reclaim and Valorization of Sea Buckthorn (*Hippophae rhamnoides*) by-Product: Antioxidant Activity and Chemical Characterization. *Foods* **2022**, *11*, 462. [CrossRef]
50. Akbas, M.; Kilmanoglu, H. Evaluation of the effects of the used of vegetable and fruit extracts on bread quality properties. *TURJAF* **2022**, *10*, 1838–1844.
51. Walker, R.; Tseng, A.; Cavender, G.; Ross, A.; Zhao, Y. Physicochemical, Nutritional, and Sensory Qualities of Wine Grape Pomace Fortified Baked Goods. *J. Food Sci.* **2014**, *79*, S1811–S1822. [CrossRef] [PubMed]
52. Stoin, D.; Dogaru, D.V.; Poiana, M.A.; Bordean, D.; Cocan, I. Use of sea buckthorn bioactive potential in obtaining of farinaceous functional products. *JAP&T* **2014**, *20*, 396–403.
53. Othman, L.; Sleiman, A.; Abdel-Massih, R.M. Antimicrobial Activity of Polyphenols and Alkaloids in Middle Eastern Plants. *Front. Microbiol.* **2019**, *10*, 911. [CrossRef] [PubMed]
54. Ou, J. Incorporation of polyphenols in baked products. *Adv. Food Nutr.* **2021**, *98*, 207–252.
55. Manso, T.; Lores, M.; de Miguel, T. Antimicrobial Activity of Polyphenols and Natural Polyphenolic Extracts on Clinical Isolates. *Antibiotics* **2022**, *11*, 46. [CrossRef] [PubMed]
56. Stamatie, G.D.; Susman, I.E.; Bobea, S.A.; Matei, E.; Duta, D.E.; Israel-Roming, F. The Influence of the Technological Process on Improving the Acceptability of Bread Enriched with Pea Protein, Hemp and Sea Buckthorn Press Cake. *Foods* **2022**, *11*, 3667. [CrossRef]

Disclaimer/Publisher's Note: The statements, opinions and data contained in all publications are solely those of the individual author(s) and contributor(s) and not of MDPI and/or the editor(s). MDPI and/or the editor(s) disclaim responsibility for any injury to people or property resulting from any ideas, methods, instructions or products referred to in the content.

Article

Insights into the Potential of Buckwheat Flour Fractions in Wheat Bread Dough

Ionica Coțovanu [1,*], Costel Mironeasa [2] and Silvia Mironeasa [1,*]

[1] Faculty of Food Engineering, Ștefan cel Mare University of Suceava, 13 Universitatii Street, 720229 Suceava, Romania
[2] Faculty of Mechanical Engineering, Automotive and Robotics, Ștefan cel Mare University of Suceava, 13 Universitatii Street, 720229 Suceava, Romania; costel.mironeasa@usm.ro
* Correspondence: ionica.cotovanu@usm.ro (I.C.); sllviam@fia.usv.ro (S.M.); Tel.: +40-740-816-370 (I.C.); +40-741-985-648 (S.M.)

Abstract: Buckwheat flour fractions with different particle sizes (PS), comprising various concentrations of valuable nutritional components, represent an opportunity to enhance refined wheat bakery products. The aim of this research was to assess the potential of buckwheat flour (BF) fractions (large, L > 300 μm, medium, 180 μm < M < 300 μm and small, S < 180 μm) to substitute refined wheat flour at 0, 5, 10, 15, and 20% in wheat bread dough and to establish the optimal amount for each fraction. The results revealed significant changes during different bread-making stages and on the finished product. A decrease in falling number index, water absorption, starch gelatinization, elastic modulus, and bread hardness with increasing PS was observed. The increase of BF amount led to an increase in dough development time, speed of protein weakening, gel starch stability, alveograph ratio, rheofermentation properties, maximum creep-recovery compliance, and bread hardness. The optimal values for falling number, mixing–heating–cooling dough parameters, dough biaxial extension, rheofermentation, storage and loss moduli, creep-recovery compliance, loaf volume, and bread hardness were obtained depending on PS based on the generation of predictive models. It was established that the best formulations, with respect to dough rheology and bread characteristics, included BF at 9.13% for large, 10.57% for medium, and 10.25% for small PS.

Keywords: particle size; refined wheat flour; Mixolab; alveograph; rheometer; bread characteristics

1. Introduction

Baking is a component of the market surrounding the demand for conventional food, in which the element of innovation is gradually taking over. Bread is generally obtained from wheat flour, but over the course of the refining process, wheat flour loses valuable nutritional substances such as minerals, vitamins, and fibers, which are found mainly in the outer layers of the grain shell [1,2]. Many manufacturers are concentrating their efforts on improving the commercial quality of baking products, often to the detriment of their nutritional value.

In recent years, various scientific research and innovation efforts have been directed towards the nutritional enrichment of refined wheat flour, especially by adding vegetables and pseudocereals [1]. One solution has been presented in the form of the use of different milling fractions of protein-rich flour, such as buckwheat flour.

Buckwheat is a pseudocereal from the genus *Fagopyrum* containing three species (*F. esculentum*, *F. tataricum*, and *F. cymosum*), but *F. esculentum* Moench is the most widely cultivated for human food [3]. It contains proteins with high nutritional value, due to its well-balanced amino acid composition, especially lysine, arginine, methionine, and tryptophan [4,5], which are only present in limited quantities in most grains. Buckwheat proteins are rich in albumin and globulin, but very poor in prolamin and gluten, and are characterized by the absence of structure-forming gluten, resulting in buckwheat dough

having poor resistance [6]. Buckwheat is considered a valuable ingredient for obtaining bakery products with high nutritional value, because it represents a valuable source of fiber, minerals, and other bioactive compounds with health benefits [7,8].

The chemical composition of buckwheat flour depends on the particle size (PS) following the milling and sieving process, which is correlated with the relative abundance of the seed tissues [9]. Buckwheat seed particles contains various proportions of central endosperm, which is rich in starch (81.2–75%), protein (5.4–6%), lipid (0.7–1%), and dietary fibers (3%), while the embryo contains high amounts of protein (12–27.7%), lipid (4–5.2%), dietary fibers (3.8–7%), and minerals (K, P, Mg, Ca) [8–11]. In a previous study, significant differences between the chemical composition of different sizes of buckwheat particle were found. The medium-sized particles were richest in protein (26.61%), lipid (5.61%), and ash (4.23%), while large and small particles contained high values of carbohydrate content (74.02% and 73.96%, respectively) [12]. Yu and collaborators (2018) [8] stated that buckwheat flour obtained from the central endosperm is rich in starch, and is used mainly in pancakes, breads, and soba noodles. The variation in the particle size distribution of cereal flours, fluctuating with respect to their chemical and physical properties, plays an essential role in the quality characteristics of their resulting products [13]. Various studies have evaluated the particle size of cereals and pseudocereals and highlighted its crucial effect on proximate composition, dough rheology, and the final baked products [14,15]. Song et al. (2016) [16] stated that particle size is one of the important factors, and can exert a remarkable effect on the characteristics of wheat flour and its baked products. In a previous study [17], it was shown that the flour fractions with a medium particle size separated by sieving offered better baking quality. On the other hand, another study [18] showed that flour fractions with a fine particle size separated by air classification offered better baking quality due to the higher protein content found in these fractions. Different baking results were found when flours with different particle sizes, but the same content of proteins were used [19]. Milling and sieving technologies can be used to obtain specific particle sizes with different structural characteristics and chemical compositions that impact dough rheology and final product quality [20,21]. Knowledge is needed regarding the simultaneous influence of PS and the amount of added BF on functional and structural properties in order to confirm the suitability of their use in the development of baked products, given that PS influences dough rheology and the quality of baked products [22,23]. Empirical dough testing methods such as falling number, Mixolab, alveograph, rheofermentometer, and, recently, the dynamic rheometer have been used in various studies to assess the rheological properties of dough, elucidating the interaction among dough components [24,25] that influence the technological process, and consequently the characteristics of the bread. Understanding the rheological properties of the dough in the formula when developing future products is important in predicting dough behavior during processing, with its quality affecting the finished loaf of bread. A dough for producing high-quality loaves must have a sufficiently high viscosity in order to prevent the ascent of gas cells, and it must remain extensible long enough during baking to avoid premature rupture of the membranes between gas cells [26]. To ensure the stability of gas cells, the dough needs to be sufficiently extensible to respond to gas pressure, but also strong enough to resist collapse [27]. The rheological properties are largely determined by the wheat gluten proteins, interactions between the gluten protein matrix and non-gluten proteins, and other flour components (e.g., flour lipids, fibers, non-gluten proteins) that may affect the rheological properties [28]. Loaf volume and texture profile analysis are used to evaluate the characteristics of the final baked products, and these are essential parameters influencing consumer purchase decisions.

In our previous research [15,29], we investigated the chemical composition, functional properties, microstructure, and molecular characteristics of buckwheat flour fractions, and the effect of their addition on the empirical rheology of wheat flour dough. Considering the factors mentioned above and the lack of studies regarding the influence of the incorporation of buckwheat flour with different particle sizes in wheat bread, this research aims to investigate how buckwheat flour at specific particle sizes and different addition levels affects the wheat flour amylase activity, dough behavior during mixing, heating–cooling stages, biaxial extension, rheofermentation, oscillatory and creep-recovery tests, and bread quality parameters in order to establish an optimal wheat–buckwheat composite flour for each particle size. The scientific research on buckwheat flour fractions addition in wheat flour is sparse; therefore, the present investigation focuses on complex rheological characterization and impact on bread characteristics.

2. Materials and Methods

2.1. Material Used

In this study, wheat flour with an extraction rate of 65% (harvest 2020) acquired from a local mill (Mopan S.R.L., Suceava County, România) was used. The samples were analyzed as described in the Romanian standard method (SR 90: 2007) [30]: gluten deformation (6.00 mm) and wet gluten (30.00%), and using the International Association for Cereal Chemistry (ICC) method [31]: moisture content (110/1), fat content (ICC 136), protein content (ICC 105/2), ash content (ICC 104/1), and falling number (ICC 107/1). The determination of total carbohydrate content was performed based on the difference of mean values: 100—(the amount of the protein, moisture content, fat, and ash) [12]. Milling fractions at three different particle sizes (large, L > 300 µm; medium, 180 µm < M < 300 µm; and small, S < 180 µm fractions) were obtained by subjecting buckwheat seeds (Sanovita, România) to a milling and sieving process, according to the methods reported in our previous research [15,29]. The approximate composition of the buckwheat grain, including moisture (13.28%), protein (13.26%), ash (2.00%), and fat (3.40%), along with the analytical characteristics of buckwheat flour (BF) fractions, was determined and reported in a previous work [29].

2.2. Composite Flour Formulations and Bread Processing

Each buckwheat milling fraction with three particle sizes (L, M, and S) was incorporated into refined wheat flour at five levels (0, 5, 10, 15, and 20%), and then mixed for 30 min (Yucebas Y21 mixer, Izmir, Turkey). Using a full factorial design, a set of 15 samples was obtained (Table 1). The experimental breads contained 300 g wheat–buckwheat flour, commercial fresh yeast of the type *Saccharomyces cerevisiae* (3% flour basis), salt (1.8% flour basis), and water up to the optimum wheat–buckwheat composite flour hydration capacity. The amount of water required to form the desired dough consistency was calculated based on the water absorption capacity of the flour, tested at Mixolab (Chopin, Tripetteet Renaud, Paris, France), ranging from 57.60 to 58.70%, depending on the level of addition and the particle size of the buckwheat flour.

The bread recipe was based on a biphasic procedure according to the method of Coțovanu et al. (2021) [12]. Half of the quantity of flour was mixed with water and yeast, and left to sit for leaven development (2 h, 30 ± 2 °C, and 85% relative humidity) inside a leavening room PL2008 (Piron, Cadoneghe, Padova, Italy). When the process was finished, the obtained leaven was mixed with the rest of the flour and salt in a laboratory mixer (Kitchen Aid, Benton Harbor, MI, USA) at 200 rpm for 10 min, and left for fermentation under the same conditions (1 h). Then, the dough samples were manually molded, placed into aluminum trays for final fermentation (1 h), and baked in an oven (Caboto PF8004D, Cadoneghe, Padova, Italy) for 25 min at 220 °C. The cooled experimental breads (2 h) were investigated with respect to their physical and textural characteristics.

Table 1. Effects of PS and BF addition level on: (**a**) falling number and dough Mixolab parameters, (**b**) alveograph and rheofermentometer parameters, (**c**) dynamic rheology and bread characteristics.

(a)

Run	Falling Number	Mixolab						
	FN (s)	WA (%)	DT (min)	ST (min)	C1-2 (N·m)	C3-2 (N·m)	C3-4 (N·m)	C5-4 (N·m)
1	307.00 ± 9.21	57.70 ± 1.04	3.68 ± 0.06	9.07 ± 0.16	0.66 ± 0.03	1.30 ± 0.05	0.14 ± 0.01	0.71 ± 0.03
2	358.50 ± 15.77	57.90 ± 0.75	1.12 ± 0.04	8.58 ± 0.09	0.65 ± 0.03	1.43 ± 0.03	0.06 ± 0.00	0.82 ± 0.02
3	299.00 ± 7.77	57.60 ± 0.81	3.80 ± 0.06	8.15 ± 0.10	0.73 ± 0.03	1.22 ± 0.05	0.30 ± 0.01	0.78 ± 0.03
4	345.50 ± 11.06	58.30 ± 1.81	1.33 ± 0.01	8.55 ± 0.12	0.65 ± 0.02	1.45 ± 0.03	0.09 ± 0.00	0.98 ± 0.02
5	312.00 ± 12.48	58.50 ± 1.11	1.69 ± 0.07	9.96 ± 0.39	0.61 ± 0.01	1.41 ± 0.02	0.05 ± 0.00	1.15 ± 0.04
6	303.00 ± 9.39	57.80 ± 0.92	4.65 ± 0.20	9.92 ± 0.22	0.64 ± 0.02	1.33 ± 0.02	0.09 ± 0.00	0.87 ± 0.02
7	317.00 ± 12.05	58.40 ± 0.82	1.62 ± 0.06	9.23 ± 0.27	0.68 ± 0.02	1.39 ± 0.02	0.07 ± 0.00	0.75 ± 0.01
8	312.00 ± 14.04	58.50 ± 1.17	1.69 ± 0.07	9.96 ± 0.40	0.61 ± 0.01	1.41 ± 0.02	0.05 ± 0.00	1.15 ± 0.02
9	349.00 ± 13.61	58.70 ± 1.47	1.38 ± 0.05	8.98 ± 0.17	0.65 ± 0.02	1.39 ± 0.02	0.07 ± 0.00	1.06 ± 0.02
10	369.50 ± 11.82	57.50 ± 0.92	0.90 ± 0.02	9.15 ± 0.16	0.70 ± 0.03	1.44 ± 0.06	0.13 ± 0.01	0.84 ± 0.03
11	314.00 ± 6.28	58.20 ± 1.16	3.40 ± 0.07	7.05 ± 0.15	0.73 ± 0.01	1.23 ± 0.02	0.20 ± 0.00	0.67 ± 0.01
12	321.00 ± 11.88	58.00 ± 1.28	1.48 ± 0.03	10.17 ± 0.32	0.58 ± 0.01	1.41 ± 0.02	0.04 ± 0.00	1.05 ± 0.01
13	299.50 ± 1.10	57.80 ± 1.10	4.35 ± 0.06	8.75 ± 0.12	0.66 ± 0.03	1.24 ± 0.05	0.21 ± 0.01	0.75 ± 0.03
14	326.00 ± 9.13	58.30 ± 1.57	3.55 ± 0.14	9.03 ± 0.15	0.65 ± 0.02	1.28 ± 0.01	0.17 ± 0.01	0.68 ± 0.01
15	312.00 ± 12.48	58.50 ± 1.05	1.69 ± 0.06	9.96 ± 0.38	0.61 ± 0.01	1.41 ± 0.03	0.05 ± 0.00	1.15 ± 0.04

(b)

Run	Alveograph				Rheofermentometer			
	P (mm H20)	L (mm)	W (×10⁻⁴) (J)	P/L (adim.)	H'm (mm)	TV (mL)	VR (mL)	CR (%)
1	93 ± 1.49	31 ± 0.56	115 ± 1.96	3.06 ± 0.05	85.40 ± 2.22	1547 ± 43.32	1533 ± 39.86	99.10 ± 2.68
2	84 ± 1.76	42 ± 0.76	131 ± 2.62	2.03 ± 0.04	71.80 ± 2.80	1311 ± 18.35	1192 ± 13.11	90.90 ± 1.18
3	89 ± 1.69	26 ± 0.55	95 ± 1.90	3.40 ± 0.03	89.80 ± 3.50	1638 ± 36.04	1628 ± 37.44	99.40 ± 2.19
4	83 ± 1.49	49 ± 0.74	146 ± 5.69	1.70 ± 0.03	77.70 ± 3.03	1390 ± 16.68	1247 ± 18.71	89.70 ± 1.36
5	87 ± 2.66	94 ± 3.67	253 ± 9.87	0.93 ± 0.03	62.00 ± 1.18	1168 ± 18.69	991 ± 15.86	84.80 ± 1.36
6	84 ± 1.85	42 ± 0.80	132 ± 5.15	2.01 ± 0.04	74.00 ± 2.89	1405 ± 15.46	1336 ± 17.76	97.20 ± 1.17
7	78 ± 1.40	50 ± 0.80	139 ± 5.42	1.56 ± 0.03	68.00 ± 2.65	1278 ± 17.89	1163 ± 18.61	91.00 ± 1.37
8	87 ± 2.44	94 ± 2.82	253 ± 0.12	0.93 ± 0.03	62.00 ± 1.12	1168 ± 23.36	991 ± 19.82	84.80 ± 1.70
9	88 ± 2.11	52 ± 1.09	169 ± 6.59	1.69 ± 0.04	70.90 ± 1.28	1278 ± 20.45	1172 ± 18.75	91.70 ± 1.47
10	83 ± 1.16	34 ± 0.27	112 ± 1.34	2.50 ± 0.03	77.50 ± 1.71	1430 ± 28.60	1320 ± 39.60	92.30 ± 2.31
11	69 ± 1.38	66 ± 1.32	130 ± 2.60	1.04 ± 0.02	71.70 ± 1.43	1329 ± 26.58	1204 ± 24.08	90.60 ± 1.81
12	86 ± 2.06	48 ± 1.06	156 ± 6.08	1.79 ± 0.04	69.30 ± 2.70	1638 ± 29.48	1217 ± 23.12	93.40 ± 1.68
13	72 ± 0.65	53 ± 0.48	107 ± 0.96	1.34 ± 0.01	77.30 ± 1.24	1413 ± 26.85	1241 ± 26.06	87.80 ± 1.76
14	81 ± 1.30	40 ± 0.40	121 ± 4.72	2.02 ± 0.03	72.20 ± 2.82	1330 ± 11.97	1201 ± 9.61	90.30 ± 1.37
15	87 ± 2.61	94 ± 3.57	253 ± 9.61	0.93 ± 0.03	62.00 ± 1.15	1168 ± 21.02	991 ± 15.86	84.80 ± 1.70

(c)

Run	Rheometer						Bread Characteristics	
	G' (Pa)	G'' (Pa)	tan δ (adim.)	T_{max} (°C)	Jc_{max} (×10⁻⁵) (Pa⁻¹)	Jr_{max} (×10⁻⁵) (Pa⁻¹)	BV (cm³)	BH (g)
1	20,080 ± 542	7413 ± 200	0.3690 ± 0.01	81.76 ± 1.55	24.47 ± 0.44	17.18 ± 0.31	350.58 ± 7.71	988.50 ± 9.89
2	30,130 ± 422	11,242 ± 180	0.3730 ± 0.01	79.36 ± 1.27	28.63 ± 0.34	18.59 ± 0.24	232.54 ± 4.88	1793.00 ± 10.76
3	25,050 ± 976	9211 ± 359	0.3150 ± 0.01	82.36 ± 3.21	32.07 ± 0.58	20.00 ± 0.32	312.70 ± 5.63	1048.00 ± 12.58
4	28,030 ± 224	11,240 ± 101	0.4010 ± 0.00	79.63 ± 1.43	21.49 ± 0.26	15.01 ± 0.24	278.93 ± 6.69	1308.00 ± 17.00
5	26,370 ± 237	9488 ± 85	0.3598 ± 0.00	83.24 ± 2.16	22.01 ± 0.42	15.34 ± 0.37	378.20 ± 14.75	786.00 ± 7.07
6	23,820 ± 143	8548 ± 77	0.3580 ± 0.00	81.48 ± 2.36	17.69 ± 0.14	12.80 ± 0.12	387.50 ± 6.20	938.00 ± 4.69
7	26,650 ± 319	10,285 ± 102	0.3850 ± 0.01	80.38 ± 1.61	23.00 ± 0.25	15.36 ± 0.18	355.03 ± 6.39	464.50 ± 4.18
8	26,370 ± 316	9488 ± 104	0.3598 ± 0.00	83.24 ± 2.00	22.01 ± 0.44	15.34 ± 0.40	378.20 ± 14.37	786.00 ± 6.29
9	23,960 ± 191	8962 ± 80	0.3740 ± 0.00	79.36 ± 1.51	15.94 ± 0.13	9.81 ± 0.09	287.54 ± 10.35	1576.00 ± 12.61
10	33,810 ± 710	11,360 ± 227	0.3400 ± 0.01	81.95 ± 1.31	30.07 ± 0.63	18.78 ± 0.41	230.90 ± 4.16	3070.00 ± 30.71
11	27,030 ± 541	10,210 ± 204	0.3700 ± 0.01	82.46 ± 1.65	25.02 ± 0.50	17.60 ± 0.35	374.14 ± 7.48	578.00 ± 4.62
12	26,980 ± 297	9394 ± 75	0.3480 ± 0.01	80.15 ± 1.76	27.10 ± 0.38	16.73 ± 0.20	348.92 ± 6.28	1083.00 ± 11.91
13	32,700 ± 621	10,650 ± 224	0.3250 ± 0.01	82.79 ± 1.49	26.09 ± 0.52	18.05 ± 0.36	293.60 ± 4.70	1471.50 ± 13.24
14	26,990 ± 297	10,210 ± 133	0.3600 ± 0.01	80.48 ± 1.45	23.61 ± 0.38	15.85 ± 0.22	368.78 ± 5.16	748.00 ± 5.32
15	26,370 ± 316	9488 ± 104	0.3598 ± 0.00	83.24 ± 2.33	22.01 ± 0.48	15.34 ± 0.46	378.20 ± 14.37	786.00 ± 6.29

(**a**) FN: falling number index; WA: water absorption; DT: development time; ST: stability; C1-2: speed of protein weakening stage; C3-2: starch gelatinization; C3-4: cooking stability; C5-4: starch retrogradation. (**b**) P, dough elasticity; L, dough extensibility; W, deformation energy; P/L, alveograph ratio; H'm: maximum height of the gas release curve; TV: total volume of gas produced; VR: volume of the gas retained in the dough; CR: retention coefficient. (**c**) G', G'': storage and viscous modulus; tan δ: loss tangent; T_{max}: maximum gelatinization temperature; Jc_{max}, Jr_{max}: maximum creep-recovery compliance; BV: bread volume, BF: bread hardness.

2.3. Empirical Dough Rheology

The falling number index (FN) of the formulated flour samples was determined using a Falling Number device (Perten Instruments AB, Stockholm, Sweden) in order to evaluate the α-amylase activity.

Dough characterization during the mixing and heating–cooling stages was performed using Mixolab equipment (Chopin Technologies, Paris, France) according to ICC standard method 173. The instrumental settings defined in the Mixolab were as follows: total analysis time: 45 min; heating rate: 4 °C/min; and mixing tank temperature of 30 °C. All the samples were made at the optimum hydration level for achieving the optimum dough consistency with respect to a target torque (C1) of 1.10 N·m. The following parameters were evaluated on the basis of the Mixolab curve: water absorption (WA), dough development time (DT), dough stability (ST), the speed of protein weakening due to heat (C1-2), starch gelatinization (C3-2), cooking stability (C3-4), and starch retrogradation during the cooling stage (C5-4).

The biaxial extension of the dough was analyzed using an alveograph (Chopin Technologies, Cedex, France) according to American Association of Cereal Chemists (AACC) International approved method 54–30.02 at constant hydration with a 14% moisture basis and 2.50% salt. The alveograph test was performed to determine the following parameters: dough tenacity (P), dough extensibility (L), deformation energy (W), and alveograph ratio (P/L).

Rheofermentation parameters were analyzed using a Rheofermentometer F4 device (Chopin Technologies, France) according to AACC method 89–01.01. The following characteristics were determined: the maximum height of the gas release curve (H'm), the total volume of gas produced (TV), the volume of the gas retained in the dough (VR), and retention coefficient (CR).

2.4. Dynamic Dough Rheology

The dynamic oscillatory measurements of the dough made with the composite flours were obtained by performing a preliminary stress sweep test to identify the limits of the linear viscoelastic region (LVR), increasing the strain from 0.01 to 1%, at a constant oscillation frequency of 1 Hz, according to Mironeasa and Mironeasa (2019) [32]. The dough samples were prepared without yeast and salt, and were placed at optimum water absorption (determined using Mixolab, until optimum consistency had been achieved) in a Mars 40 rheometer (Thermo-Haake, Karlsruhe, Germany) coupled with a Peltier temperature control unit, using a parallel plate–plate geometry (40 mm diameter) and a gap of 4 mm, chosen based on the viscosity range of dough. The samples were left to rest for 5 min before testing to allow relaxation and to stabilize the temperature, according to settings from other studies; all measurements were performed at 20.0 ± 0.1 °C [25,32,33]. The excess dough was trimmed, and a layer of Vaseline was applied to the exposed edge of the sample to avoid the evaporation of moisture during testing. Frequency sweep tests were performed in an oscillation frequency range from 0.01 to 20 Hz, with a constant stress of 10 Pa, which was previously established in the LVR test. In the oscillatory test, the elastic or storage modulus, G', viscous or loss modulus, G'', and loss tangent, $\tan \delta$ (G''/G'), were acquired as a function of frequency using the Rheowin Job software (v.4.86, Haake). For the temperature sweep test, the dough samples were heated from 20 to 100 °C at a rate of 4.0 ± 0.1 °C/min, at a constant strain of 0.10%, and a frequency of 1 Hz, to determine the maximum gelatinization temperature (T_{max}) [34,35] corresponding to the maximum value of loss modulus.

To simulate different stresses during the breadmaking process, a creep-recovery test with small forces was applied. The dough was kept under a constant stress of 25 Pa for 60 s, and after removing the shear stress, there was a relaxation time of 180 s [25,35]. The maximum compliance (Jc_{max}) value reached in the creep phase after 60 s, corresponding to the maximum deformation, was reported, as well as the maximum compliance value at the end of the recovery phase (Jr_{max}), related to partial reformation after stress removal.

2.5. Physical and Textural Characteristics of Bread

Two hours after baking, when the samples were cooled, bread volume (BV) was analyzed in accordance with the Romanian standard procedure SR 90:2007 [30]. Loaf volume (cm^3) was obtained by employing the rapeseed displacement method. The bread hardness (BH) was determined using a TVT-6700 texture analyzer (Perten Instruments, Hägersten, Sweden). Bread slices were compressed twice with a 2.5-cm stainless-steel cylinder, with a penetration depth of 20%, a speed of 1.0 mm/s, an auto-trigger force of 5.0 g, and a time between compressions of 15 s.

2.6. Statistical Analysis

Data modeling was carried out using a trial version of the Design-Expert 12.0 software (Stat-Ease, Inc., Minneapolis, MN, USA). The values of the determined parameters for all formulations are shown in Table 1a–c.

The influence of the factors PS (L, M, and S) and BF addition level (0, 5, 10, 15, and 20%) on empirical and dynamic rheological properties of the dough, as well as on some of the technical characteristics of the bread, were evaluated using a full factorial design () and Response Surface Methodology. Analysis of variance (ANOVA) was applied in order to evaluate the influence of factors and their interactions (at a confidence level of 95%) on the following responses: FN, WA, DT, ST, C1-2, C3-2, C3-4, C5-4, P, L, W, P/L, H'm, TV, VR, CR, G', G'', tan δ, T_{max}, $J_{c\,max}$, $J_{r\,max}$, BV, and BH. The most adequate model for predicting experimental data variation for each response was chosen by taking into account F-test values, coefficient of determination (R^2), and adjusted coefficients of determination ($Adj.\text{-}R^2$).

Optimization of the BF addition level for each PS was performed by applying multiple response analysis to the predictive models generated, in conjunction with the desirability function approach. In this process, each predicted response is transformed into an individual desirability function, d_n, which includes the researcher's desired priorities when building the optimization procedure for each of the factors. The individual desirability functions are then combined into an objective function, named the overall desirability function, D, computed as the geometric mean of values of the individual desirability function, d_n, which varies from 0 to 1 [36,37]. For this purpose, for each response, the desired goals were: BF addition level, ST, C3-4, H'm, TV, VR, CR, W, Jr_{max}, and BV set at maximum value; minimization of C1-2, C5-4, P/L; and the level of all remaining responses maintained within range.

3. Results

3.1. Effects of Formulated Factors on FN, Mixolab, and Alveographic Parameters

The quadratic model adequately represents the data for falling number, water absorption, development time, speed protein weakening, starch gelatinization, hot starch stability, starch retrogradation, dough tenacity, dough extensibility, deformation energy, and alveograph ratio, explaining between 66 and 93% of the variation, whereas the two-factor interaction (2FI) model obtained for dough stability explained 51% of the variation (Table 2).

Table 2. ANOVA results of the models fitted for falling number and dough rheological properties determined using Mixolab and alveograph.

Factors	Falling Number	Mixolab							Alveograph			
	FN (s)	WA (%)	DT (min)	ST (min)	C1-2 (N·m)	C3-2 (N·m)	C3-4 (N·m)	C5-4 (N·m)	P (mm H$_2$O)	L (mm)	W (×10^{-4}) (J)	P/L
Constant	318.82	58.28	3.09	9.10	0.66	1.31	0.122	0.73	76.64	49.22	125.81	1.50
A	−19.25 ***	−0.13 *	0.89 **	0.20	−0.00	−0.04 **	0.02	−0.02	1.40	−3.00	−6.00	0.23
B	3.77	−0.43 ***	0.78 *	−0.75 **	0.04 **	−0.05 **	0.07 **	−0.18 ***	−2.67	−23.27 **	−65.20 ***	0.66 **
A × B	−16.45 **	0.07	0.83 *	−0.27	0.01.	−0.05 **	0.04 *	−0.02	2.30	−2.30	−3.70	0.27
A^2	13.95 *	−0.19	−0.75	-	−0.01	0.07 **	−0.03	0.14 **	9.00 **	−9.40	6.20	0.62 *
B^2	−10.24	−0.09	−0.33	-	0.00	−0.00	0.03	0.14 *	1.52	22.76 *	48.38 **	−0.24
					Model Assession							
R^2	0.88	0.86	0.76	0.51	0.66	0.87	0.86	0.88	0.66	0.81	0.93	0.86
$Adj.\text{-}R^2$	0.82	0.79	0.62	0.38	0.47	0.79	0.79	0.82	0.48	0.70	0.89	0.79
p-value	0.0005	0.0011	0.0130	0.0411	<0.0001	0.0010	0.0011	0.0006	0.0461	0.0048	<0.0001	0.0225

FN: falling number index; WA: water absorption; DT: development time; ST: stability; C1-2: speed of protein weakening stage; C3-2: starch gelatinization; C3-4: cooking stability; C5-4: starch retrogradation; P: dough tenacity; L: dough extensibility; W: deformation energy; P/L: alveograph ratio; A: particle size (µm); B: buckwheat flour addition level (%); R^2, $Adj.\text{-}R^2$: measures of model fit. ***, **, * indicated significance at $p < 0.0001$, $p < 0.001$, and $p < 0.05$, respectively.

The falling number was negatively influenced by particle size, and a decrease in this parameter was observed with increasing PS (Figure 1). The linear and quadratic effect of the level of buckwheat addition in wheat flour had no significant ($p > 0.05$) effect on FN, while the interaction between factors led to a considerably lower FN with increasing interaction between the factors.

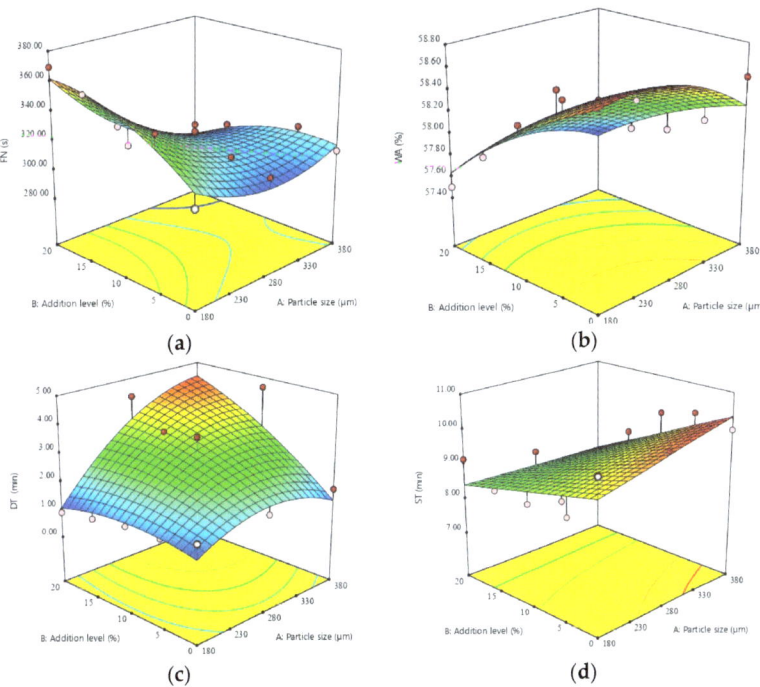

Figure 1. Response surface graphic of (**a**) falling number (FN), (**b**) water absorption (WA), (**c**) development time (DT), and (**d**) dough stability (ST) as a function of buckwheat flour fraction and addition level.

During Mixolab measurements the mixing and baking processes were simulated, because the protein and starch behavior are key factors in predicting technical properties in baking. As can be seen from Table 2, the PS and addition level influenced were remarkably negatively correlated with water absorption (WA).

When BF particle size and addition level decreased, WA increased (Figure 1). Particle size and the addition level of buckwheat flour, as well as the interaction between these factors, exerted a considerable influence on dough development time (DT) (Table 2). The variations of DT with buckwheat flour PS and addition levels presented an increase in DT with an increase of both factors (Figure 1). Buckwheat flour addition level showed a positive significant ($p < 0.05$) influence on dough stability (ST), while PS had no significant effect (Table 2), with DT decreasing proportionally with increasing BF addition level (Figure 1). The rate of protein thermal weakening (C1-2) was significantly influenced by the BF addition level, while the PS did not influence this parameter ($p > 0.05$) (Table 2).

The response surface plot (Figure 2) reveals an increase of C1-2 with increasing BF addition level. The degree of starch gelatinization (C3-2) is substantially affected by the linear terms of both factors, PS and addition level, as well as by the interaction between them, and by the quadratic term of PS (Table 2). Increases in PS and addition level were correlated with a decrease in C3-2 (Figure 2). The hot-starch gel stability (C3-4) was found to be significantly ($p < 0.05$) influenced by BF addition level and by the interaction between PS and BF addition level, while the linear term of PS did not affect gel stability (Table 2).

Hot-gel stability (or cooking stability) increases with increasing BF addition level, as can be observed from the surface response graph (Figure 2). Starch retrogradation (C5-4) of composite flour dough was significantly influenced by the amount of BF added (Table 2) and quadratic term of PS. Figure 2 shows a decrease in C5-4 values with increasing BF addition level, revealing a low starch retrogradation.

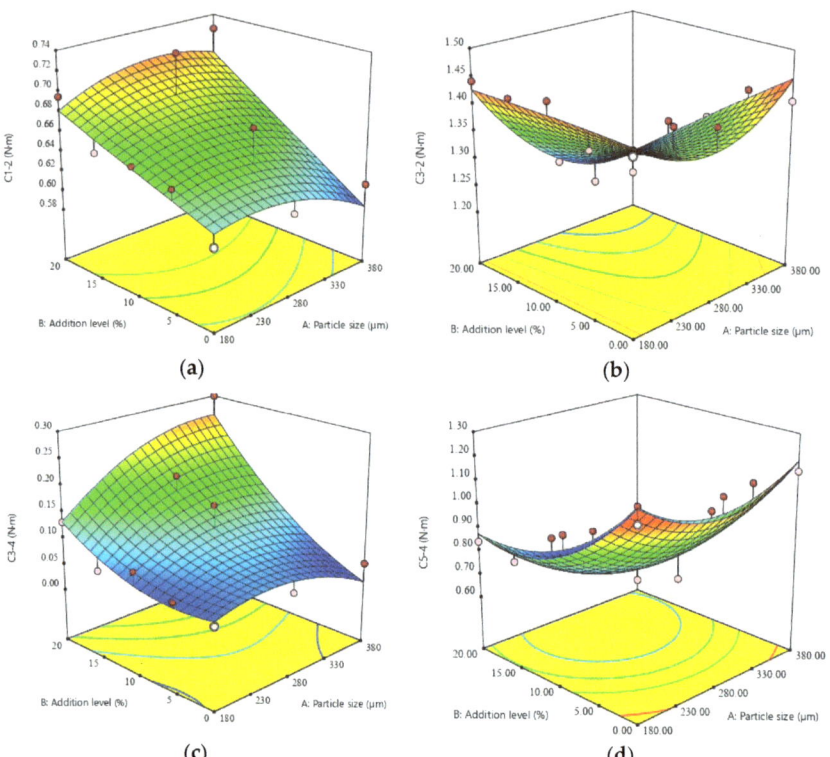

Figure 2. Response surface graphic of (**a**) protein weakening (C1-2), (**b**) starch gelatinization (C3-2), (**c**) cooking stability (C3-4), and (**d**) starch retrogradation (C5-4) as a function of buckwheat flour fraction and addition level.

Regarding the alveograph characteristics, the linear term of PS and the BF addition level did not significantly affect dough tenacity ($p > 0.05$), while the quadratic term of PS was correlated with an increase in this parameter with increasing PS. BF addition level showed a considerable influence on the dough extensibility, deformation energy, and alveograph ratio parameters (Table 2). The response surface plots shown in Figure 3 for dough tenacity, extensibility and deformation energy are characteristic for the quadratic model, and show a decrease in these parameters with increasing addition level. On the other hand, the alveograph ratio exhibits an increasing trend with increasing BF addition level and PS.

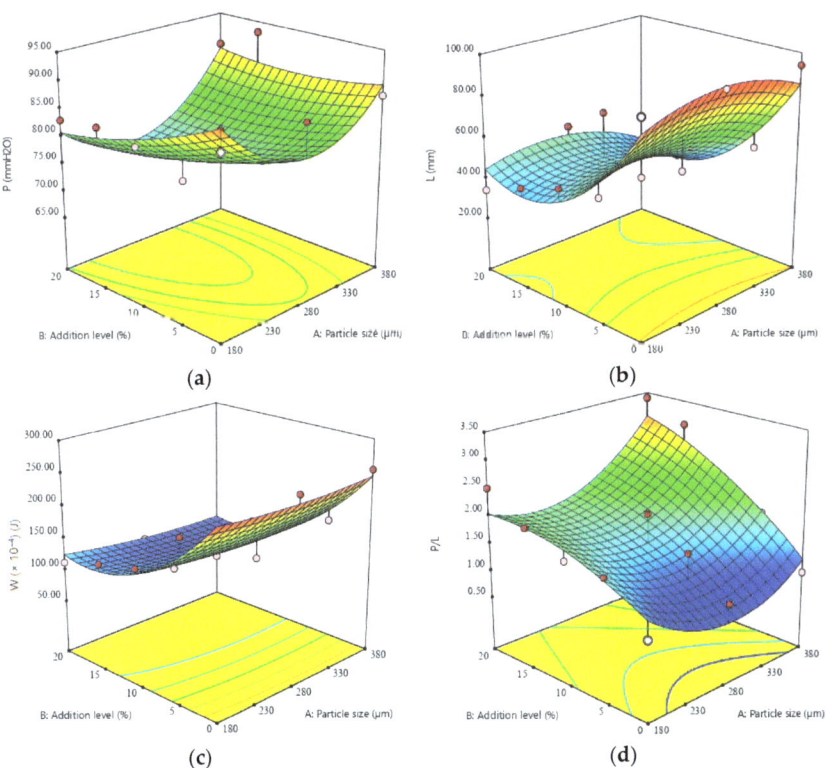

Figure 3. Response surface graphic of (**a**) dough tenacity (P), (**b**) dough extensibility (L), (**c**) deformation energy (W), and (**d**) alveograph ratio (P/L) as a function of buckwheat flour fractions and addition level.

3.2. Effects of Formulated Factors on Dough Reofermentation, Dynamic Rheology and Bread Characteristics

The maximum height of the gas release curve (H′m), the total volume of gas produced (TV), the volume of gas retained in the dough (VR), the retention coefficient (CR), storage modulus (G′), loss tangent (tan δ), maximum gelatinization temperature (T_{max}), bread volume (BV), and the bread hardness (BH) were successfully fitted to the quadratic model, explaining 72 to 92% of the data variation (Table 3). Of the data variation in loss modulus (G″) 69% was explained by the 2FI model, while for maximum creep compliance (Jc_{max}) and maximum recovery compliance (Jr_{max}), only 42–43% of the data variation was explained by the linear model (Table 3).

The maximum height of the gas release curve (H′m) showed a significant increase with increasing BF addition level and interaction between factors, while PS had no significant effects. The total volume of gas produced (TV) was considerably influenced by PS and addition levels (Table 3). Similarly, the volume of gas retained in the dough (VR) and the retention coefficient (CR) were substantial impacted by the PS and addition level factors. Figure 4 shows the positive influence of both of the studied factors on all rheofermentografic parameters, showing an increasing trend with increasing BF addition level and PS.

Table 3. ANOVA results of the models fitted for dough rheofermentation, and dynamic rheological and bread characteristics.

Factors	Rheofermentometer				Rheometer						Bread Parameters	
	H'm (mm)	TV (mL)	VR (mL)	CR (%)	G' (Pa)	G" (Pa)	tan δ (adim.)	T_{max} (°C)	Jc_{max} ($\times 10^{-5}$) (Pa^{-1})	Jr_{max} ($\times 10^{-5}$) (Pa^{-1})	BV (cm^3)	BH (g)
Constant	71.37	1335.08	1197.38	90.96	26,575.14	9812.60	0.3760	80.58	24.08	16.12	357.55	646.50
A	2.06	81.90 *	81.30 **	2.45 **	−2000.00 **	823.80 **	−0.0093	0.54	0.52	0.45	36.98 **	−369.00 **
B	9.19 ***	129.80 **	187.30 ***	3.65 **	1636.67 *	382.53	−0.0130 *	−0.10	3.77 *	2.21 *	−41.94 **	446.67 **
A × B	3.98 *	29.20	91.20 *	2.07	−3059.00 **	−855.90 *	−0.0001	0.24	-	-	22.03	−435.65 *
A^2	3.80	93.70	105.70 *	3.43 *	1488.00	-	−0.0007	−0.61	-	-	−0.35 *	528.10 *
B^2	−2.26	−62.95	−74.76	−4.12 *	2745.71 *	-	−0.0330 **	2.57 **	-	-	−0.72	326.19
					Model Assession							
R^2	0.90	0.72	0.92	0.86	0.83	0.69	0.75	0.75	0.42	0.43	0.83	0.86
Adj.-R^2	0.85	0.57	087	0.77	0.73	0.60	0.61	0.61	0.32	0.33	0.74	0.78
p-value	0.0003	0.0216	0.0001	0.0014	0.0031	0.0036	0.0143	0.0153	0.04	0.0029	0.0014	0.0014

H'm: maximum height of the gas release curve; TV: total volume of gas produced; VR: volume of the gas retained in the dough; CR: retention coefficient; G': storage modulus; G": loss modulus; tan δ: loss tangent; T_{max}: maximum gelatinization temperature; Jc_{max}, Jr_{max}: maximum creep-recovery compliance; BV: bread volume, BF: bread hardness. A: Particle size (μm); B: Buckwheat flour addition level (%); R^2, Adj.-R^2: measures of model fit. ***, **, * indicate significance at $p < 0.0001$, $p < 0.001$, and $p < 0.05$, respectively.

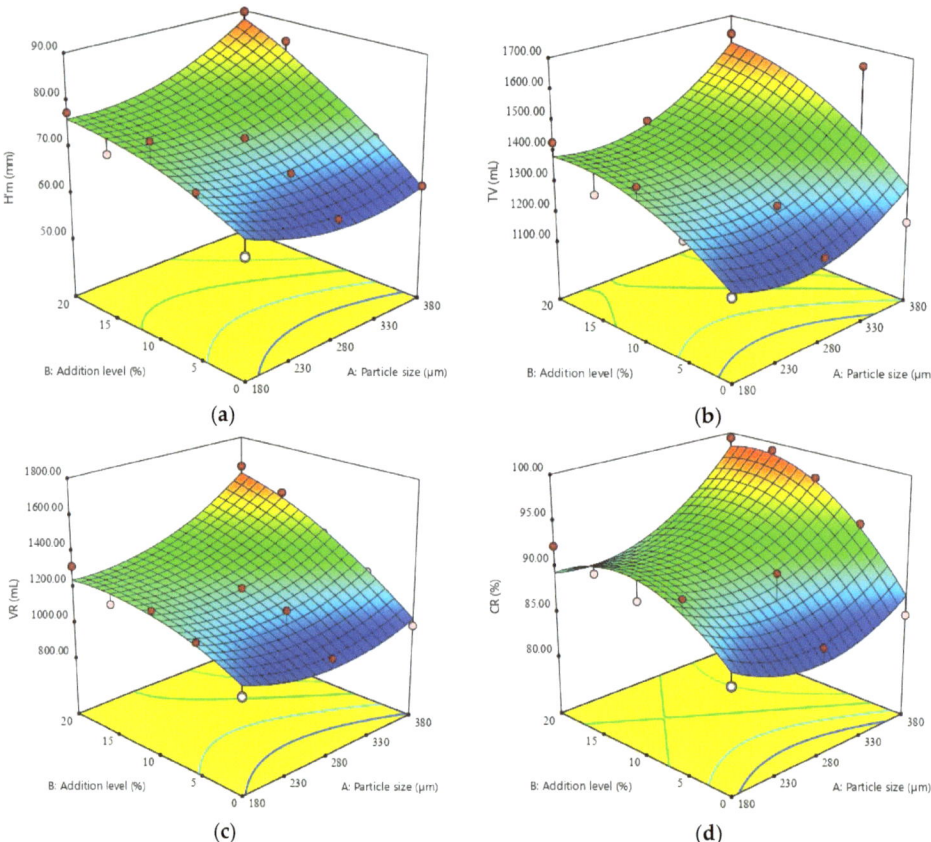

Figure 4. Response surface graphic of (**a**) maximum height of the gas release curve (H'm), (**b**) total volume of gas produced (TV), (**c**) volume of gas retained in the dough (VR), and (**d**) retention coefficient (CR) as a function of buckwheat flour fractions and addition level.

The dynamic rheological properties, in terms of storage modulus, loss modulus, loss tangent, and maximum gelatinization temperature, were influenced by PS and BF addition level, but with different trends. G′ decreased significantly with increasing PS and interactions between PS and BF addition level (Figure 5a), while BF addition level exerted a significant positive effect on G′ (Table 3). Loss modulus (G″) was considerably affected only by PS and by the interaction between PS and addition level. An increasing trend for G″ was noted with increasing PS (Figure 5). The quadratic model obtained for loss tangent adequately represented the experimental data. Loss tangent was significantly affected by the linear and quadratic terms of BF addition levels in a negative mode (Table 3). The maximum gelatinization temperature (T_{max}) was significantly affected by the quadratic term of BF addition level, while PS had a non-significant effect (Table 3).

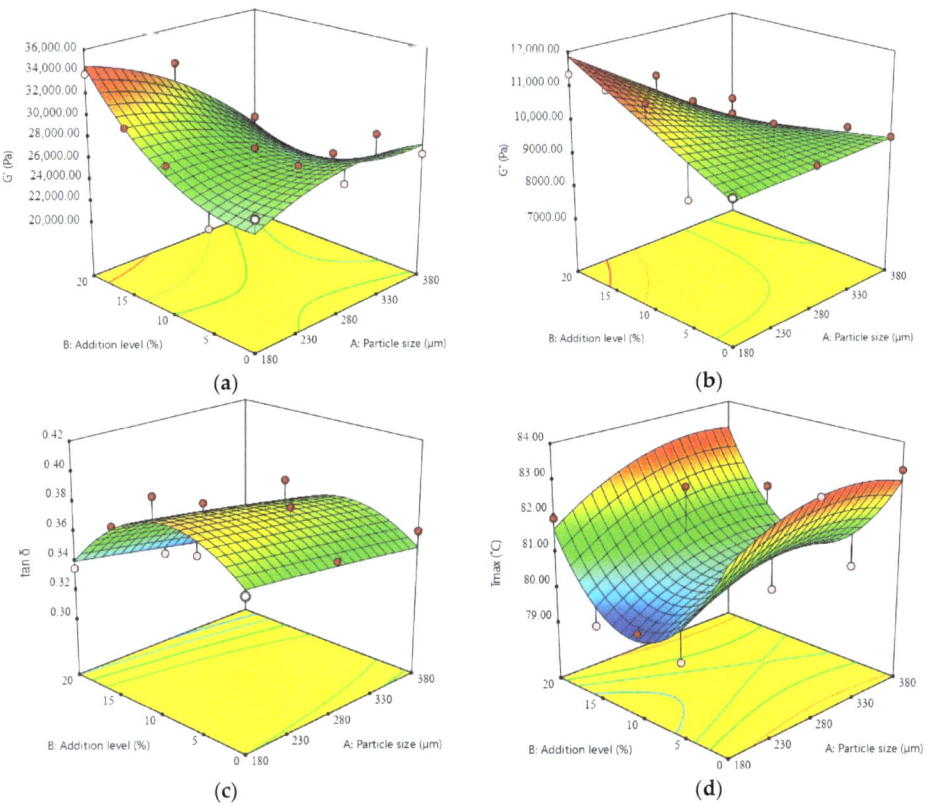

Figure 5. Response surface graphic of (**a**) storage modulus (G′), (**b**) loss modulus (G″), (**c**) loss tangent (tan δ), and (**d**) maximum gelatinization temperature (T_{max}) as a function of buckwheat flour fractions and addition level.

The amount of buckwheat flour added had a significant influence on maximum creep compliance, while PS did not exert a significant influence (Table 3). An increase in the maximum creep compliance was observed with increasing BF addition level (Figure 6). The maximum recovery compliance was influenced by the amount of BF, while the PS had a non-significant effect. The amount of BF had a considerable positive influence on Jr_{max}, presenting an increase in Jr_{max} with increasing BF addition level (Figure 6).

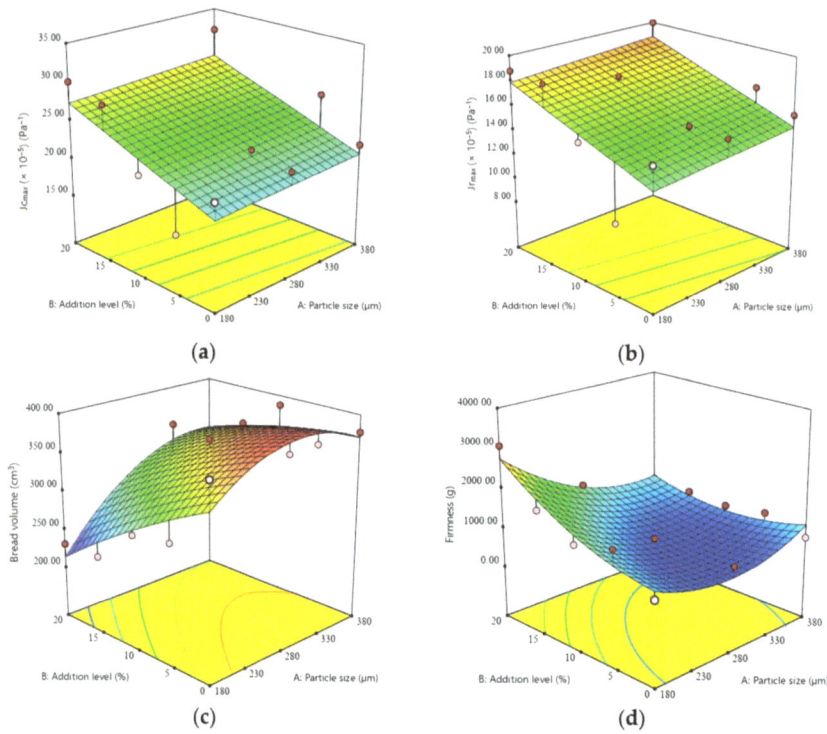

Figure 6. Response surface graphic of (**a**) maximum creep compliance (Jc$_{max}$), (**b**) maximum recovery compliance (Jr$_{max}$), (**c**) bread volume (BV), and (**d**) bread hardness (BH) as a function of buckwheat flour fractions and addition level.

Bread volume (BV) was significantly affected by the PS and the amount added to the wheat flour, as shown by the ANOVA test (Table 3). It can be seen that the BF addition level had a negative influence on the BV, and it can be observed in Figure 6 that this bread technical parameter decreased with increasing BF addition level. Rather, the particle size significantly affected the bread volume in a positive way, with an increase in BV being apparent with increasing PS (Figure 6). The bread hardness (BH) was significantly influenced by the PS, the BF addition level, and the interactions between factors, with an increase being observed with decreasing PS (Figure 6).

3.3. Optimal Amount of Buckwheat Flour Fraction

The parameters of the samples with the optimal addition levels of buckwheat flour for each PS compared to the control are presented in Table 4. The optimum values of the variables with the highest desirability, obtained through the numerical optimization analysis, were selected.

Table 4. Wheat flour dough and optimized composite flour for each buckwheat flour particle size.

Variable	Control	O_BL	O_BM	O_BS
Adition Level	100% WF	9.13%	10.57%	10.25%
FN (s)	312.00	314.54	319.00	352.51
WA (%)	58.50	57.99	58.25	58.21
DT (min)	1.69	3.08	3.13	1.44
ST (min)	9.96	9.39	9.05	8.88
C1-2 (N·m)	0.61	0.63	0.66	0.65
C3-2 (N·m)	1.41	1.34	1.30	1.42
C3-4 (N·m)	0.05	0.09	0.12	0.06
C5-4 (N·m)	1.15	0.85	0.71	0.89
P (mm H_2O)	87.00	87.08	76.49	84.11
L (mm)	91.00	39.20	47.95	42.30
W ($\times 10^{-4}$) (J)	253.00	132.35	122.22	136.50
P/L (adim.)	0.95	2.27	1.53	1.90
H'$_m$ (mm)	62.00	76.08	71.88	73.23
TV (mL)	1168.00	1496.39	1342.32	1349.35
VR (mL)	991.00	1359.63	1207.89	1224.13
CR (%)	84.80	96.31	91.15	91.97
G' (Pa)	26,370.00	23,231.33	26,678.24	27,206.25
G''(Pa)	9488.00	9029.90	9834.58	10,667.36
tan δ (adim.)	0.360	0.368	0.376	0.385
T_{max} (°C)	83.24	80.52	80.58	79.41
Jc_{max} ($\times 10^{-5}$) (Pa^{-1})	24.50	24.27	24.29	23.65
Jr_{max} ($\times 10^{-5}$) (Pa^{-1})	16.62	16.37	16.24	15.72
BV (cm^3)	372.20	360.85	355.11	283.61
BH (g)	786.00	807.10	673.24	1565.86

O_BL, O_BM, O_BS: optimal samples with fractions of buckwheat flour with large, medium, and small particle sizes; FN: falling number; WA: water absorption; DT: development time; ST: dough stability; C1-2: speed of protein weakening; C3-2: starch gelatinization; C3-4: cooking stability; C5-4: starch retrogradation; P: dough tenacity; L: dough extensibility; W: deformation energy; P/L: alveograph ratio; H'm: maximum height of the gas release curve; TV: total volume of gas produced; VR: volume of the gas retained in the dough; CR: retention coefficient; G': storage modulus; G'': loss modulus; tan δ: loss tangent; T_{max}: maximum gelatinization temperature; Jc_{max}, Jr_{max}: maximum creep-recovery compliance; BV: bread volume; BH: bread hardness.

4. Discussion

This research presents for the first time a complex investigation into the influence on the empirical-dynamic rheological properties of dough and bread characteristics of buckwheat flour PS and addition level when incorporated into wheat flour.

Wheat–buckwheat composite flour presented a higher α-amylase activity, since this is inversely correlated with FN, which exhibited a decreasing value with increasing PS. This effect could be related to the intake of calcium, which stabilizes α-amylase, producing a favorable effect on bread quality [29]. BF with small particle sizes possesses a higher quantity of phenolic compounds [38], which form bonds with α-amylase, leading to a decrease in α-amylase in composite flours containing small PS.

Water absorption increased with decreasing PS and BF addition level, which means that composite flours require a higher quantity of water to swell the starch and to achieve the optimum consistency. This trend could be due to the damaged granules of starch that result during the milling process, which affect the behavior of starch-containing systems, facilitating the penetration of water for hydration [39]. A similar effect of small PS on WA was also observed by other authors [40,41]. The phenolic compounds (tyrosol, lkylphenol, and phenolic acids) in buckwheat flour tend to absorb water, and can form non-covalent bonds with starch, decreasing the pH of the dough, and modifying the water absorption of starch granules, starch gelatinization, pasting properties, and starch retrogradation [42]. The increase of PS and BF addition levels led to an increase in DT that can be attributed the content of dietary crude fibers (pentosane), which require a longer period to absorb water, including a longer mixing time [43]. Other authors who have studied the incorporation of buckwheat flour into wheat flour have observed the same decreasing trend in dough development time [43–47]. Dough stability, which provides information on dough strength, decreases with increasing BF addition level, a phonemonen that can be attributed to BF proteins, which mainly consist of albumin and globulin [47], and the lack of structure-forming capacity of the non-gluten proteins in buckwheat. Dough stability is influenced by diluted gluten content and gluten quality. With the addition of buckwheat flour, which contains non-gluten proteins, a weaker structure-forming ability and viscoelastic behavior were reported [48]. The protein from buckwheat flour contributes to the replacement of gluten, which also influences the dough's rheological behavior, similarly to protein from amaranth [49]. The interactions between the different types of starch and protein impact the rheological characteristics of the dough, also affecting the pasting properties of the starch [50]. The protein weakening due to heat (C1-2) increased with BF addition. This effect can presumably be attributed to the gluten dilution effect, with the protein network becoming less compact during heating, a fact that favors enzymatic attacking points. Similar findings with respect to the effect of different PS on protein weakening in wheat flour dough were observed by Torbica et al. (2010) [45]. The starch gelatinization (C3-2) of composite flour, describing the transition of semicrystalline starch granules into an amorphous structure, presented a decreasing trend when both factors, PS and BF addition level, increased, probably due to the formation of buckwheat amylose with complex lipid compounds [51]. The lower degree of swelling and gelatinization of the starch granules can probably be attributed to the presence of phenolic components from BF, such as rutin, which acts as a rigid filler in the starch matrix [52,53]. Our results are in accordance with the findings reported by Filipèev et al. (2015) and Sciarini et al. (2019) [54,55]. Decreasing gel stability (C3-4) with increasing BF addition level in composite flour can be associated with the fraction components, most probably with the soluble fiber and phenolic compounds, which bind water through hydrogen bonds, limiting the availability of water for the starch [48]. Similar results were obtained by Filipèev et al. (2015) [54]. Starch retrogradation (C5-4), which describes the changes in viscosity during starch re-association during the cooling phase, indicates cooking stability, which is correlated with gel firmness. The decrease in starch recrystallization occurring in the composite flour samples with increasing amounts of BF could be due to the effects of the content of leached amylose and long-chain amylopectin [56]. This decrease could be also associated with increased amounts of lipid from BF [12,43], because the amylose chain cannot bind and reassociate in a free manner like in lower lipid systems with low lipid content, which will improve the staling of the formulated bread. These results are in line with those reported by other researchers [43,57].

With respect to dough biaxial extension parameters, it can be stated that dough tenacity increased with increasing PS decreasing BF addition level, most likely because of the fiber content of BF with large PS, which can form strong interactions with wheat flour proteins. Instead, dough extensibility decreased remarkably with the addition of BF. This decrease in dough extensibility may be due to the fact that BF is gluten-free, which is the main cause of modifications in the rheological properties of wheat dough, whose protein components create a three-dimensional sponge-mesh structure in the dough, determining the physical properties of the dough. Dough deformation energy presented significant decreases with increasing BF addition level, revealing a weakening effect of the addition of BF on the formulated dough. This consequence could derive from the BF protein content substituting for gluten proteins, causing a dilution effect. Similar results regarding dough behavior during biaxial extension were found in other studies when different amounts of BF were used, but they did not take into account the PS [58].

The incorporation of buckwheat flour PS at different amounts into wheat flour resulted in an increase in the dough rheofermentation parameters assessed. Buckwheat differs from wheat flour in its starch, monosaccharide, and disaccharide contents, which represent the substrate for the future fermentation process. The fermentation resulted in significant hydrolysis of 13S globulin, the main storage protein in buckwheat [59], with the intensification of enzymatic hydrolysis of the polymers resulting in the release of small polypeptides and an increase in the amount of nutrients for the yeast [60]. Additionally, the proteolytic activities during fermentation release amino acids and peptides, improving this process, and strongly enhancing the taste and the flavor of the bread [61]. Similar results were found by Mariotti et al. (2008) [58] when substituting wheat flour with 20–40% wholemeal buckwheat flour, reporting that replacement of up to 20% did not significantly lower the maximum dough development and but that a higher volume of released gas was obtained during leavening.

The dynamic modulus provides information about the three-dimensional conformation of the gluten matrix. An increase in storage modulus (G') with the addition of BF and the increase of loss modulus (G'') with increasing PS may be linked with the presence of phenolic compounds and damaged starch from BF and the attractive forces between buckwheat dietary fibers and starch granules, indicating that the WF–BF dough possesses a solid, elastic-like behavior. The increase in both of the dynamic modulus values depending on these factors is presumably due to the interactions between the proteins and starch. Some studies have reported that the damaged starch that results during the milling process affects dough visco-elasticity, resulting in stiffer doughs [62]. The addition buckwheat flour with large PS to the dough resulted in a decrease in storage modulus (G'), suggesting that this fraction decreases the strength and elasticity of the dough. This phenomenon could be due to lipid–starch interactions, suggesting a shift in the relaxation time of the dough cross-links to shorter times, causing decreased rigidity with increasing PS [60]. Similar findings were reported in another study [63]. Loss tangent (tan δ) values decreased with increasing BF addition levels, resulting in an elastic and firm dough. This trend could be due to the damaged starch present in the buckwheat fractions. Partial replacement of wheat flour with buckwheat flour was conducted to increase Jc_{max} and Jr_{max} values, leading to higher dough extensibility. Maximum creep compliance data exhibited a similar trend to that of the oscillation data, which also revealed a decrease in dough strength with increasing BF addition level. Data on the elastic properties of the dough with respect to tan δ were also in accordance. These changes in gel strength could potentially be due to modification of the macromolecular organization with respect to the interactions between components of the system [45], because it is well known that protein–polysaccharide interactions modify the visco-elastic profile [64]. Our results are in agreement with previous reports studying gluten-free bread based on buckwheat flour [45,57].

Bread volume increased when the dough contained BF with large particle sizes, but decreased when the amount of BF added increased. These changes could be linked to the decrease in the wheat gluten-forming protein content (prolamin represented by gliadin and glutelin, by glutenin), which lead to the dough having diminished gas-retention capacity. In any case, some authors have demonstrated that gluten dilution has only a secondary physical effect on the decrease in loaf volume, because the interaction between gluten proteins and ferulic acid monomers, glutathione, and phytate present in the fractions with large particle size have an evident positive impact [65]. This fact can be demonstrated by the difference between bread with large particle sizes (which have fibers and other compounds from the outer layers of buckwheat) and that containing the fractions with smaller particle sizes [65]. Our results fall in line with those reported in other studies, revealing that a low volume and high crumb texture occurred in bread with increasing buckwheat flour replacement [45,66]. Furthermore, the bread quality parameters, volume, and hardness can be affected by the phenolic compounds in buckwheat flour that are able to bind the gluten thiol groups from wheat flour doughs, leading to a weak gluten network [67]. The decrease of crumb hardness with increasing particle size may be due to the fiber content of buckwheat endosperm (which is rich in starch). The fibers may retard starch retrogradation during storage, thus having a positive effect on the crumb structure [68].

The optimization of BF addition level in wheat flour for each PS revealed that the best rheological properties were obtained when the BF addition level was 9.13% BF for large PS, 10.57% BF for medium PS, and 10.25% BF for small PS. Compared to the control, the optimized samples were characterized by lower values of dough stability, hot starch stability, starch retrogradation, dough extensibility, deformation energy, storage modulus, maximum gelatinization temperature, and bread volume. Conversely, development time, protein weakening, alveograph ratio, the maximum height of the gas release curve, total volume gas produced, the volume of the gas retained in the dough, retention coefficient, and bread hardness were higher in the optimal samples compared to the control. These differences could be due to the addition of non-gluten proteins, lipids, minerals, and carbohydrates in the BF fractions and their interactions within the dough matrix [29].

5. Conclusions

Buckwheat flour fractions present good potential for achieving nutritious baking products, while minimally diminishing their quality characteristics. An evaluation of the effect of PS and BF addition level, along with their interactions, on composite flour, dough properties, and bread characteristics was performed on the basis of mathematical data modeling. With increasing BF particle size, falling number, water absorption, starch gelatinization, storage modulus, and bread hardness decreased, while dough development time, rheofermentation parameters, loss modulus, and bread volume increased. Increasing BF addition level was correlated with lower values of water absorption, dough stability, starch gelatinization, retrogradation, dough extensibility, deformation energy, loss tangent, and bread volume, while dough development, speed of protein weakening, stability of starch gel, alveograph ratio, rheofermentation parameters, maximum creep and recovery compliance, and bread hardness increased.

The results of the optimization process indicated a decrease in consistency during the starch retrogradation stage, extensibility, deformation energy, maximum gelatinization temperature, and bread volume, and an increase in the maximum height of the gas release curve, total volume of gas produced, volume of the gas retained in the dough at the end of the test, retention coefficient, and loss tangent compared to the control. On the basis of these results, new bread formulations with the desired characteristics of higher volume, porosity, and elasticity, as well as improved nutritional profiles in comparison with wheat bread, can be introduced to the market. Further research regarding the influence of the optimal amount of buckwheat flour corresponding to each fraction substituting wheat flour on the sensory bread characteristics would be needed.

Author Contributions: I.C., C.M. and S.M. contributed equally to the study design, collection of data, development of the sampling, analyses, interpretation of results, and preparation of the paper. All authors have read and agreed to the published version of the manuscript.

Funding: This work was funded by the Ministry of Research, Innovation and Digitalization within Program 1—Development of national research and development system, Subprogram 1.2—Institutional Performance—RDI excellence funding projects, under contract no. 10PFE/2021.

Institutional Review Board Statement: Not applicable.

Informed Consent Statement: Not applicable.

Data Availability Statement: Not applicable.

Conflicts of Interest: The authors declare no conflict of interest.

References

1. Dewettinck, K.; Van Bockstaele, F.; Kühne, B.; Van de Walle, D.; Courtens, T.M.; Gellynck, X. Nutritional value of bread: Influence of processing, food interaction and consumer perception. *J. Cereal Sci.* **2008**, *48*, 243–257. [CrossRef]
2. Dall' Asta, M.; Dodi, R.; Di Pede, G.; Marchini, M.; Spaggiari, M.; Gallo, A.; Righetti, L.; Brighenti, F.; Galaverna, G.; Dall'Asta, C.; et al. Postprandial blood glucose and insulin responses to breads formulated with different wheat evolutionary populations (*Triticum aestivum* L.): A randomized controlled trial on healthy subjects. *Nutrition* **2021**, 111533. [CrossRef]

3. Yasui, Y.; Hirakawa, H.; Ueno, M.; Matsui, K.; Katsube-Tanaka, T.; Yang, S.J.; Aii, J.; Sato, S.; Mori, M. Assembly of the draft genome of buckwheat and its applications in identifying agronomically useful genes. *DNA Res.* **2016**, *23*, 215–224. [CrossRef] [PubMed]
4. Alonso-Miravalles, L.; O'Mahony, J.A. Composition, protein profile and rheological properties of pseudocereal-based protein-rich ingredients. *Foods* **2018**, *7*, 73. [CrossRef]
5. Martínez-Villaluenga, C.; Penas, E.; Hernández-Ledesma, B. Pseudocereal grains: Nutritional value, health benefits and current applications for the development of gluten-free foods. *Food Chem. Toxicol.* **2020**, 111178. [CrossRef]
6. Gimenez-Bastida, J.A.; Zielinski, H. Buckwheat as a functional food and its effects on health. *J. Agric. Food Chem.* **2015**, *63*, 7896–7913. [CrossRef]
7. Gao, L.; Xia, M.; Li, Z.; Wang, M.; Wang, P.; Yang, P.; Gao, J. Common buckwheat-resistant starch as a suitable raw material for food production: A structural and physicochemical investigation. *Int. J. Biol. Macromol.* **2020**, *145*, 145–153. [CrossRef]
8. Yu, D.; Chen, J.; Ma, J.; Sun, H.; Yuan, Y.; Ju, Q.; Luan, G. Effects of different milling methods on physicochemical properties of common buckwheat flour. *LWT* **2018**, *92*, 220–226. [CrossRef]
9. Steadman, K.J.; Burgoon, M.S.; Lewis, B.A.; Edwardson, S.E.; Obendorf, R.L. Buckwheat seed milling fractions: Description, macronutrient composition and dietary fibre. *J. Cereal Sci.* **2001**, *33*, 271–278. [CrossRef]
10. Sinkovič, L.; Sinkovič, D.K.; Meglič, V. Milling fractions composition of common (*Fagopyrum esculentum* Moench) and Tartary (*Fagopyrum tataricum* (L.) Gaertn.) buckwheat. *Food Chem.* **2021**, *365*, 130459. [CrossRef]
11. Horbowicz, M.; Obendorf, R.L. Changes in sterols and fatty acids of buckwheat endosperm and embryo during seed development. *J. Agric. Food Chem.* **1992**, *40*, 745–750. [CrossRef]
12. Coțovanu, I.; Mironeasa, S. Impact of different amaranth particle size addition level on wheat flour dough rheology and bread features. *Foods* **2021**, *10*, 1539. [CrossRef] [PubMed]
13. Ma, S.; Wang, C.; Li, L.; Wang, X. Effects of particle size on the quality attributes of wheat flour made by the milling process. *Cereal Chem.* **2020**, *97*, 172–182. [CrossRef]
14. Coțovanu, I.; Ungureanu-Iuga, M.; Mironeasa, S. Investigation of Quinoa Seeds Fractions and Their Application in Wheat Bread Production. *Plants* **2021**, *10*, 2150. [CrossRef] [PubMed]
15. Coțovanu, I.; Mironeasa, S. Influence of Buckwheat Seed Fractions on Dough and Baking Performance of Wheat Bread. *Agronomy* **2022**, *12*, 137. [CrossRef]
16. Song, Y.Y.; Wang, Y.H.; Chen, J. Physicochemical properties of wheat flour with different particle size ranges. *Mod. Food Sci.* **2016**, *32*, 116–120.
17. Choi, H.; Baik, B. Significance of wheat flour particle size on sponge cake baking quality. *Cereal Chem.* **2013**, *90*, 150–156. [CrossRef]
18. Sakhare, S.D.; Inamdar, A.A.; Soumya, C.; Indrani, D.; Rao, G.V. Effect of flour particle size on microstructural, rheological and physico-sensory characteristics of bread and south Indian parotta. *J. Food Sci. Technol.* **2014**, *51*, 4108–4113. [CrossRef]
19. Wang, N.; Hou, G.G.; Kweon, M.; Lee, B. Effects of particle size on the properties of whole-grain soft wheat flour and its cracker baking performance. *J. Cereal Sci.* **2016**, *69*, 187–193. [CrossRef]
20. Niu, M.; Zhang, B.; Jia, C.; Zhao, S. Multi-scale structures and pasting characteristics of starch in whole-wheat flour treated by superfine grinding. *Int. J. Biol. Macromol.* **2017**, *104*, 837–845. [CrossRef]
21. Marchini, M.; Carini, E.; Cataldi, N.; Boukid, F.; Blandino, M.; Ganino, T.; Vittadini, E.; Pellegrini, N. The use of red lentil flour in bakery products: How do particle size and substitution level affect rheological properties of wheat bread dough? *LWT* **2021**, *136*, 110299. [CrossRef]
22. Ahmed, J.; Ptaszek, P.; Basu, S. Influence of fibers and particle size distribution on Food Rheology. In *Advances in Food Rheology and Its Applications*; Woodhead Publishing: Cambridge, UK, 2016.
23. Mironeasa, S.; Iuga, M.; Zaharia, D.; Mironeasa, C. Optimization of grape peels particle size and flour substitution in white wheat flour dough. *Sci. Study Res. Chem. Eng. Biotechnol. Food Ind.* **2019**, *20*, 29–42.
24. Mironeasa, S.; Iuga, M.; Zaharia, D.; Mironeasa, C. Optimization of white wheat flour dough rheological properties with different levels of grape peels flour. *Bull. UASVM Food Sci. Technol.* **2019**, *76*, 27–39. [CrossRef]
25. Iuga, M.; Mironeasa, C.; Mironeasa, S. Oscillatory rheology and creep-recovery behaviour of grape seed-wheat flour dough: Effect of grape seed particle size, variety and addition level. *Bull. UASVM Food Sci. Technol.* **2019**, *76*, 40–51. [CrossRef]
26. Bloksma, A.H. Rheology of the breadmaking process. *Cereal Foods World* **1990**, *35*, 228–236.
27. Sroan, B.S.; Bean, S.R.; MacRitchie, F. Mechanism of gas cell stabilization in bread making. I. The primary gluten–starch matrix. *J. Cereal Sci.* **2009**, *49*, 32–40. [CrossRef]
28. Veraverbeke, W.S.; Delcour, J.A. Wheat protein composition and properties of wheat glutenin in relation to breadmaking functionality. *Crit. Rev. Food Sci. Nutr.* **2002**, *42*, 179–208. [CrossRef] [PubMed]
29. Coțovanu, I.; Mironeasa, S. Buckwheat Seeds: Impact of Milling Fractions and Addition Level on Wheat Bread Dough Rheology. *Appl. Sci.* **2021**, *11*, 1731. [CrossRef]
30. *Romanian Standard SR 90:2007*; Wheat Flour. Analysis Method. Romanian Standards Association: Bucharest, Romania, 2007.
31. ICC. *Standard Methods of the International Association for Cereal Chemistry. Methods 104/1, 110/1, 136, 105/2, 171, 121, 107/1, 173*; International Association for Cereal Chemistry: Vienna, Austria, 2010.
32. Mironeasa, S.; Mironeasa, C. Dough bread from refined wheat flour partially replaced by grape peels: Optimizing the rheological properties. *J. Food Process Eng.* **2019**, *42*, e13207. [CrossRef]

33. Sanz, T.; Salvador, A.; Hernández, M.J. Creep–recovery and oscillatory rheology of flour-based systems. In *Advances in Food Rheology and Its Applications*; Woodhead Publishing: Sawston, UK, 2017; pp. 277–295. [CrossRef]
34. Moreira, R.; Chenlo, F.; Torres, M.D.; Rama, B. Fine particle size chestnut flour doughs rheology: Influence of additives. *J. Food Eng.* 2014, 120, 94–99. [CrossRef]
35. Coțovanu, I.; Mironeasa, S. Features of Bread Made from Different Amaranth Flour Fractions Partially Substituting Wheat Flour. *Appl. Sci.* 2022, 12, 897. [CrossRef]
36. Wu, C.J.; Hamada, M.S. *Experiments: Planning, Analysis, and Optimization*; John Wiley & Sons: Hoboken, NJ, USA, 2011; p. 552.
37. Coțovanu, I.; Stoenescu, G.; Mironeasa, S. Amaranth influence on wheat flour dough rheology: Optimal particle size and amount of flour replacement. *J. Microbiol. Biotechnol. Food Sci.* 2021, 10, 366–373. [CrossRef]
38. Skrabanja, V.; Kreft, I.; Golob, T.; Modic, M.; Ikeda, S.; Ikeda, K.; Kosmelj, K. Nutrient Content in Buckwheat Milling Fractions. *Cereal Chem.* 2004, 81, 172–176. [CrossRef]
39. Zhu, F. Impact of ultrasound on structure, physicochemical properties, modifications, and applications of starch. *Trends Food Sci. Technol.* 2015, 43, 1–17. [CrossRef]
40. Sapirstein, H.; Wu, Y.; Koksel, F.; Graf, R.J. A study of factors influencing the water absorption capacity of Canadian hard red winter wheat flour. *J. Cereal Sci.* 2018, 81, 52–59. [CrossRef]
41. Bressiani, J.; Oro, T.; Da Silva, P.M.L.; Montenegro, F.M.; Bertolin, T.E.; Gutkoski, L.C.; Gularte, M.A. Influence of milling whole wheat grains and particle size on thermo-mechanical properties of flour using Mixolab. *Czech J. Food Sci.* 2019, 37, 276–284. [CrossRef]
42. Xu, J.; Wang, W.; Li, Y. Dough properties, bread quality, and associated interactions with added phenol compounds: A review. *J. Funct. Foods* 2019, 52, 629–639. [CrossRef]
43. Sedej, I.; Sakač, M.; Mandić, A.; Mišan, A.; Tumbas, V.; Hadnađev, M. Assessment of antioxidant activity and rheological properties of wheat and buckwheat milling fractions. *J. Cereal Sci.* 2011, 54, 347–353. [CrossRef]
44. Nikolić, N.; Sakač, M.; Mastilović, J. Effect of buckwheat flour addition to wheat flour on acylglycerols and fatty acids composition and rheology properties. *LWT Food Sci. Technol.* 2011, 44, 650–655. [CrossRef]
45. Torbica, A.; Hadnađev, M.; Dapčević, T. Rheological, textural and sensory properties of gluten-free bread formulations based on rice and buckwheat flour. *Food Hydrocoll.* 2010, 24, 626–632. [CrossRef]
46. Yıldız, G.; Bilgiçli, N. Utilisation of buckwheat flour in leavened and unleavened Turkish flat breads. *Qual. Assur. Saf. Crop.* 2015, 7, 207–215. [CrossRef]
47. Alvarez-Jubete, L.; Wijngaard, H.; Arendt, E.K.; Gallagher, E. Polyphenol composition and in vitro antioxidant activity of amaranth, quinoa buckwheat and wheat as affected by sprouting and baking. *Food Chem.* 2010, 119, 770–778. [CrossRef]
48. Lazaridou, A.; Duta, D.; Papageorgiou, M.; Belc, N.; Biliaderis, C.G. Effects of hydrocolloids on dough rheology and bread quality parameters in gluten-free formulations. *J. Food Eng.* 2007, 79, 1033–1047. [CrossRef]
49. Matos, M.E.; Rosell, C.M. Understanding gluten-free dough for reaching breads with physical quality and nutritional balance. *J. Sci. Food Agric.* 2015, 95, 653–661. [CrossRef]
50. Schirmer, M.; Jekle, M.; Becker, T. Starch gelatinization and its complexity for analysis. *Starch-Stärke* 2015, 67, 30–41. [CrossRef]
51. Qian, J.; Rayas-Duarte, P.; Grant, L. Partial characterization of buckwheat (*Fagopyrum esculentum*) starch. *Cereal Chem.* 1998, 75, 365–373. [CrossRef]
52. Wang, N.; Hou, G.G.; Dubat, A. Effects of flour particle size on the quality attributes of reconstituted whole-wheat flour and Chinese southern-type steamed bread. *LWT Food Sci. Technol.* 2017, 82, 147–153. [CrossRef]
53. Wu, K.; Gan, R.; Dai, S.; Cai, Y.Z.; Corke, H.; Zhu, F. Buckwheat and millet affect thermal, rheological, and gelling properties of wheat flour. *J. Food Sci.* 2016, 81, E627–E636. [CrossRef]
54. Filipčev, B.; Šimurina, O.; Bodroža-Solarov, M. Impact of buckwheat flour granulation and supplementation level on the quality of composite wheat/buckwheat ginger-nut-type biscuits. *Ital. J. Food Sci.* 2015, 27, 495–504.
55. Sciarini, L.S.; Steffolani, M.E.; Fernández, A.; Paesani, C.; Pérez, G.T. Gluten-free breadmaking affected by the particle size and chemical composition of quinoa and buckwheat flour fractions. *Food Sci. Technol. Int.* 2020, 26, 321–332. [CrossRef]
56. Xu, M.; Saleh, A.S.; Liu, Y.; Jing, L.; Zhao, K.; Wu, H.; Li, W. The changes in structural, physicochemical, and digestive properties of red adzuki bean starch after repeated and continuous annealing treatments. *Starch-Stärke* 2018, 70, 1700322. [CrossRef]
57. Hadnađev, T.D.; Torbica, A.; Hadnađev, M. Rheological properties of wheat flour substitutes/alternative crops assessed by Mixolab. *Procedia Food Sci.* 2011, 1, 328–334. [CrossRef]
58. Mariotti, M.; Lucisano, M.; Pagani, M.A.; Iametti, S. Macromolecular interactions and rheological properties of buckwheat-based dough obtained from differently processed grains. *J. Agric. Food Chem.* 2008, 56, 4258–4267. [CrossRef] [PubMed]
59. Radovic, S.R.; Maksimovic, V.R.; Varkonji-Gasic, E.I. Characterization of buckwheat seed storage proteins. *J. Agric. Food Chem.* 1996, 44, e972–e974. [CrossRef]
60. Moroni, A.V.; Dal Bello, F.; Zannini, E.; Arendt, E.K. Impact of sourdough on buckwheat flour, batter and bread: Biochemical, rheological and textural insights. *J. Cereal Sci.* 2011, 54, 195–202. [CrossRef]
61. Gänzle, M.G.; Loponen, J.; Gobbetti, M. Proteolysis in sourdough fermentations: Mechanisms and potential for improved bread quality. *Trends Food Sci. Technol.* 2008, 19, e513–e521. [CrossRef]
62. Hatcher, D.W.; Anderson, M.J.; Desjardins, R.G.; Edwards, N.M.; Dexter, J.E. Effects of flour particle size and starch damage on processing and quality of white salted noodles. *Cereal Chem.* 2002, 79, 64–71. [CrossRef]

63. Hadnađev, T.R.D.; Torbica, A.M.; Hadnađev, M.S. Influence of buckwheat flour and carboxymethyl cellulose on rheological behaviour and baking performance of gluten-free cookie dough. *Food Bioproc. Tech.* **2013**, *6*, 1770–1781. [CrossRef]
64. Fitzsimons, S.M.; Mulvihill, D.M.; Morris, E.R. Large enhancements in thermogelation of whey protein isolate by incorporation of very low concentrations of guar gum. *Food Hydrocoll.* **2008**, *22*, 576–586. [CrossRef]
65. Noort, M.W.J.; van Haaster, D.; Hemery, Y.; Schols, H.A.; Hamer, R.J. The effect of particle size of wheat bran fractions on bread quality—Evidence for fibre–protein interactions. *J. Cereal Sci.* **2010**, *52*, 59–64. [CrossRef]
66. Drobot, V.; Semenova, A.; Smirnova, J.; Mykhonik, L. Effect of buckwheat processing products on dough and bread quality made from whole-wheat flour. *Int. J. Food Stud.* **2014**, *3*. [CrossRef]
67. Koh, B.-K.; Ng, P.K.W. Effects of Ferulic Acid and Transglutaminase on Hard Wheat Flour Dough and Bread. *Cereal Chem. J.* **2009**, *86*, 18–22. [CrossRef]
68. Schmiele, M.; Jaekel, L.Z.; Patricio, S.M.C.; Steel, C.J.; Chang, Y.K. Rheological properties of wheat flour and quality characteristics of pan bread as modified by partial additions of wheat bran or whole grain wheat flour. *Int. J. Food Sci.* **2012**, *47*, 2141–2150. [CrossRef]

Article

Characterization of Sorghum Processed through Dry Heat Treatment and Milling

Ana Batariuc [1], Mădălina Ungureanu-Iuga [1,2,*] and Silvia Mironeasa [1,*]

[1] Faculty of Food Engineering, "Ştefan cel Mare" University of Suceava, 13 Universitatii Street, 720229 Suceava, Romania; ana.batariuc@usm.ro

[2] Integrated Center for Research, Development and Innovation in Advanced Materials, Nanotechnologies, and Distributed Systems for Fabrication and Control (MANSiD), "Ştefan cel Mare" University of Suceava, 13th University Street, 720229 Suceava, Romania

* Correspondence: madalina.iuga@usm.ro (M.U.-I.); silviam@fia.usv.ro (S.M.)

Abstract: Sorghum grain nutritional quality can be enhanced by applying dry heat treatments. The purpose of this study was to investigate the effects of dry heat treatment at two temperatures (121 and 140 °C) with three fractionation factors (S fraction < 200 μm, M fraction 200–250 μm and, L fraction > 300 μm) on sorghum flour chemical and functional properties, to optimize processes by means of a desirability function, and to characterize the optimal products. Treatment temperature negatively affected oil- and water-absorption capacity, protein and moisture contents, while the opposite trend was obtained for hydration capacity, swelling power, emulsifying properties, fat, ash, and carbohydrate content. Sorghum flour fractions positively influenced the hydration and water-retention capacities, emulsifying properties, and protein and carbohydrate content, while oil absorption, swelling power, fat, ash, and moisture were negatively affected. The optimal processing determined for each fraction was heat treatment at 121.00 °C for S fraction, 132.11 °C for M, and 139.47 °C for L. Optimal product characterization revealed that the color, bioactive properties, and protein and starch structures of the optimal samples had changed after heat treatment, depending on the fraction. These findings could be helpful for the cereal industry, since sorghum flour could be an alternative for conventional crops for the development of new products, such as snacks, baked goods, and pasta.

Keywords: sorghum grains; fractionation; molecular characteristics; heat processing; functional properties; nutritional composition

Citation: Batariuc, A.; Ungureanu-Iuga, M.; Mironeasa, S. Characterization of Sorghum Processed through Dry Heat Treatment and Milling. *Appl. Sci.* 2022, 12, 7630. https://doi.org/10.3390/app12157630

Academic Editors: Massimo Lucarini and Anabela Raymundo

Received: 15 June 2022
Accepted: 25 July 2022
Published: 28 July 2022

Publisher's Note: MDPI stays neutral with regard to jurisdictional claims in published maps and institutional affiliations.

Copyright: © 2022 by the authors. Licensee MDPI, Basel, Switzerland. This article is an open access article distributed under the terms and conditions of the Creative Commons Attribution (CC BY) license (https://creativecommons.org/licenses/by/4.0/).

1. Introduction

Cereal-processing industry dynamics and the continuous growth in consumer demand for special purpose foods, such as those with lower glycemic index, higher fiber content, gluten free products, etc. are the key factors that drive researchers and producers to find solutions to diversify the variety of cereal-based products. Sorghum is the sixth most cultivated cereal in the world [1]. The biggest producer of sorghum in the world is the United States [2]. The surface area cultivated with sorghum in Romania in 2020 was 8.4 million ha, with a production of 26.6 thousand tons [3]. A gluten-free cereal, sorghum flour is usually employed for the manufacturing of bread, porridges, fermented beverages [4], tortilla, cookies, and pasta [5].

The results presented in the literature show that sorghum is generally rich in phenolic phytochemicals that may exert important health-promoting properties [6]. It has been stated that sorghum has low protein quality due to the structural property of its predominant protein (kafirin, Figure 1), to deficiency in amino acids, such as threonine, tryptophan, and lysine, and to interactions with phenol molecules, such as condensed tannins, which form complexes with proteins and reduce their digestion [7].

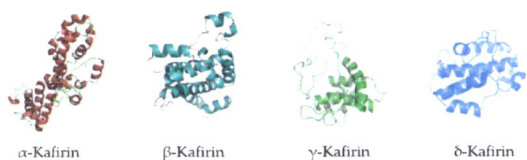

α-Kafirin   β-Kafirin   γ-Kafirin   δ-Kafirin

Figure 1. Kafirin structure. Image adapted from Zhu et al. [8] with permission from Elsevier.

Sorghum grain chemical composition (Table 1) is influenced by genetic and environmental factors. A significant and inverse correlation has been observed for protein with grain weight and starch content, while the ash content was positively correlated with protein content [9]. The main carbohydrates found in sorghum grains are starch, soluble sugar, and fiber. Sorghum is a rich source of slowly digestible starch (SDS), which is beneficial for the digestion and intestinal absorption of carbohydrates in intestine, while the raised amount of dietary fiber (9.7–14.3 g) contributes to digestion enhancement and carbohydrate and cholesterol metabolization [9,10]. Sorghum grain contains low amounts of fat, which is mainly found in the scutellar zone of the germ [9]. Kim et al. [10] reported that sorghum grain contains compounds that could be introduced in foods or dietary supplements to regulate cholesterol levels. Sorghum is rich in minerals and vitamins, which are mainly found in the aleurone coating and germ, being an important fount of B vitamins, except for vitamin B_{12} [11].

Table 1. Sorghum grain fraction chemical composition, adapted from Taylor and Duodu [12] with permission from Elsevier.

Anatomic Part	Whole Kernel (%)	Protein (%)	Fat (%)	Ash (%)	Starch (%)
Whole grain	100	11.5–12.3	3.6	1.6–1.7	72.3–75.1
Endosperm	81.7–86.5	8.7–13.0	0.4–0.8	0.3–0.7	81.3–83.0
Germ	8.0–10.9	17.8–19.2	26.9–30.6	3.9–10.4	-
Pericarp	4.3–8.7	5.2–7.6	3.7–6.0	2.0–3.8	-

The main structural components of sorghum grain are the pericarp, germ, and endosperm (Figure 2). Sorghum grains are of the caryopsis type, with the pericarp fully merged with the endosperm. Component weights are pericarp 6%, endosperm 84%, and germ 10% [9]. Sorghum grain processing implies the partial removal and/or change of the three most important components (germ, endosperm, and pericarp), some of the techniques applied being decortication, malting, fermentation, roasting, flaking, and grinding [9].

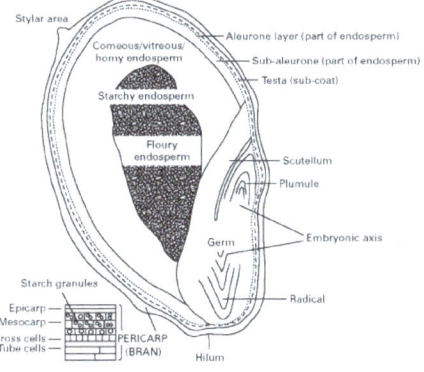

Figure 2. Sorghum grain structure. Image reproduced after Taylor and Emmambux [13] with permission from Elsevier.

The processing of sorghum grains by milling promotes lipase enzyme release, which results in triglyceride transformation, rise of free fatty acid content, and consequently diminishes the shelf life of the flour [14]. Heat-treatment processing of the grains may be considered a suitable practice for overcoming these disadvantages, causing at the same time changes in final product texture, proximate composition, tannin and antinutrient content, antioxidant properties, digestibility, etc. [15]. Dry heat treatment is a practice in snack manufacturing that has been proved to enhance product edibility, digestibility, and organoleptic characteristics [16]. Schlörmann et al. [17] demonstrated that dry heat treatment of oat up to 160 °C generated an improved sensory profile of the final product and enhanced nutritional value, with lipid, protein, starch, and β-glucan content not being influenced by the treatment, whereas the dietary fiber content was slightly affected. Dry heat treatment of little millet generated the increase of the total phenolic, flavonoid, and tannin content, and antioxidant activity [18].

Milling has a great influence on starch digestibility of cereal-based products, higher digestibility being obtained for lower particle dimensions [19]. On the other hand, obtaining fine particles implies higher energy costs and might not be recommended due to their ability to increase the erosion of the gastrointestinal wall [19]. Furthermore, fine particles may diminish enzyme flow to substrates caused by the release of viscosity, promoting such components as β-glucans, which may influence gastrointestinal residence time, distribution, and digestion [20]. Thus, while it may be preferable to decrease sorghum flour particle-size dimensions for enhanced digestibility, other factors must also be considered to achieve an effective size for a desired product quality. Fractions should be selected according to the purpose of the flour. Alvarenga et al. [21] stated that the fraction containing more sorghum bran resulted in lower expansion and stability when extruded than flour with smaller particles. The chemical and physical properties of sorghum flour fractionated by two milling methods were distinguished depending on the particle size, indicating that sorghum grains were separated into different anatomical parts [22].

The transformation of sorghum grain components into forms that can be readily used in food products without generating negative effects on food sensory characteristics are of interest and under continuous development. To our knowledge, no paper has been published to date regarding the combined effect of sorghum grain dry heat treatment and grain fractionation. The purpose of this paper was: (1) to study the effects of dry heat treatment, temperature, and fractionation on the functional properties and proximate composition of sorghum flour, (2) to optimize the processing conditions for each fraction and compare the characteristics of the optimal samples with untreated ones, and (3) to characterize optimal products regarding molecular, physical characteristics, and bioactive compounds.

2. Materials and Methods

2.1. Materials

White sorghum (*ES Albanus* hybrid) was acquired from the Secuieni Agricultural Development Research Station (Neamț, Romania).

The dry heat treatment of sorghum grains at two temperatures (121 °C and 140 °C) was performed in a Binder ED53 L convection oven (Binder, Tuttlingen, Germany) for 15 min. The grains were ground in a laboratory mill (grain mill, KitchenAid, model 106 5 KGM, Italy) and sieved in a Retsch AS 200 basic vibratory sieve shaker (Haan, Germany) to get three different fractions: large (L > 300 μm), medium (200 μm < M < 250 μm), and small (S < 200 μm). Optimization of treatment temperature for each fraction considering the functional and proximate composition of the flours was done, then the optimal samples were compared to the control (untreated) fractions and characterized.

2.2. Methods
2.2.1. Functional Properties
Hydration Capacity (HC)

HC was evaluated in duplicate, following the protocol presented by Bordei et al. [23]. An amount of 5 g sorghum flour was weighed and mixed with 30 mL tap water by using a rod for 30 s every 10 min for 1 h. The rod was washed with 10 mL water at the end and the suspension was centrifuged for 20 min at 2300 rpm. The supernatant was discarded and the sample kept at 50 °C for 25 min, then it was weighed after reaching room temperature. The HC was calculated according to Equation (1):

$$\text{HC }(\%) = \frac{(w_2 - w_0) - w_1}{w_1} \times 100 \quad (1)$$

where w_0 = tube weight, w_1 = sample weight before analysis, and w_2 = sample weight after water absorption.

Oil-Absorption Capacity (OAC)

OAC was evaluated in duplicate using a protocol adapted from the one described by Elkhalifa and Bernhardt [24]. For this purpose, 3 g sorghum flour was mixed with sunflower oil (30 mL) and stirred for 1 min at 10 min intervals for 30 min. Then, the mix was centrifuged at 3000 rpm for 15 min, the supernatant discarded, and the pellet weighed after 5 min of draining. OAC was calculated using Equation (2):

$$\text{OAC }(\%) = \frac{w_1}{w_0} \times 100 \quad (2)$$

where w_0 = sample weight before analysis and w_1 = sample weight after supernatant discard.

Water-Retention Capacity (WRC)

WRC was evaluated in duplicate using the method presented by Zhu et al. [25]. An amount of 1 g sorghum flour was mixed with 30 mL distilled water and left for 18 h at room temperature in a sealed tube. After centrifugation at 3000 rpm for 20 min, the supernatant was discarded, the sample dried for 2 h at 105 °C in a convection oven, and weighed. WRC was calculated with Equation (3):

$$\text{WRC}\left(\frac{g}{g}\right) = \frac{w_1 - w_2}{w_2} \quad (3)$$

where w_1 = sample weight before drying an w_2 = sample weight after drying.

Swelling Power (SP)

SP was evaluated in duplicate following the protocol described by Elkhalifa and Bernhardt [24], with some modifications. The sorghum flour (0.5 g) was mixed with 15 mL distilled water and heated in a water bath at 90 °C for 30 min. After cooling at room temperature (15 min) and centrifugation at 3000 rpm for 25 min, the supernatant was discarded and the swollen sorghum flour sediment was weighed.

Emulsion Activity (EA) and Stability (ES)

The emulsifying characteristics of sorghum flour fractions were achieved in duplicate following the protocol presented by Elkhalifa and Bernhardt [24]. For EA determination, 2 g sorghum flour was mixed with 20 mL distilled water at 4 °C and 20 mL of sunflower oil. After stirring for 20 min and centrifugation at 4000 rpm for 10 min, the height of the emulsion layer formed was measured. The EA was calculated with Equation (4):

$$\text{EA }(\%) = \frac{H_e}{H_w} \times 100 \quad (4)$$

where H_e = height of the emulsion layer and H_w = height of the whole layer.

For ES evaluation, the emulsion formed in the previous step was heated in a water bath at 80 °C for 30 min, then it was cooled at room temperature for 20 min. After another centrifugation at 4000 rpm for 10 min, the height of the emulsified layer was measured and ES calculated with Equation (5):

$$ES\ (\%) = \frac{H_{eh}}{H_w} \times 100 \quad (5)$$

where H_{eh} = height of the emulsion layer after heating and H_w = height of the whole layer.

2.2.2. Proximate Composition

The proximate composition in terms of moisture, protein, fat, and ash of sorghum flour fractions was determined by following the ICC standard protocols: moisture (101/1), fat (104/1), protein (105/2), and ash (105/1). The content of carbohydrates was calculated by difference [26].

2.2.3. Optimization of Sorghum Dry Heat Treatment

The response surface methodology (RSM) and D-optimal design (from Design Expert software, Stat-Ease, Minneapolis, MN, USA, trial version) were employed to evaluate the influence of two factors: treatment temperature (121 and 140 °C) and fractionation (L > 300, M 200–300 and S < 200 μm) on sorghum flour functional (HC—hydration capacity, OAC—oil-absorption capacity, SP—swelling power, WRC—water-retention capacity, EA—emulsion activity and ES—emulsion stability), and proximate composition (protein, ash, fat, moisture and carbohydrates). The maximum and minimum values of the responses used in the data matrix are listed in Table 2.

Table 2. Minimum and maximum values of the responses used in the experimental design.

Variable	Minimum Value	Maximum Value
HC (%)	88.80	102.59
OAC (%)	150.66	177.33
SP (g/g)	3.28	4.35
WRC (g/g)	0.77	1.23
EA (%)	38.00	55.00
ES (%)	56.00	66.00
Protein (%)	8.57	11.17
Fat (%)	1.58	6.35
Ash (%)	0.29	2.16
Moisture (%)	8.72	10.02
Carbohydrates (%)	67.00	71.58

HC—hydration capacity, OAC—oil-absorption capacity, SP—swelling power, WRC—water-retention capacity, EA—emulsion activity, ES—emulsion stability.

The effects of factors and their interaction on the considered responses were evaluated through analysis of variance (ANOVA) for the mathematical model fitted to each property. The mathematical model suitability was determined by F sequential test, coefficient of determination (R^2), and adjusted coefficients of determination ($Adj.\text{-}R^2$). Thus, the 2 FI model (Equation (6)) was selected for all the responses based on the highest R^2 and $Adj.\text{-}R^2$ values.

$$Y = x_0 + x_1 A + x_2 B + x_3 AB \quad (6)$$

where Y = response, x_{0-3} regression coefficients, and A and B factors.

Desirability function was considered to optimize the treatment temperature for each of the three fractions. The constraints applied consisted of the maximization of HC, SP, WRC, EA, ES, protein, fat, and ash, minimization of carbohydrates, and maintaining the OAC and moisture content within the range. All the factors and responses received the same importance. Model validation was done by verifying the optimal values through experimental determinations, and the obtained values were compared to the fractions of the control sample. Optimal and control samples were used for further characterization.

2.2.4. Characterization of Optimal Samples

Color Evaluation

The color parameters of sorghum flour fractions were measured by reflectance using the CIE Lab system on a Konica Minolta CR-400 (Konica Minolta, Tokyo, Japan) device. The parameters recorded in triplicate were: L^* (luminosity), a^* (describing red nuance if positive or green nuance if negative), and b^* (describing yellow nuance if positive or blue nuance if negative).

Starch Digestibility

Starch fractions, such as rapidly digestible starch (RDS), slowly digestible starch (SDS), resistant starch (RS), and total digestible starch (TDS), and total starch contents of sorghum flour fractions were determined in triplicate following the international AOAC 2017.16 protocol using a Megazyme kit (K-DSTRS; Megazyme, Bray, Ireland). The principle consists of spectrophotometric measurement (at 510 nm) of the glucose released after sample digestion with α-amylase and amyloglucosidase for 20 min (RDS), 120 min (SDS), or 240 min (RS, TDS) using GOPOD reagent.

Total Polyphenols and Antiradical Activity

The extract was prepared by mixing 1 g sorghum flour with methanol 99.9% (1:20 w/v), then the mix was sonicated at 50 °C, at 40 kHz for 30 min.

For total polyphenol content (TPC) determination, 0.2 mL extract was mixed with 2 mL Folin–Ciocâlteu reagent and 1.8 mL sodium carbonate (7.5%). After resting at room temperature in the darkness for 30 min, the absorbance was read at 750 nm [27]. The calibration curve made with gallic acid had $R^2 = 0.99$ and the equation y = 0.00949x + 0.02950.

For antiradical activity (AA) of sorghum flour evaluation, the 2,2 diphenyl-1-picrylhydrazyl (DPPH) method was used. The extract (2 mL) was mixed with DPPH reagent (2 mL) and the absorbance read at 517 nm after 30 min of resting in the darkness at room temperature. The DPPH AA was calculated by using Equation (7).

$$\text{DPPH AA (\%)} = \left(1 - \frac{A_{sample}}{A_{blank}}\right) \cdot 100 \qquad (7)$$

Starch and Protein Molecular Characteristics

Sorghum flour starch and protein characteristics were evaluated by ATR-FTIR analysis, on a Thermo Scientific Nicolet iS20 (Waltham, MA, USA) device. Three spectra were recorded in the range of 650 cm^{-1} to 4000 cm^{-1}, with a resolution of 4 cm^{-1} and 32 scans. After ATR correction of all spectra, the fractions of amide I (1652 cm^{-1} for α-helix structure and 1624 cm^{-1} for β-sheets) and starch (1049 cm^{-1} for crystalline area and 1022 cm^{-1} for the amorphous region) were evaluated after applying Fourier self-deconvolution on the average spectra. The ratio of structures was determined by reporting the corresponding peak areas [28,29]. Background was collected after each sorghum flour sample.

2.2.5. Statistics

The differences among the experimental and predicted values of the optimal samples were evaluated trough Student's t test. The differences among the optimal and control samples were checked trough ANOVA and Tukey's test. The significance level was 95%. All statistical tests and Pearson correlations were done using XLSTAT for Excel 2022 version (Addinsoft, New York, NY, USA) software.

3. Results and Discussion

3.1. Functional Properties and Proximate Composition

3.1.1. Influence of Factors

The experimental data obtained for sorghum flour fraction functional properties and chemical properties were successfully fitted to the 2 FI model (Table 3). The ANOVA results revealed that the model proposed was significant for all the responses, since the F-values were significant at $p < 0.05$, while R^2 values were 0.58–0.97.

Table 3. ANOVA results for the 2 FI model fitted for the functional properties of sorghum flour.

Factor	HC (%)	OAC (%)	SP (g/g)	WRC (g/g)	EA (%)	ES (%)	Protein (%)	Fat (%)	Ash (%)	Moisture (%)	Carbohydrates (%)
Constant	96.17	164.78	3.85	0.90	45.36	61.67	9.85	4.09	1.39	9.28	68.80
A	1.77	−1.86	0.14 **	−0.07 **	1.53 **	0.50	−0.24 **	0.28 *	0.06 *	−0.41 **	0.30
B	4.16 *	−9.73 **	−0.41 **	0.11 **	5.63 **	1.25 *	1.01 **	−1.84 **	−0.77 **	−0.17 **	1.44 **
A × B	1.37	−0.21	−0.05	−0.07 *	2.12 **	2.75 **	−0.22 *	−0.08	0.16 **	0.08 *	−0.30
						Model fitting					
p-value	<0.05	<0.01	<0.01	<0.01	<0.01	<0.01	<0.01	<0.01	<0.01	<0.01	<0.05
R^2	0.58	0.69	0.90	0.73	0.93	0.68	0.90	0.93	0.97	0.95	0.58
Adj.-R^2	0.49	0.62	0.88	0.67	0.91	0.61	0.88	0.91	0.96	0.94	0.49

* $p < 0.05$, ** $p < 0.01$, HC—hydration capacity, OAC—oil-absorption capacity, SP—swelling power, WRC—water-retention capacity, EA—emulsion activity, ES—emulsion stability, A—treatment temperature factor, B—fractions factor.

HC was positively influenced by both factors—treatment temperature and fraction dimension—and their interaction (Table 3), but only in the case of the fraction factor was the effect significant ($p < 0.05$). The increase in HC with temperature and fraction-dimension increase (Figure 3a) could be related to the increase of damaged starch induced by gelatinization during heat treatment [15]. As the determination method implied the application of an external force to the sample, as stated by Jacobs et al. [30], the nanopores caused by fractionation present on the surface of the sorghum flour particles could play an essential role. A similar trend of HC was reported by these authors for wheat bran with different particle sizes.

A negative effect of fractionation was observed for sorghum flour OAC (Table 3). OAC decreased with treatment temperature and fraction dimensions decreased, as can be seen in Figure 3d. Almaiman et al. [31] also reported a decrease in OAC of sorghum flour when microwave treatment was applied, probably as a result of the changes in protein hydrophobic properties, oil-absorption capacity being affected by starch–protein–lipid linkages, sequence of polypeptides, various conformational properties of macromolecules, and quantity of apolar amino acids [32]. Furthermore, OAC has been proved to be influenced also by the ability of oil physical entrapment of the sample, surface area, size of macromolecules, charge, and hydrophobicity [33].

Sorghum flour SP was positively influenced by treatment temperature and negatively by fractionation ($p < 0.01$), as shown in Table 3. A similar increasing trend of SP was obtained by Zou et al. [34] for thermally treated maize starch, which could be explained by the modification of starch granule surface, which determines higher water absorption, SP also being influenced by amylose/amylopectin ratio, molecular weight, and starch–lipid complexes. The milling process affects starch, protein, and cell wall components of sorghum grains, leading to changes in their capacity to bind water and release soluble components [20].

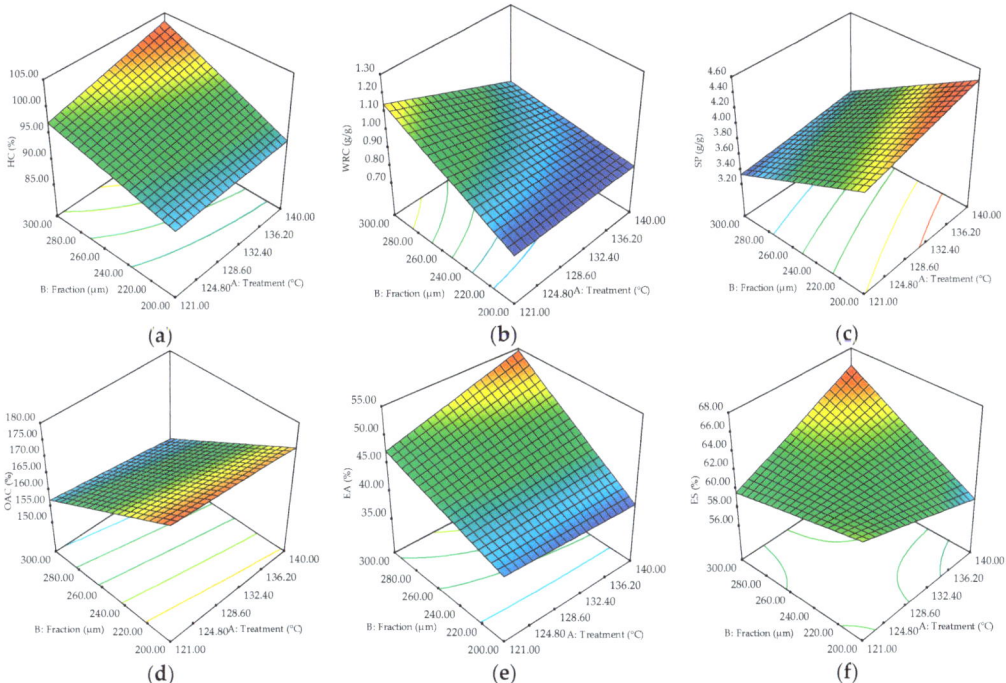

Figure 3. Three-dimensional graphic showing the combined effects of factors (treatment temperature and fractions) on the sorghum flour: (**a**) hydration capacity (HC), (**b**) water-retention capacity (WRC), (**c**) swelling power (SP), (**d**) oil-absorption capacity (OAC), (**e**) emulsion activity (EA), and (**f**) emulsion stability (ES).

Fractionation exhibited significant positive ($p < 0.01$) effects (Table 3) on sorghum flour WRC, while treatment temperature and the interaction between factors had a negative influence ($p < 0.05$). The increase in fraction dimensions resulted in proportional rise of flour WRC (Figure 3b), while an opposite trend was obtained when temperature was increased. The reduction in WRC at the higher treatment temperature could be due to physical damage to the fiber structure at high temperatures, which may lead to changes in water-retention ability [35]. Heat treatment causes a rupture of weak linkages between polysaccharide chains, high temperatures also being able to break the glycosidic bonds in the polysaccharides [35]. The structure of sorghum flour particles comprising nanopores could be responsible for the water binding of sorghum flour, since this water is likely to be retained even when subjected to external forces, such as centrifugation due to capillary forces and the cell walls' mechanical strength [30].

The EA and ES parameters were positively influenced by treatment temperature and fractionation ($p < 0.05$), while the interaction between factors showed significant positive influence ($p < 0.05$) (Table 3). The EA is mainly dependent on the dispersion of proteins at the surface tension of water and air by unfolding its structure, while ES is determined by the creation of a dense cohesive layer around the air bubble [36]. Similar enhancement of the emulsifying properties was obtained by Hassan et al. [36] when radio-frequency heat treatment was applied to maize grains. The increase in protein content as the fraction dimension was higher could represent an explanation for the enhanced EA and ES of sorghum flour with greater fraction size. The different nutritional composition could be explained by the different fractions obtained from sieving.

Sorghum flour protein content was positively influenced (at $p < 0.01$) by fractionation (Figure 4a), while treatment temperature and the interaction between factors had a significant negative effect. Treatment temperature significantly influenced ($p < 0.05$) in a positive way fat and ash content, while the fractionation negatively affected ($p < 0.01$) these parameters (Table 3).

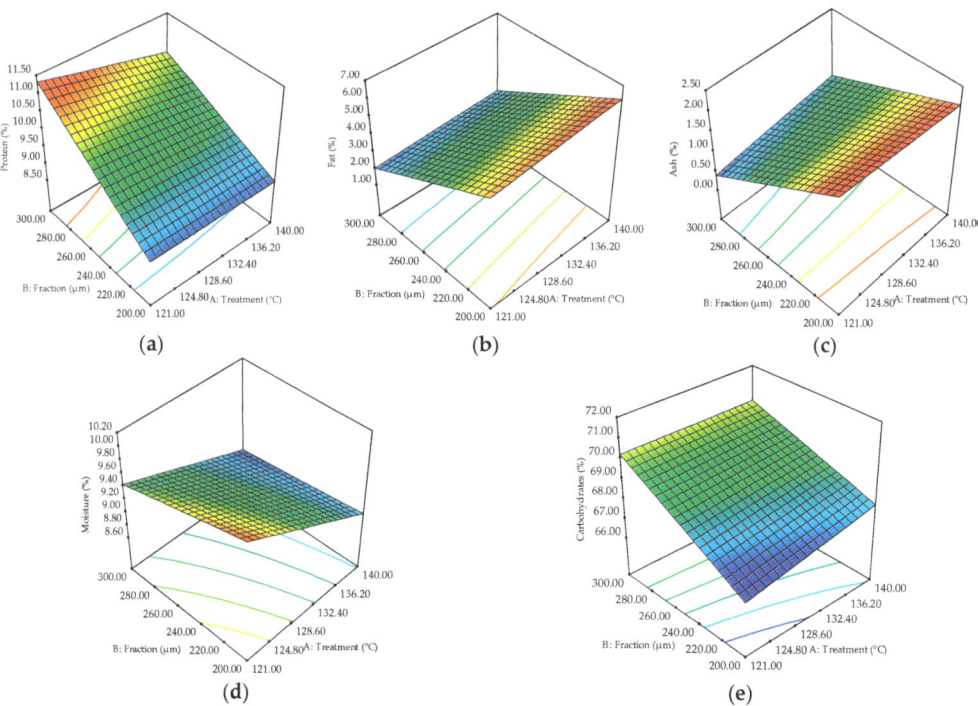

Figure 4. Three-dimensional graphic showing the combined effects of factors (treatment temperature and fractionation) on the sorghum flour content of: (**a**) protein, (**b**) fat, (**c**) ash, (**d**) moisture, and (**e**) carbohydrates.

Sorghum flour moisture content was negatively affected ($p < 0.01$) by both factors, while their interaction positively influenced this parameter (Table 3). Positive effects of both factors on the carbohydrate content of sorghum flour were observed, but only the fractionation exhibited a significant effect ($p < 0.01$). Sorghum fractions can determine the contents of macronutrients and digestion rate [37]. Increased protein (Figure 4a) and carbohydrate (Figure 4e) contents as the fraction size raised was observed. Alvarenga et al. [21] stated that the proteins were most abundant in the fibrous mill-feed fraction because the aleurone layer in the pericarp and the peripheral endosperm tissue, which would be found in the sorghum mill-feed, are rich in proteins. On the other hand, the ash (Figure 4c) and moisture (Figure 4d) contents decreased with fraction dimension increase, which is explainable since the endosperm, which is mostly found in smaller fractions, is abundant in minerals [38] and moisture loss can occur during milling [21]. A reduction trend with particle-size increase was also observed for fat (Figure 4c) content, which seems to be more abundant in the germ [22]. The decrease in protein and moisture content with treatment temperature could be due to the denaturation of proteins and moisture loss induced by heating, similar results being reported by Mahajan and Gupta [39] for roasted sorghum. Higher amounts of fat, ash, and carbohydrates were obtained as the treatment temperature raised, which could be due to changes in lipid complexes with other molecules and partial starch gelatinization during heating, which also affects nutrient digestibility [40]. Dharmaraj and Malleshi [40] stated that the extractability of bound lipids increased after hydrothermal processing of millet, changing the fatty acid profile due to the formation of complexes between linoleic acid and amylose.

3.1.2. Pearson Correlations between Variables

Pearson correlation coefficients are presented in Table 4. The protein content was negatively correlated ($p < 0.05$) with OAC ($r = -0.52$) and SP ($r = -0.75$), while with WRC ($r = 0.50$), HC ($r = 0.55$), and EA ($r = 0.60$), significant positive correlations were obtained. The correlations of SP with protein and carbohydrates support the fact that polysaccharide–protein complexes have good emulsifying activity, their capacity to stabilize interfaces being led to the isoelectric point of proteins [41].

Table 4. Pearson correlation coefficients.

Variables	Protein	Fat	Ash	Moisture	Carbohydrates	OAC	WRC	HC	SP	EA	ES
Protein	1.00										
Fat	−0.77 **	1.00									
Ash	−0.86	0.93 **	1.00								
Moisture	−0.12	0.19	0.23	1.00							
Carbohydrates	0.51 *	−0.89 **	−0.82 **	−0.53 **	1.00						
OAC	−0.52 *	0.82 **	0.84 **	0.26	−0.85 **	1.00					
WRC	0.50 *	−0.69 **	−0.66 **	−0.28	0.68 **	−0.50 *	1.00				
HC	0.55 **	−0.44 *	−0.41 *	−0.35	0.38	−0.21	−0.11	1.00			
SP	−0.75 **	0.80 **	0.84 **	0.16	−0.68 **	0.71 **	−0.70 **	−0.14	1.00		
EA	0.60 **	−0.83 **	−0.75 **	−0.51 *	0.87 **	−0.80 **	0.44 *	0.59 **	−0.66	1.00	
ES	0.20	−0.34	−0.11	−0.11	0.28	−0.15	−0.15	0.69 **	−0.07	0.60 **	1.00

* $p < 0.05$, ** $p < 0.01$, HC—hydration capacity, OAC—oil-absorption capacity, SP—swelling power, WRC—water-retention capacity, EA—emulsion activity, ES—emulsion stability.

Sorghum flour fat content was negatively correlated with WRC ($r = -0.69$), HC ($r = -0.44$), and EA ($r = -0.83$), while with OAC ($r = 0.82$) and SP ($r = 0.80$), positive relationships were observed. These correlations support the observation made by Joshi et al. [42] that different amounts and types of lipids, proteins, and carbohydrates can impact the WRC and HC because they may have distinct polarity and thus different water-binding and -retention capacities. Similar to the fat content variable, the ash content was positively correlated with OAC and SP ($r = 0.84$, $p < 0.05$) and negatively with WRC, HC, and EA ($-0.41 > r > -0.75$, $p < 0.05$). The moisture content and EA were negatively correlated ($r = -0.51$, $p < 0.05$), a negative correlation also being observed between OAC and EA ($r = -0.80$, $p < 0.05$).

Significant negative ($p < 0.01$) correlations were obtained for carbohydrates with OAC ($r = -0.85$) and SP ($r = -0.68$) and positive with WRC ($r = 0.68$) and EA ($r = 0.87$). WRC and HC are strongly influenced by carbohydrate content and structure, damaged starch, and fibers playing an essential role [22]. The ES parameter was significantly positively ($p < 0.01$) correlated with HC ($r = 0.69$) and EA ($r = 0.60$). OAC is determined by the physical oil entrapment and fat binding to apolar protein molecules, and it is directly dependent on the lipophilicity, amino acid composition, and surface polarity [14].

3.1.3. Optimization and Model Validation

The optimization process for each sorghum flour fraction revealed that dry treatment at 121.00 °C would be recommended for S fraction, 132.11 °C would be appropriate for the M fraction, while for L fraction a temperature of 139.47 °C (Table 5) can be applied to obtain the desired functional and nutritional properties.

Table 5. Model validation. Properties of the optimal samples compared to the control samples.

Property	Optimal S Predicted	Optimal S Experimental	Optimal M Predicted	Optimal M Experimental	Optimal L Predicted	Optimal L Experimental	Control S	Control M	Control L
Treatment (°C)	121.00	121.00	132.11	132.11	139.47	139.47	-	-	-
HC (%)	91.60 ± 3.86 [a]	96.55 ± 0.11 [ax]	96.47 ± 3.86 [a]	101.74 ± 0.54 [aw]	103.29 ± 3.86 [a]	110.84 ± 0.08 [aj]	91.30 ± 1.50 [y]	97.80 ± 1.40 [z]	98.11 ± 0.90 [k]
OAC (%)	176.16 ± 6.28 [a]	179.71 ± 0.06 [ax]	164.47 ± 6.28 [a]	174.77 ± 0.15 [aw]	153.10 ± 6.28 [a]	158.94 ± 0.38 [aj]	171.33 ± 0.43 [y]	166.71 ± 0.39 [z]	152.36 ± 0.54 [k]
SP (g/g)	4.08 ± 0.14 [a]	4.32 ± 0.06 [ax]	3.87 ± 0.14 [b]	4.13 ± 0.04 [aw]	3.52 ± 0.14 [a]	3.74 ± 0.06 [aj]	4.31 ± 0.01 [x]	3.98 ± 0.01 [z]	3.51 ± 0.01 [k]
WRC (g/g)	0.79 ± 0.09 [a]	0.83 ± 0.02 [ay]	0.88 ± 0.09 [a]	0.94 ± 0.01 [az]	0.87 ± 0.09 [a]	0.93 ± 0.01 [ak]	1.32 ± 0.02 [x]	1.06 ± 0.03 [w]	1.31 ± 0.08 [j]
EA (%)	40.33 ± 1.65 [a]	41.75 ± 1.06 [ax]	45.62 ± 1.65 [a]	46.75 ± 1.06 [aw]	54.43 ± 1.65 [a]	57.75 ± 1.06 [aj]	41.50 ± 0.50 [x]	45.50 ± 0.50 [w]	56.50 ± 0.50 [j]
ES (%)	62.67 ± 1.95 [a]	64.50 ± 0.72 [ax]	61.75 ± 1.95 [a]	63.50 ± 0.72 [aw]	65.98 ± 1.95 [a]	66.50 ± 0.72 [aj]	62.50 ± 0.50 [y]	54.50 ± 0.50 [z]	66.50 ± 0.50 [j]
Protein (%)	8.86 ± 0.34 [a]	8.87 ± 0.06 [ay]	9.81 ± 0.34 [a]	8.96 ± 0.06 [bw]	10.42 ± 0.34 [a]	11.3 ± 0.06 [ak]	10.62 ± 0.14 [x]	8.83 ± 0.02 [z]	11.85 ± 0.05 [j]
Fat (%)	5.56 ± 0.48 [a]	5.69 ± 0.01 [ax]	4.13 ± 0.48 [a]	4.22 ± 0.03 [aw]	2.44 ± 0.48 [a]	2.53 ± 0.04 [aj]	3.20 ± 0.02 [y]	3.15 ± 0.02 [z]	3.12 ± 0.01 [k]
Ash (%)	2.26 ± 0.13 [a]	2.29 ± 0.01 [ax]	1.40 ± 0.13 [a]	1.47 ± 0.01 [aw]	0.83 ± 0.13 [a]	0.86 ± 0.01 [aj]	2.26 ± 0.02 [x]	1.15 ± 0.01 [z]	0.76 ± 0.01 [k]
Moisture (%)	9.93 ± 0.11 [a]	10.13 ± 0.01 [ay]	9.21 ± 0.11 [b]	9.89 ± 0.02 [az]	8.80 ± 0.11 [a]	8.87 ± 0.01 [ak]	10.98 ± 0.01 [x]	11.20 ± 0.02 [w]	11.30 ± 0.03 [j]
Carbohyd. (%)	66.76 ± 1.19 [a]	73.01 ± 0.08 [bx]	68.85 ± 1.19 [b]	75.45 ± 0.00 [az]	70.23 ± 1.19 [b]	76.41 ± 0.01 [aj]	72.93 ± 0.15 [x]	75.65 ± 0.02 [w]	72.98 ± 0.06 [k]

(a,b) different letters in the same row for each sample mean significant differences between predicted and experimental values ($p < 0.05$), different letters in the same row indicate significant differences among optimal and control sample (x,y for S particle size, w,z for M fraction and j,k for L fraction) ($p < 0.05$), HC—hydration capacity, OAC—oil-absorption capacity, SP—swelling power, WRC—water-retention capacity, EA—emulsion activity, ES—emulsion stability.

Mathematical model validation revealed that the differences between the predicted and experimental values were less than 10%. Compared to Control S, the Optimal S sample presented higher OAC, SP, ES, fat, and ash contents, while the WRC, EA, and carbohydrates were lower. The Optimal M sample exhibited higher EA, ES, protein, fat, and ash contents and lower SP, WRC, and carbohydrates than Control M. Raised HC and ash content and smaller WRC, EA, ES, protein, fat, and carbohydrates values were observed for the Optimal L sample than Control L. Fractionation can lead to nutrient structure changes, mainly due to photo-oxidation, and to the decrease of some compounds' bioavailability [12]. Previous research revealed that sorghum grain processing through thermal treatment, soaking, and steaming can significantly influence physical tissue structure, nutrient levels, and functional properties of grains [43].

3.2. Characterization of the Optimal Samples

3.2.1. Total Polyphenols, DPPH Antiradical Activity (AA), and Starch Digestibility

Dry heat treatment and fractionation significantly influenced TPC and DPPH AA (Table 6). TPC of the optimal samples was higher compared to the corresponding controls, except for the L fraction. The main polyphenols of sorghum are represented by phenolic acid, flavonoid, procyanidin, and stilbenoids [44]. Woo et al. [45] reported higher flavonoid content in sorghum bran compared to other fractions. Cardoso et al. [46] demonstrated that flavonoids were more sensitive to heat than other phenolics. Considering that, maybe the decrease of TPC in the L fraction could be related to the thermal degradation of the sensitive phenolics. Heat treatment determined a decrease of DPPH AA for all the fractions. Both TPC and DPPH AA reduced as the fraction size increased, the fraction containing the highest amount of bran registering the lowest values. Sharanagat et al. [14] also reported raised levels of DPPH free radical-scavenging activity in native sorghum flour compared to the roasted samples, which may be due to the diminishing of flavonoids that are the most active phenolic compounds after treatment. Zhu et al. [25] stated that the superfine grinding of hull-less barley led to an increase of AA DPPH and TPC, probably because the fiber matrix was damaged, thus causing some phenolic compounds to be released or exposed. The data obtained by Almaiman et al. [31] showed that microwave treatment of sorghum grains raised the TPC in a power-dependent manner, which could lead to the release of phenolic compounds from glycosidic components and to the disintegration of larger phenolic compounds into smaller ones.

Table 6. Total polyphenols, antiradical activity, and starch digestibility of optimal and control samples.

Property	Sample					
	Optimal S	Control S	Optimal M	Control M	Optimal L	Control L
TPC (mg GAE/g)	21.76 ± 0.00 ^{xB}	21.06 ± 0.08 ^{yG}	25.80 ± 0.06 ^{wA}	21.74 ± 0.00 ^{zF}	11.01 ± 0.00 ^{kC}	16.10 ± 0.06 ^{jH}
DPPH AA (%)	89.22 ± 0.14 ^{yB}	99.81 ± 0.00 ^{xF}	97.57 ± 0.14 ^{zA}	99.03 ± 0.00 ^{wG}	77.57 ± 0.14 ^{kC}	85.34 ± 0.14 ^{jH}
RDS (g/100 g)	9.38 ± 0.04 ^{xB}	8.45 ± 0.04 ^{yG}	10.02 ± 0.02 ^{wA}	9.75 ± 0.04 ^{zF}	4.57 ± 0.06 ^{jC}	4.29 ± 0.04 ^{kH}
SDS (g/100 g)	0.66 ± 0.04 ^{xB}	0.46 ± 0.08 ^{yH}	1.02 ± 0.02 ^{wA}	1.00 ± 0.00 ^{wF}	0.29 ± 0.02 ^{kC}	0.77 ± 0.08 ^{jG}
TDS (g/100 g)	12.68 ± 0.02 ^{xB}	11.63 ± 0.02 ^{yG}	13.53 ± 0.02 ^{wA}	13.13 ± 0.06 ^{zF}	11.38 ± 0.04 ^{jC}	10.97 ± 0.02 ^{kH}
RS (g/100 g)	4.04 ± 0.05 ^{xB}	3.65 ± 0.02 ^{yG}	8.45 ± 0.05 ^{wA}	8.33 ± 0.07 ^{wF}	3.90 ± 0.05 ^{jC}	3.55 ± 0.05 ^{kG}
Total Starch (g/100 g)	16.73 ± 0.03 ^{xB}	15.28 ± 0.04 ^{yG}	21.98 ± 0.07 ^{wA}	21.46 ± 0.13 ^{zF}	15.29 ± 0.01 ^{jC}	14.51 ± 0.07 ^{kH}

TPC—total polyphenol content, DPPH AA—DPPH antiradical activity, RS—resistant starch, SDS—slowly digestible starch, RDS—rapidly digestible starch, TDS—total digestible starch, different letters in the same row indicate significant differences among optimal and control sample (x,y for S fraction, w,z for M fraction, and j,k for L fraction), different capital letters in the same row indicate significant differences among control samples (A–C) and among optimal samples (F–H) with various particle sizes ($p < 0.05$).

Starch digestibility was affected by heat treatment, the magnitude of changes depending on the fraction. The content of SDS, TDS, RS, total starch, and RDS increased after heat treatment compared to the control samples, except for SDS in the L fraction sample. SDS formation can lead to some interactions that occur between starch and fat during thermal treatment [47], a fact that may possibly explain the lower SDS in the L optimal sample compared to the control, since it had the lowest lipid content (Table 5). Furthermore, the formation of polyphenol–starch complexes during treatment could also be related to this decrease, it being known that the bran is richer in phenolic compounds compared to other fractions [44,48]. The results for RS and total starch were comparable with those previously reported [37]. The increase in digestible starch content with fraction dimension reduction is the result of the higher substrate–enzyme contact surface and/or of the different composition of the fractions. Starch digestibility is directly influenced by granule size, structure, crystalline pattern, degree of crystallinity, presence of pores or channels on the granule surface, degree of polymerization, nonstarch components, and their interactions with starch [20]. Kanagaraj et al. [49] reported that dry heat treatment methods were found to increase RS content in both rice and barnyard millet, depending on the temperature and time. These results could be due to the reorganization of amylose and amylopectin chains of starch during heating [50].

3.2.2. Color and Molecular Characteristics

The color properties of optimal sorghum flours compared to the controls are presented in Table 7. No significant differences regarding L^*, a^* and b^* values were recorded between Optimal S and Control S, while in the case of M and L fractions the optimal samples exhibited higher L^* and lower a^* and b^* compared to the controls. All the studied samples showed red nuances indicated by the positive values of a^* and yellow nuances suggested by the positive values of b^*. Fraction dimension increase determined a decrease of L^*, while the values of the a^* parameter increased. The color of sorghum products is influenced by the quantity and type of phenols and metal ions found in the grains [51]. The differences between the three fractions in terms of color and luminosity may be attributed to the distribution of these phenolics in the sorghum grain, it being known that the color of the pericarp seems to be generated by a combination of primarily anthocyanin and anthocyanidin pigments and other flavonoid compounds [52]. Winger et al. [51] reported an increase in redness of tortillas as the fraction size was higher due to the presence of larger quantities of bran.

Table 7. Color and molecular characteristics of optimal and control samples.

Property	Sample					
	Optimal S	Control S	Optimal M	Control M	Optimal L	Control L
L^*	85.23 ± 0.04 [xF]	85.51 ± 0.03 [xA]	84.54 ± 0.04 [wG]	83.25 ± 0.07 [zB]	78.88 ± 0.12 [jH]	77.08 ± 0.17 [kC]
a^*	0.97 ± 0.01 [xF]	0.78 ± 0.03 [yC]	1.11 ± 0.02 [zF]	1.23 ± 0.03 [wB]	1.11 ± 0.05 [kG]	1.93 ± 0.03 [jA]
b^*	11.68 ± 0.07 [xF]	11.69 ± 0.01 [xC]	12.00 ± 0.05 [zG]	12.63 ± 0.01 [wB]	15.24 ± 0.04 [kH]	15.80 ± 0.11 [jA]
α-helix/β sheets	3.80 ± 0.37 [xF]	2.11 ± 0.15 [yA]	2.46 ± 0.09 [wG]	3.01 ± 0.69 [wA]	1.66 ± 0.21 [kH]	2.36 ± 0.02 [jA]
Crystallin/amorphous	1.96 ± 0.00 [xG]	1.06 ± 0.01 [yB]	1.79 ± 0.02 [wH]	1.05 ± 0.01 [zB]	2.03 ± 0.03 [jF]	2.06 ± 0.16 [jA]

Different lowercase letters in the same row indicate significant differences among optimal and control sample (x,y for S fraction, w,z for M fraction and j,k for L fraction), different capital letters in the same row indicate significant differences among control samples (A–C) and among optimal samples (F–H) of various fractions ($p < 0.05$).

FT-IR spectra (Figure 5) interpretation allowed the identification of the main molecular characteristics of the optimal and control samples. The results obtained showed that for L fraction, the FT-IR spectra of the optimal sample registered some differences among peaks in the regions 1545, 1746 and 2338–2367 cm^{-1} (Figure 5a). The peaks found at 1200 to 1900 cm^{-1} indicate some functional groups and compounds, such as amides, amino acids, –C=O in aldehydes, C–O in esters, C^O in anhydrides, =O in lactones, t-butyl groups, N–O pyridine groups, esters, lactones, etc. [14]. Changes in the amide II fraction after dry heat treatment can be observed at 1545 cm^{-1}, which is given by the joining of N–H bond vibrations with CN stretching [53]. A prominent peak was observed at 1746 cm^{-1} for the Optimal L sample, suggesting the changes of amide I fractions caused by the heat treatment that determined denaturation of sorghum proteins. The differences in the 2338–2367 cm^{-1} regions could be related to the browning reactions that may occur during heat treatment that could lead to the increase in unsaturated carbonyl groups, the degradation of amino acids to aldehydes and their condensation with carbohydrate fractions, furfurals, and other species to set up chromophores and off-flavors [14].

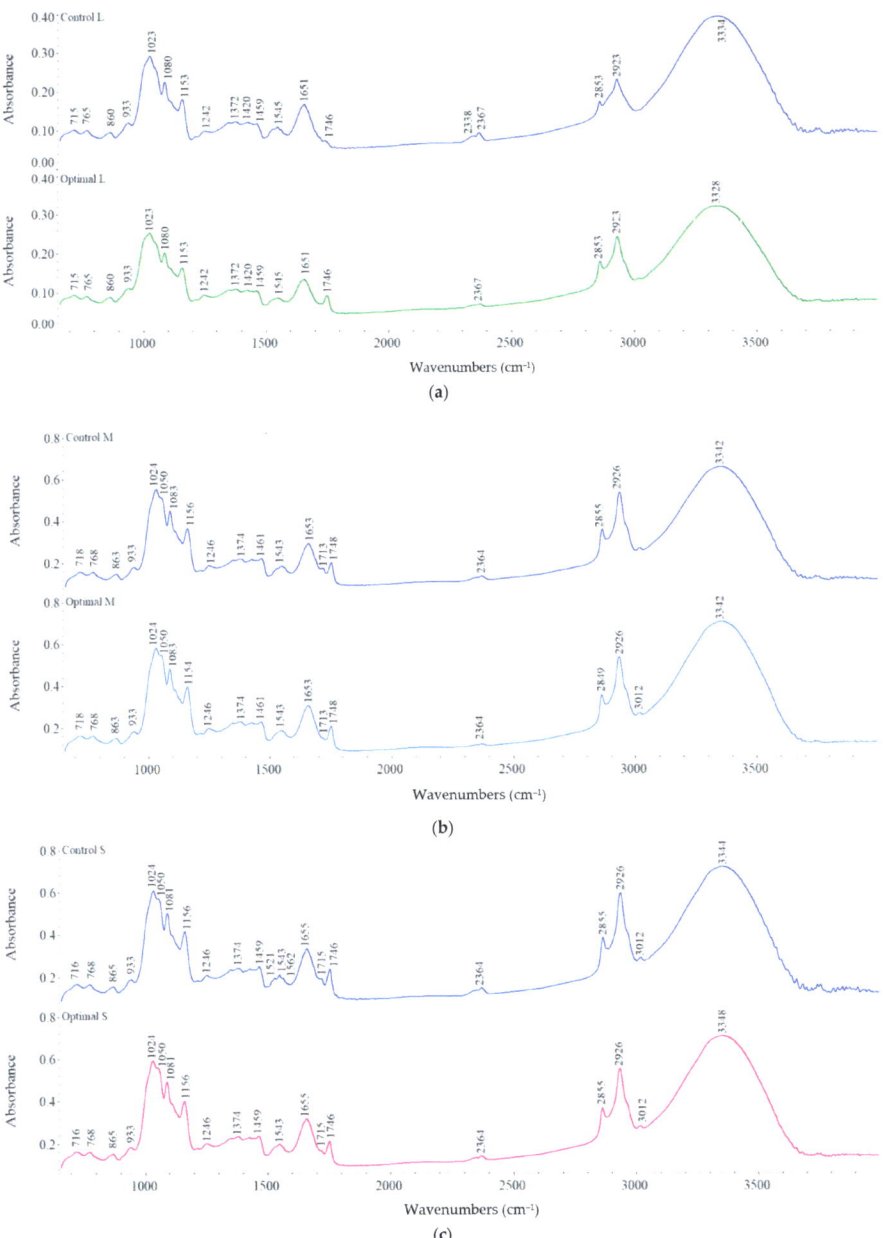

Figure 5. FT-IR spectra of control and optimal sorghum flour with: (**a**) L, (**b**) M, (**c**) S fraction.

For M and S fractions, no significant changes between optimal and control samples were observed regarding peak occurrence (Figure 5b,c). Peak intensities exhibited differences among fractions, the smallest values being obtained for L fraction. The band appeared at around 3328–3342 cm^{-1} suggesting the O–H bond stretching from starch and protein–starch interaction and also from water molecules, alcohol, phenolic compounds, or carboxylic acid, while the modification in peak intensity indicates the degree of formation of hydrogen bonds (inter- and intramolecular) [54]. The lower intensities of peaks at 3328–3342 cm^{-1} for all the optimal samples compared to the corresponding controls could lead to a rise in dehydroxylation reactions upon dry heat treatment [14]. The modification of this band position towards a lower peak intensity suggests an enhancement in the bond strength caused by the interaction within starch molecules and starch with other compounds [55]. The bands observed at about 1156, 1081, and 865 cm^{-1} for all the studied samples suggest the presence of alcohol (C–O), anhydrides (C=O), and alkane (C–H) groups, respectively [54]. The modification in the intensities of these bands in the optimal samples could be related to the deformation of angular C–H bond, the vibration of—1–4 glycosidic bonds (C–O–C) and C–O linkages in the esters (developed between the –COOH and O–H group of protein and starch, respectively), and could have led to the appearance of new functional groups [14,54].

The amide I structure study revealed that α-helix/β sheet ratio was higher in Optimal S than Control S, while an opposite trend was observed for L and M fractions (Table 7). All the studied samples showed a predominant α–helix protein conformation that increased with fraction size decrease and was reduced after heat treatment of sorghum grains. This means that heat treatment promoted the rise in β-sheet formation, which could indicate protein aggregation [56]. Regarding starch structure, the crystalline areas were more abundant compared to the amorphous ones in all samples and increased after dry heat treatment. Higher crystalline/amorphous ratios were obtained for treated samples than controls, except for L fraction (Table 7), indicating the increase in crystalline molecular structures, a similar trend being reported by González et al. [57] for wheat starch and flour treated at temperatures up to 150 °C. The loss of water molecules during treatment seems to have played an essential role in the reorganization and recrystallization of starch chains.

4. Conclusions

Dry heat treatment of sorghum grains and fractionation significantly impacted flour properties in terms of functional and nutritional characteristics. Treatment temperature and fraction dimension rises determined the increase of sorghum flour hydration capacity, emulsifying properties, and carbohydrate content and the decrease of oil-absorption capacity and moisture content. Positive effects were observed for swelling power, fat, and ash contents when temperature increased gradually, while fractionation affected these parameters. The optimal conditions for the dry heat treatment of sorghum grains for each fraction were 121.00 °C for S fraction, 132.11 °C for M fraction, while for L fraction a temperature of 139.47 °C would be indicated. The optimal samples showed lower red and yellow nuances compared to the untreated samples, while fraction dimension increase determined lightness reduction. TPC increased after dry heat treatment, while antiradical activity slightly decreased. Starch digestibility was affected by heat treatment depending on the fraction type, higher amount of resistant starch, total digestible starch, rapid digestible starch, and total starch being obtained. Molecular characteristic analysis revealed that dry heat treatment induced structure changes in proteins and starch. These results may help producers to select the appropriate processing conditions of sorghum grains depending on the final product type in order to obtain the best quality. Further studies regarding the effects of dry heat treatment and fractionation on dough rheology and final product characteristics are needed. The main limitations of this study are related to the use of a single variety of sorghum grains and the temperature range, which was restrained. The main advantages of using dry heat treatment and milling for sorghum processing led to the eco-friendly character of these techniques and the possibility to enhance the nutritional value of the flour by reducing non-nutritious factors. The sorghum flours after processing could be used in the production of gluten-free baked goods, such as bread, biscuits, and cakes, but also for pasta production. The fraction needed should be selected according to the desired characteristics of the final product.

Author Contributions: A.B., M.U.-I. and S.M. contributed equally to the experimental design, data collection and modeling, data interpretation, and writing of the manuscript. All authors have read and agreed to the published version of the manuscript.

Funding: This work received no external funding.

Institutional Review Board Statement: Not applicable.

Informed Consent Statement: Not applicable.

Data Availability Statement: Data are available on request at the corresponding authors.

Acknowledgments: This work was supported by Romania National Council for Higher Education Funding, CNFIS, project number CNFIS-FDI-2022-0259.

Conflicts of Interest: The authors declare no conflict of interest.

References

1. Tasie, M.M.; Gebreyes, B.G. Characterization of Nutritional, Antinutritional, and Mineral Contents of Thirty-Five Sorghum Varieties Grown in Ethiopia. *Int. J. Food Sci.* **2020**, *2020*, 8243617. [CrossRef] [PubMed]
2. World Agricultural Production World Sorghum Production 2021/2022. Available online: http://www.worldagriculturalproduction.com/crops/sorghum.aspx (accessed on 10 May 2022).
3. National Institute of Statistics. *Romanian Statistical Yearbook*; National Institute of Statistics: Bucharest, Romania, 2022.
4. Popescu, A.; Condei, R. Some considerations on the prospects of sorghum crop. *Sci. Pap. Ser. Manag. Econ. Eng. Agric. Rural Dev.* **2014**, *14*, 295–304.
5. Liu, L.; Herald, T.J.; Wang, D.; Wilson, J.D.; Bean, S.R.; Aramouni, F.M. Characterization of sorghum grain and evaluation of sorghum flour in a Chinese egg noodle system. *J. Cereal Sci.* **2012**, *55*, 31–36. [CrossRef]
6. Taylor, J.R.N.; Belton, P.S.; Beta, T.; Duodu, K.G. Increasing the utilisation of sorghum, millets and pseudocereals: Developments in the science of their phenolic phytochemicals, biofortification and protein functionality. *J. Cereal Sci.* **2014**, *59*, 257–275. [CrossRef]
7. Medina Martinez, O.D.; Lopes Toledo, R.C.; Vieira Queiroz, V.A.; Pirozi, M.R.; Duarte Martino, H.S.; Ribeiro de Barros, F.A. Mixed sorghum and quinoa flour improves protein quality and increases antioxidant capacity in vivo. *LWT* **2020**, *129*, 109597. [CrossRef]
8. Zhu, L.; Song, X.; Pan, F.; Tuersuntuoheti, T.; Zheng, F.; Li, Q.; Hu, S.; Zhao, F.; Sun, J.; Sun, B. Interaction mechanism of kafirin with ferulic acid and tetramethyl pyrazine: Multiple spectroscopic and molecular modeling studies. *Food Chem.* **2021**, *363*, 130298. [CrossRef]
9. Mir, S.A.; Manickavasagan, A.; Shah, M.A. *Whole Grains: Processing, Product Development, and Nutritional Aspects*; CRC Press: Boca Raton, FL, USA, 2019; ISBN 1351104756.
10. Kim, E.; Kim, S.; Park, Y. Sorghum extract exerts cholesterol-lowering effects through the regulation of hepatic cholesterol metabolism in hypercholesterolemic mice. *Int. J. Food Sci. Nutr.* **2015**, *66*, 308–313. [CrossRef]
11. Kulamarva, A.G.; Sosle, V.R.; Raghavan, G.S.V. Nutritional and rheological properties of sorghum. *Int. J. Food Prop.* **2009**, *12*, 55–69. [CrossRef]
12. Taylor, J.; Duodu, K.G. *Sorghum and Millets: Chemistry, Technology, and Nutritional Attributes*; Elsevier: Duxford, UK, 2018; ISBN 0128115289.
13. Taylor, J.R.N.; Emmambux, M.N. 13—Products containing other speciality grains: Sorghum, the millets and pseudocereals. In *Technology of Functional Cereal Products*; Hamaker, B.R., Ed.; Woodhead Publishing Series in Food Science, Technology and Nutrition; Woodhead Publishing: Sawston, UK, 2008; pp. 281–335. ISBN 978-1-84569-177-6.
14. Sharanagat, V.S.; Suhag, R.; Anand, P.; Deswal, G.; Kumar, R.; Chaudhary, A.; Singh, L.; Singh Kushwah, O.; Mani, S.; Kumar, Y.; et al. Physico-functional, thermo-pasting and antioxidant properties of microwave roasted sorghum [*Sorghum bicolor* (L.) Moench]. *J. Cereal Sci.* **2019**, *85*, 111–119. [CrossRef]
15. Meera, M.S.; Bhashyam, M.K.; Ali, S.Z. Effect of heat treatment of sorghum grains on storage stability of flour. *LWT Food Sci. Technol.* **2011**, *44*, 2199–2204. [CrossRef]
16. Sharma, P.; Gujral, H.S. Effect of sand roasting and microwave cooking on antioxidant activity of barley. *Food Res. Int.* **2011**, *44*, 235–240. [CrossRef]
17. Schlörmann, W.; Zetzmann, S.; Wiege, B.; Haase, N.U.; Greiling, A.; Lorkowski, S.; Dawczynski, C.; Glei, M. Impact of different roasting conditions on sensory properties and health-related compounds of oat products. *Food Chem.* **2020**, *307*, 125548. [CrossRef] [PubMed]
18. Pradeep, S.R.; Guha, M. Effect of processing methods on the nutraceutical and antioxidant properties of little millet (*Panicum sumatrense*) extracts. *Food Chem.* **2011**, *126*, 1643–1647. [CrossRef] [PubMed]

19. Kim, J.C.; Mullan, B.P.; Pluske, J.R. A comparison of waxy versus non-waxy wheats in diets for weaner pigs: Effects of particle size, enzyme supplementation, and collection day on total tract apparent digestibility and pig performance. *Anim. Feed Sci. Technol.* **2005**, *120*, 51–65. [CrossRef]
20. Mahasukhonthachat, K.; Sopade, P.A.; Gidley, M.J. Kinetics of starch digestion in sorghum as affected by particle size. *J. Food Eng.* **2010**, *96*, 18–28. [CrossRef]
21. Alvarenga, I.C.; Ou, Z.; Thiele, S.; Alavi, S.; Aldrich, C.G. Effects of milling sorghum into fractions on yield, nutrient composition, and their performance in extrusion of dog food. *J. Cereal Sci.* **2018**, *82*, 121–128. [CrossRef]
22. Rumler, R.; Bender, D.; Speranza, S.; Frauenlob, J.; Gamper, L.; Hoek, J.; Jäger, H.; Schönlechner, R. Chemical and physical characterization of sorghum milling fractions and sorghum whole meal flours obtained via stone or roller milling. *Foods* **2021**, *10*, 870. [CrossRef]
23. Bordei, D.; Bahrim, G.; Pâslaru, V.; Gasparotti, C.; Elisei, A.; Banu, I.; Ionescu, L.; Codină, G. Quality Control in the Bakery Industry-Analysis Methods. *Galați Acad.* **2007**, *1*, 203–212.
24. Elkhalifa, A.E.O.; Bernhardt, R. Combination Effect of Germination and Fermentation on Functional Properties of Sorghum Flour. *Curr. J. Appl. Sci. Technol.* **2018**, *30*, 1–12. [CrossRef]
25. Zhu, F.; Du, B.; Xu, B. Super fine grinding improves functional properties and antioxidant capacities of bran dietary fibre from Qingke (hull-less barley) grown in Qinghai-Tibet Plateau, China. *J. Cereal Sci.* **2015**, *65*, 43–47. [CrossRef]
26. FAO. Chapter 4: Summary—Integration of Analytical Methods and Food Energy Conversion Factors. Available online: https://www.fao.org/3/y5022e/y5022e05.htm#bm5 (accessed on 7 April 2022).
27. Singleton, V.L.; Rossi, J.A. Colorimetry of total phenolics with phosphomolybdic-phosphotungstic acid reagents. *Am. J. Enol. Vitic.* **1965**, *16*, 144–158.
28. Gao, X.; Tong, J.; Guo, L.; Yu, L.; Li, S.; Yang, B.; Wang, L.; Liu, Y.; Li, F.; Guo, J.; et al. Influence of gluten and starch granules interactions on dough mixing properties in wheat (*Triticum aestivum* L.). *Food Hydrocoll.* **2020**, *106*, 105885. [CrossRef]
29. Iuga, M.; Mironeasa, S. Application of heat moisture treatment in wheat pasta production. *Food Control* **2021**, *128*, 108176. [CrossRef]
30. Jacobs, P.J.; Hemdane, S.; Dornez, E.; Delcour, J.A.; Courtin, C.M. Study of hydration properties of wheat bran as a function of particle size. *Food Chem.* **2015**, *179*, 296–304. [CrossRef]
31. Almaiman, S.A.; Albadr, N.A.; Alsulaim, S.; Alhuthayli, H.F.; Osman, M.A.; Hassan, A.B. Effects of microwave heat treatment on fungal growth, functional properties, total phenolic content, and antioxidant activity of sorghum (*Sorghum bicolor* L.) grain. *Food Chem.* **2021**, *348*, 128979. [CrossRef]
32. Sharma, P.; Gujral, H.S. Extrusion of Hulled Barley Affecting β-Glucan and Properties of Extrudates. *Food Bioprocess Technol.* **2013**, *6*, 1374–1389. [CrossRef]
33. Wani, I.A.; Sogi, D.S.; Shivhare, U.S.; Gill, B.S. Physico-chemical and functional properties of native and hydrolyzed kidney bean (*Phaseolus vulgaris* L.) protein isolates. *Food Res. Int.* **2015**, *76*, 11–18. [CrossRef]
34. Zou, J.; Xu, M.; Tang, W.; Wen, L.; Yang, B. Modification of structural, physicochemical and digestive properties of normal maize starch by thermal treatment. *Food Chem.* **2020**, *309*, 125733. [CrossRef]
35. Caprita, R.; Caprita, A.; Cretescu, I. Effect of Heat Treatment and Digestive Enzymes on Cereal Water-Retention Capacity. *Sci. Pap. Anim. Sci. Biotechnol.* **2015**, *48*, 94–96.
36. Hassan, A.B.; von Hoersten, D.; Mohamed Ahmed, I.A. Effect of radio frequency heat treatment on protein profile and functional properties of maize grain. *Food Chem.* **2019**, *271*, 142–147. [CrossRef]
37. Moraes, É.A.; Marineli, R.D.S.; Lenquiste, S.A.; Steel, C.J.; De Menezes, C.B.; Queiroz, V.A.V.; Maróstica Júnior, M.R. Sorghum flour fractions: Correlations among polysaccharides, phenolic compounds, antioxidant activity and glycemic index. *Food Chem.* **2015**, *180*, 116–123. [CrossRef] [PubMed]
38. Xiong, Y.; Zhang, P.; Warner, R.D.; Fang, Z. Sorghum Grain: From Genotype, Nutrition, and Phenolic Profile to Its Health Benefits and Food Applications. *Compr. Rev. Food Sci. Food Saf.* **2019**, *18*, 2025–2046. [CrossRef] [PubMed]
39. Mahajan, H.; Gupta, M. Nutritional, functional and rheological properties of processed sorghum and ragi grains. *Cogent Food Agric.* **2015**, *1*, 1109495. [CrossRef]
40. Dharmaraj, U.; Malleshi, N.G. Changes in carbohydrates, proteins and lipids of finger millet after hydrothermal processing. *LWT Food Sci. Technol.* **2011**, *44*, 1636–1642. [CrossRef]
41. Lucas-González, R.; Viuda-Martos, M.; Pérez-Álvarez, J.Á.; Fernández-López, J. Evaluation of Particle Size Influence on Proximate Composition, Physicochemical, Techno-Functional and Physio-Functional Properties of Flours Obtained from Persimmon (*Diospyros kaki* Trumb.) Coproducts. *Plant Foods Hum. Nutr.* **2017**, *72*, 67–73. [CrossRef] [PubMed]
42. Joshi, A.U.; Liu, C.; Sathe, S.K. Functional properties of select seed flours. *LWT* **2015**, *60*, 325–331. [CrossRef]
43. Wu, L.; Huang, Z.; Qin, P.; Ren, G. Effects of processing on phytochemical profiles and biological activities for production of sorghum tea. *Food Res. Int.* **2013**, *53*, 678–685. [CrossRef]
44. Li, Z.; Zhao, X.; Zhang, X.; Liu, H. The Effects of Processing on Bioactive Compounds and Biological Activities of Sorghum Grains. *Molecules* **2022**, *27*, 3286. [CrossRef] [PubMed]
45. Woo, K.S.; Seo, M.C.; Kang, J.R.; Ko, J.Y.; Song, S.B.; Lee, J.S.; Oh, B.G.; Do Park, G.; Lee, Y.H.; Nam, M.H.; et al. Antioxidant compounds and antioxidant activities of the methanolic extracts from milling fractions of sorghum (*Sorghum Bicolor* L. Moench). *J. Korean Soc. Food Sci. Nutr.* **2010**, *39*, 1695–1699. [CrossRef]

46. Cardoso, L.d.M.; Pinheiro, S.S.; de Carvalho, C.W.P.; Queiroz, V.A.V.; de Menezes, C.B.; Moreira, A.V.B.; de Barros, F.A.R.; Awika, J.M.; Martino, H.S.D.; Pinheiro-Sant'Ana, H.M. Phenolic compounds profile in sorghum processed by extrusion cooking and dry heat in a conventional oven. *J. Cereal Sci.* **2015**, *65*, 220–226. [CrossRef]
47. Silva, W.M.F.; Biduski, B.; Lima, K.O.; Pinto, V.Z.; Hoffmann, J.F.; Vanier, N.L.; Dias, A.R.G. Starch digestibility and molecular weight distribution of proteins in rice grains subjected to heat-moisture treatment. *Food Chem.* **2017**, *219*, 260–267. [CrossRef] [PubMed]
48. Liu, B.; Zhong, F.; Yokoyama, W.; Huang, D.; Zhu, S.; Li, Y. Interactions in starch co-gelatinized with phenolic compound systems: Effect of complexity of phenolic compounds and amylose content of starch. *Carbohydr. Polym.* **2020**, *247*, 116667. [CrossRef] [PubMed]
49. Kanagaraj, S.P.; Ponnambalam, D.; Antony, U. Effect of dry heat treatment on the development of resistant starch in rice (Oryza sativa) and barnyard millet (Echinochloa furmantacea). *J. Food Process. Preserv.* **2019**, *43*, e13965. [CrossRef]
50. Adebowale, K.O.; Olu-Owolabi, B.I.; Olayinka, O.O.; Lawal, O.S. Effect of heat moisture treatment and annealing on physicochemical properties of red sorghum starch. *African J. Biotechnol.* **2005**, *4*, 928–933. [CrossRef]
51. Winger, M.; Khouryieh, H.; Aramouni, F.; Herald, T.; Al, M.W.E.T. Sorghum Flour characterization and evaluation in gluten free flour tortilla. *J. Food Qual.* **2014**, *37*, 95–106. [CrossRef]
52. Ratnavathi, C.; Patil, L.V.; Chavan, U. *Sorghum Biochemistry: An Industrial Perspective*; Academic Press: Oxford, UK, 2016; ISBN 9780415475976.
53. Kamble, D.B.; Singh, R.; Rani, S.; Kaur, B.P.; Upadhyay, A.; Kumar, N. Optimization and characterization of antioxidant potential, in vitro protein digestion and structural attributes of microwave processed multigrain pasta. *J. Food Process. Preserv.* **2019**, *43*, e14125. [CrossRef]
54. Navyashree, N.; Singh Sengar, A.; Sunil, C.K.; Venkatachalapathy, N. White Finger Millet (KMR-340): A comparative study to determine the effect of processing and their characterisation. *Food Chem.* **2022**, *374*, 131665. [CrossRef]
55. Sun, Q.; Dai, L.; Nan, C.; Xiong, L. Effect of heat moisture treatment on physicochemical and morphological properties of wheat starch and xylitol mixture. *Food Chem.* **2014**, *143*, 54–59. [CrossRef]
56. Oladiran, D.A.; Emmambux, N.M. Nutritional and Functional Properties of Extruded Cassava-Soy Composite with Grape Pomace. *Starch* **2018**, *70*, 1700298. [CrossRef]
57. González, M.; Vernon-Carter, E.J.; Alvarez-Ramirez, J.; Carrera-Tarela, Y. Effects of dry heat treatment temperature on the structure of wheat flour and starch in vitro digestibility of bread. *Int. J. Biol. Macromol.* **2021**, *166*, 1439–1447. [CrossRef]

Article

Response-Surface-Methodology-Based Optimization of High-Quality *Salvia hispanica* L. Seed Oil Extraction: A Pilot Study

Manee Saelee [1], Bhagavathi Sundaram Sivamaruthi [2,3], Periyanaina Kesika [2,3], Sartjin Peerajan [4], Chawin Tansrisook [2], Chaiyavat Chaiyasut [2,*] and Phakkharawat Sittiprapaporn [1,*]

[1] Neuropsychological Research Laboratory, Neuroscience Research Center, School of Anti-Aging and Regenerative Medicine, Mae Fah Luang University, Bangkok 10110, Thailand; maneenana17@gmail.com
[2] Innovation Center for Holistic Health, Nutraceuticals, and Cosmeceuticals, Faculty of Pharmacy, Chiang Mai University, Chiang Mai 50200, Thailand; sivamaruthi.b@cmu.ac.th (B.S.S.); kesika.p@cmu.ac.th (P.K.)
[3] Office of Research Administration, Chiang Mai University, Chiang Mai 50200, Thailand
[4] Health Innovation Institute, Chiang Mai 50200, Thailand; s.peerajan@gmail.com
* Correspondence: chaiyavat@gmail.com (C.C.); wichian.sit@mfu.ac.th (P.S.)

Featured Application: The present results provide the optimal conditions (size of the restriction die, pressing temperature, and duration of dry heat exposure) to extract chia seed oil via the screw press method.

Abstract: Chia seeds play an important role in human health and nutrition since they contain dietary fiber, lipids, protein, polyphenolic compounds, and polyunsaturated fatty acids. The present study aimed to evaluate the yield and quality of chia seed oil (extracted using the screw press method) in terms of total phenolic content, acid, and peroxide levels. A central composite design was used to optimize the extraction procedure, and the response surface methodology was used to assess the results. The restriction die size of 1 cm, pressing temperature of 53 °C, and no dry heat were the optimal conditions for extracting the desired quality of chia seed oil according to the predicted response surface methodology model. The conditions were evaluated and a 29.47% yield was achieved, with a TPC of 2.20 µg GAE/g of oil, acid content of 0.96 mg KOH/g of oil, and peroxide content of 2.87 mEq/Kg of oil. The extraction process exceeded 45.10 min. Antioxidant activities of 19.21 µg TE/g of oil (ABTS radical scavenging activity), 5.69 µg TE/g of oil (DPPH radical scavenging activity), and 186.68 µg CE/g of oil (nitric oxide free radical scavenging activity) were observed. The fatty acid composition of the chia seed oil samples is also reported herein. We report the optimal conditions for extracting oil from local cultivar chia seeds, thus helping to analyze changes in the composition and impact due to geographical differences in oil quality. The extracted chia seed oil could be utilized for functional foods, cosmetics, and pharmaceutical applications.

Keywords: chia seed oil; response surface methodology; phenolic content; antioxidants; fatty acids

Citation: Saelee, M.; Sivamaruthi, B.S.; Kesika, P.; Peerajan, S.; Tansrisook, C.; Chaiyasut, C.; Sittiprapaporn, P. Response-Surface-Methodology-Based Optimization of High-Quality *Salvia hispanica* L. Seed Oil Extraction: A Pilot Study. *Appl. Sci.* **2023**, *13*, 6600. https://doi.org/10.3390/app13116600

Academic Editors: Georgiana Gabriela Codină and Adriana Dabija

Received: 30 March 2023
Revised: 26 May 2023
Accepted: 27 May 2023
Published: 29 May 2023

Copyright: © 2023 by the authors. Licensee MDPI, Basel, Switzerland. This article is an open access article distributed under the terms and conditions of the Creative Commons Attribution (CC BY) license (https://creativecommons.org/licenses/by/4.0/).

1. Introduction

Salvia hispanica L., popularly known as chia, is a member of the Lamiaceae family [1]. Chia is an annual plant widely cultivated because of its culinary and medicinal uses [2]. This plant can be grown in various climates, from tropical to subtropical [3]. The chia plant can reach a height of 1 m and has oppositely arranged leaves with small white or purple hermaphrodites. The oval, smooth, and shiny seeds can be gray, black, black spotted, or white, with sizes ranging from 1 to 2 mm [4]. When soaked in water, chia seeds become gelatinous and absorb much water [5].

An early study reported the composition of chia seeds grown in several geographical locations under different climatic conditions [6]. Chia seeds have high antioxidant potential [7] and are widely recognized for their high amounts of alpha-linolenic acid (ALA), dietary fiber, minerals, omega 3 (n − 3), proteins, phytochemicals such as phenolic compounds [8] (such as caffeic acid, daidzin, gallic acid, protocatechuic ethyl ester, and rosmarinic acid) [7], and vitamins [8]. Quercetin and kaempferol are the main substances in hydrolyzed and crude chia seed extracts, but caffeic and chlorogenic acids are only found in trace amounts [9]. Fatty acids, including alpha-linolenic acid, linoleic acid (L), oleic acid, palmitic acid, and stearic acid, are predominantly observed in chia seed oil (CSO) [3,6].

CSO is extracted using various methods, resulting in varying quality of the oil in terms of fatty acid content, antioxidant activity, and functional properties. Conventional solvent extraction, the cold-pressing method, the ultrasound-assisted method, and the supercritical fluid extraction are techniques used to extract CSO [10]. The cold-press extraction method is a mechanical extraction technique using a screw press without organic treatment or heat application on the expeller. This technique is a cheap, nontoxic, environmentally friendly, and green method. Some parameters such as barrel temperature, restriction die, screw press speed, and seed moisture content can affect CSO extraction [11].

The biochemical components of chia seeds increase the satiety index, improve serum lipid levels, prevent inflammation and cardiovascular diseases, and decrease the risk of chronic diseases due to chia's antioxidant properties [2,12,13]. Chia has also been studied for its cosmetic benefits. Preliminary findings indicate that chia seed exhibits biological functions in the skin, including maintenance of the stratum corneum epidermal barrier, the prevention of transepidermal water loss, and disruption of melanogenesis in epidermal melanocytes [13].

However, many studies have also been conducted to assess the influencing factors of the quality and quantity of CSO. The results could be more consistent due to the variety of chia seeds and culture location. Currently, chia is popular in Thailand due to its health benefits, but there is no experimental design to extract the oil from local chia seeds. Thus, the purpose of this study was to extract CSO using the screw-pressing method and evaluate the quality of the oil. The response surface methodology (RSM) and central composite design (CCD) were used to optimize the conditions to achieve high-quality CSO.

2. Materials and Methods

2.1. Materials

Chia seeds were purchased from a local market in Chiang Mai province. The following chemicals were used in this study: Folin–Ciocalteu reagent, sulfanilamide, K_2HPO_4, and KH_2PO_4 were purchased from Loba Chemie (Maharashtra, India). Naphthyl ethylenediamine dihydrochloride and Na_2CO_3 were bought from HiMedia (Maharashtra, India) and RCI Labscan (Bangkok, Thailand), respectively. Merck (Darmstadt, Germany) supplied gallic acid, methanol, and phosphoric acid. Acetic acid, hexanes, sodium thiosulfate pentahydrate, NaCl, KI, ethanol, and KOH were purchased from RCI Labscan (Bangkok, Thailand). Lastly, 2,2′-azinobis (3-ethylbenzothiazoline-6-sulfonic acid) di-ammonium salt and 2,2-diphenyl-1-picrylhydrazyl (±)-6-hydroxy-2,5,7,8-tetramethyl were purchased from Sigma-Aldrich (Oakville, Canada).

2.2. Moisture Content, Extraction, and Variables

The moisture content of the chia seeds was analyzed before the experiments were designed. Seeds were incubated at 100 °C for 0, 15, and 30 min [14], and the moisture content was measured using a Moisture Analyzer (Moisture HC103, Mettler Toledo, Switzerland).

Chia seed oil (CSO) was obtained via the screw press extraction method (FEA-101ss-M-H-Tc-2015, Energy Friend Ltd., Chiang Mai, Thailand) [15].

The extraction conditions were optimized through a central composite design (CCD) with three independent variables: the restriction die size, the pressing temperature, and the duration of dry heat exposure at 100 °C. The influence of the variables on CSO extraction was studied through response surface methodology (RSM).

The levels of the duration of dry heat exposure at 100 °C (0, 15, and 30 min), size of the restriction die (1.0, 1.2, and 1.4 cm), and pressing temperature (40, 50, and 60 °C) were chosen as variable factors to attain a high yield and high quality of CSO. An experimental diagram of the chia seed oil extraction is shown in Figure 1. About 600 g of chia seeds were used for the extraction process.

Figure 1. The procedure followed for chia seed oil extraction.

The evaluated responses presented the yield and quality of the oil and time. The antioxidant activities (ABTS, DPPH, and nitric oxide radical scavenging assay) were evaluated by selecting a treatment that affected the total phenolic content (TPC).

2.3. RSM and CCD

To achieve the highest yield and quality of chia seed oil, RSM and CCD were used to optimize the conditions. Design Expert, version 10.0 (Stat-Ease Inc., Minneapolis, MN, USA), was used for the statistical analysis. The objective outcomes were the oil yield and quality. As mentioned previously, the duration of dry heat exposure at 100 °C (0, 15, and 30 min), the size of the restriction die (1.0, 1.2, and 1.4 cm), and the pressing temperature were chosen as the variable factors.

Seventeen tests were carried out, including 14 combinations and 3 center-point replicates (Table 1). The variation in the response values (Y) versus the independent variables was fit into a response surface model and presented (Equation (1)):

$$Y = \beta_0 + \sum \beta_i X_i + \sum \beta_{ii} X_i^2 + \sum_{i \neq j} \beta_{ij} X_i X_j \tag{1}$$

where Y is the dependent response; β_0, β_i, β_{ii}, and β_{ij} are the regression coefficients for the intercept, linear, quadratic, and interaction terms, respectively; and X_i and X_j are independent variables.

Table 1. Details of the standard orders (STDs), size of the restriction die, pressing temperature, and duration of dry heat exposure at 100 °C.

STD	Size of the Restriction Die (cm)	Temperature (°C)	Duration of Dry Heat Exposure at 100 °C (min)
1	1	40	0
2	1.4	40	0
3	1	60	0
4	1.4	60	0
5	1	40	30
6	1.4	40	30
7	1	60	30
8	1.4	60	30
9	1	50	15
10	1.4	50	15
11	1.2	40	15
12	1.2	60	15
13	1.2	50	0
14	1.2	50	30
15	1.2	50	15
16	1.2	50	15
17	1.2	50	15

2.4. Extraction and Determination of Total Phenolic Content (TPC)

The phenolic compounds in CSO were extracted [16]. In brief, 2.5 g of CSO was mixed with 5 mL hexane and 3 mL of methanol/water (60:40, vol/vol) and vortexed for 2 min. The samples were centrifuged at $3500 \times g$ rpm for 10 min, and the methanolic phase was collected. The samples were further extracted twice with a methanol/water solution, and the methanolic phase was collected to determine the TPC [16].

The Folin–Ciocalteu colorimetric technique [17] was used to evaluate the TPC of CSO, and results are expressed as the mg gallic acid equivalent (mg GAE)/g of oil [17].

2.5. Acid and Peroxide Values of CSO

CSO's acid and peroxide values were determined based on US Pharmacopeia 37, as detailed previously [14]. The acid values are expressed as the mg KOH equivalent/g of oil (Equation (2)) [14]:

$$\text{mg KOH/g oil} = V \times 5.61/W \quad (2)$$

where V and W represent the volume of KOH (mL) and weight of oil (g), respectively, and 5.61 is a constant value (equivalent to a mass of 0.1 M KOH).

The peroxide value is expressed in milliequivalents of oxygen per kg of CSO (mEq/Kg) (Equation (3)) [14]:

$$\text{mEq/kg} = 2 \times (A - B)/\text{CSO sample (g)}. \quad (3)$$

where A and B represent the sodium thiosulfate (mL) volume in the test and blank, respectively.

2.6. Antioxidant Capacity of CSO

The free radical scavenging activities of CSO were determined using 1,1-diphenyl-2picrylhydrazyl (DPPH), 2,2-azino-bis-3-ethylbenzothiazoline-6-sulfonic acid (ABTS) [18], and a nitric oxide (NO) radical scavenging assay [19], as detailed in our previous studies.

The DPPH and ABTS radical scavenging activity of CSO is expressed as μg Trolox equivalent antioxidant capacity (TEAC)/g of oil. CSO's nitric oxide scavenging activity is expressed in terms of the μg curcumin equivalent (CE)/g of oil.

2.7. Fatty Acid Content of CSO

As reported in our previous study, the fatty acid content of CSO was analyzed via gas chromatography [14]. The analysis was performed at the Halal Science Center of Chulalongkorn University, Bangkok, Thailand.

2.8. Statistics

The quality evaluation of CSO was performed in duplicate. All values were recorded as the mean ± standard deviation (SD). One-way ANOVA was used to evaluate the differentiation between groups. The differences were considered significant when $p < 0.05$.

3. Results

The results obtained for the extraction of CSO using the screw press method with different conditions such as the size of the restriction die, pressing temperature, and duration of dry heat exposure are shown in Table 2. Outlier trials were excluded from the analysis. The oil yield varied from 23.86% to 30.15%, and the maximum yield was observed in STD 17. However, when both the quality and amount of oil were considered, STD 13 was found to be suitable.

Table 2. The observed and predicted oil yields, acid and peroxide values, total phenolic content, and time.

STD	Yield (%)		Acid Value (mg KOH/g of Oil)		Peroxide Value (mEq/Kg of Oil)		TPC (µg GAE/g of Oil)		Pressing Time (min)	
	Observed	Predicted	Observed	Predicted	Observed	Predicted	Observed	Predicted	Observed	Predicted
1	29.86	29.88	1.02	1.02	3.58	3.54	1.60	1.87	50.00	51.71
2	27.74	27.81	0.85	0.86	3.71	3.47	1.61	1.71	69.00	69.31
3	27.59	27.66	0.84	0.84	3.53	3.35	1.72	1.52	58.00	59.11
4	28.44	28.46	0.91	0.91	4.38	4.26	1.39	1.38	60.00	62.21
5	26.70	26.77	0.85	0.50	3.89	3.96	1.94	1.52	74.00	72.10
6	29.16	29.17	0.91	0.91	3.94	3.87	1.29	1.38	105.00	104.20
7	23.86	23.87	0.85	0.84	4.23	4.50	0.97	1.22	80.00	46.00
8	28.52	28.60	0.80	1.48	6.22	5.83	1.21	1.10	65.00	63.60
9	28.82	28.63	0.80	0.80	5.97	5.79	1.23	2.44	58.00	57.23
10	29.11	28.92	1.06	1.04	5.47	6.55	2.29	2.24	75.00	74.83
11	28.24	28.06	0.85	0.82	8.08	8.71	1.66	1.61	68.00	74.33
12	29.08	28.89	0.90	1.02	4.20	4.47	1.29	1.30	54.00	57.73
13	28.35	28.84	0.88	0.91	3.12	3.64	2.60	2.57	70.00	60.58
14	29.83	30.35	0.88	0.93	4.48	4.46	1.87	2.11	59.00	71.48
15	29.48	29.58	1.18	0.92	6.37	6.15	2.86	2.33	80.00	66.03
16	29.75	29.58	0.93	0.92	6.64	6.15	2.09	2.33	51.00	66.03
17	30.15	29.58	0.95	0.92	7.39	6.15	2.31	2.33	68.00	66.03

3.1. Moisture Content

The moisture content of chia seeds was determined before the variables for RSM were designed. The seeds were dried at 100 °C for different durations. About 4.64 ± 0.17, 4.49 ± 0.18, and 4.01 ± 0.50 % MC was observed in the seeds dried for 0, 15, and 30 min, respectively. There were no significant changes observed in the moisture content of the seeds processed for 0 and 15 min, whereas seeds dried for 30 min showed a significant difference in their moisture content (Table 3).

Table 3. The average moisture content (% MC) of chia seeds.

Duration (min)	Moisture Content (% MC)
0	4.64 ± 0.17 [a]
15	4.49 ± 0.18 [a]
30	4.01 ± 0.50 [b]

[a,b] indicates a significant difference (at $p < 0.05$) between times analyzed using Duncan's multiple range test.

3.2. Yield of CSO

The predicted yield of CSO in the three center-point STDs (size of the restriction die: 1.2 cm; pressing temperature: 50 °C; and duration of dry heat exposure at 100 °C: 15 min) was 29.58%. The observed yield of the three center-point STDs was 29.48, 29.75, and 30.15% (Table 2). The highest predicted yield was observed in STD 14, followed by STD 1. However, the observed maximum yield was obtained in STD 17, followed by STD 1. The changes between STDs 1 and 14 were insignificant. The lowest yield was observed (23.86%) and predicted (23.87%) in STD 7 (size of the restriction die: 1 cm; pressing temperature: 60 °C; and duration of dry heat exposure at 100 °C: 30 min) (Table 2). A reduced cubic model used to evaluate variance in the yield of CSO is presented in Table 4.

Table 4. Analysis of variance for the studied variables.

Response	Models	Model (p-Value)	Lack of Fit (p-Value)	R^2	Adjusted R^2	Predicted R^2	Adequate Precision
Yield (%)	Reduced Cubic	0.0021	0.2967	0.9771	0.9268	0.7595	18.9895
Acid value	2FI	0.0056	0.2351	0.9139	0.8279	0.6874	10.5796
Peroxide value	Reduced Cubic	0.0005	0.2604	0.9094	0.8389	0.7110	11.7801
TPC	Reduced Quadratic	0.0004	0.7247	0.8205	0.7553	0.5608	10.2480
Pressing time	2FI	0.0233	0.9798	0.7814	0.6175	0.5119	8.9976

The reduced cubic model for oil extraction was significant ($p = 0.0021$) with adjusted R^2 and predicted R^2 values of 0.9268 and 0.7595, respectively, and a nonsignificant lack of fit ($p = 0.2967$). These results indicated that the reduced cubic model equation was appropriate for the prediction of oil extraction.

The CCD-generated reduced cubic model equation for the yield of CSO (%) was as follows (Equation (4)):

$$\text{Yield} = \frac{1}{\sqrt[0.37]{\begin{array}{c}(-0.1545 + 1.1648A - 0.0031B + 0.0144C - 0.0134AB - 0.0235AC \\ + 7.6562 \times 10 - 6BC - 0.5828A^2 + 0.0002B^2 + 0.0104A^2B + 0.0092A^2C - 0.0001AB^2)\end{array}}} \quad (4)$$

where A is the size of the restriction die, B is the pressing temperature, and C is the duration of dry heat exposure.

The size of the restriction die and temperature ($p = 0.0059$), size of the restriction die and dry heat exposure time ($p = 0.0008$), size of the restriction die^2 ($p = 0.0265$), temperature^2 ($p = 0.0076$), size of the restriction die^2 and temperature ($p = 0.0244$), and size of the restriction die^2 and dry heat exposure time ($p = 0.0082$) significantly affected the yield of CSO (Table 5). However, the restriction die, temperature, dry heat exposure time, and interaction between temperature and dry heat exposure time did not significantly affect the yield of CSO (Figure 2).

Table 5. Estimated coefficients of coded factors for CSO extraction and other desirable factors.

Variable	Yield	p-Value	Acid Value	p-Value	Peroxide Value	p-Value	TPC	p-Value	Time	p-Value
A	−0.0005	0.6673	0.1187	0.0007	−0.0098	0.1463	−0.0102	0.3742	8.7998	0.0258
B	−0.0016	0.2378	0.0968	0.0032	0.0528	0.0039	−0.0233	0.0465	−8.2998	0.0328
C	−0.0027	0.0680	0.0121	0.3457	−0.0181	0.0161	−0.0232	0.0470	5.4478	0.1623
AB	−0.0027	0.0059	0.0573	0.0058	−0.0117	0.1223			−3.6253	0.3600
AC	−0.0042	0.0008	0.1400	0.0008					3.6247	0.3601
BC	0.0011	0.1056	0.1297	0.0012	−0.0081	0.2680			−8.3747	0.0552
A^2	0.0029	0.0265								
B^2	0.0041	0.0076					−0.1082	<0.0001		
C^2					0.0712	<0.0001				
A^2B	0.0041	0.0244								
A^2C	0.0055	0.0082								
AB^2	−0.0026	0.1070								
BC^2					−0.0676	0.0017				

A: size of the restriction die; B: pressing temperature; C: dry heat exposure time; TPC: total phenolic content.

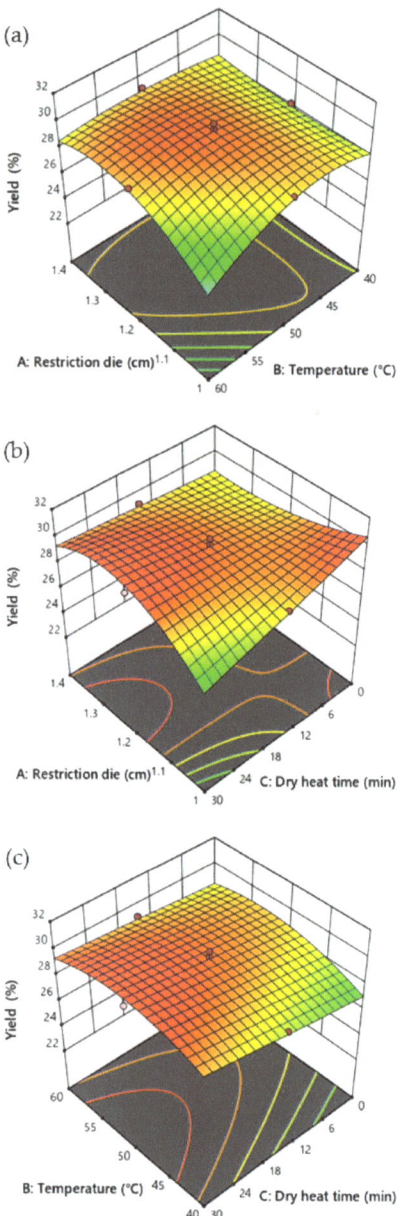

Figure 2. Response surface plot illustrating the influence of restriction die size, temperature, and duration of dry heat time on the yield of CSO. (**a**) Influence of restriction die and temperature (**b**) restriction die and dry heat time, and (**c**) temperature and dry heat time on the yield of the CSO.

3.3. Total Phenolic Content of CSO

The predicted TPC of CSO in the three center-point STDs (size of the restriction die: 1.2 cm; temperature: 50 °C; and duration of dry heat exposure at 100 °C: 15 min) was 2.33 µg GAE/g of oil. The observed TPC of CSO center-point STDs was 2.86, 2.09, and 2.31 µg GAE/g of oil (Table 2). The maximum TPC was recorded in the predicted (2.57 µg GAE/g of oil) and observed values (2.60 µg GAE/g of oil) for STD 13 (size of the restriction die: 1.2 cm; temperature: 50 °C; and duration of dry heat exposure at 100 °C: 0 min) (Table 2). The reduced quadratic model used to evaluate variances in the TPC of CSO is presented in Table 4.

The reduced quadratic model for the TPC of CSO was significant ($p = 0.0004$) with adjusted R^2 and predicted R^2 values of 0.7553 and 0.5608, respectively, and a nonsignificant lack of fit ($p = 0.7247$). The results indicated that the reduced quadratic model equation was suitable for predicting the TPC of CSO.

The CCD-generated reduced quadratic model equation used for the TPC of CSO (μg GAE/g of oil) was as follows (Equation (5)):

$$\text{Total phenolic acid} = \frac{1}{\sqrt[-0.2]{(-1.3199 - 0.0512A + 0.1060B - 0.0015C - 0.0011B2)}} \quad (5)$$

where A is the size of the restriction die, B is the pressing temperature, C is the duration of the dry heat exposure.

The temperature ($p = 0.0465$), dry heat exposure time ($p = 0.0470$), and temperature^2 ($p < 0.0001$) significantly affected the TPC of CSO (Table 5). An increase in the duration of dry heat exposure reduced the TPC of CSO. The low (40 °C) and high (60 °C) temperatures affected the TPC of CSO. High TPC was observed in the samples treated at 50 °C. The size of the restriction die did not influence the TPC of CSO (Figure 3).

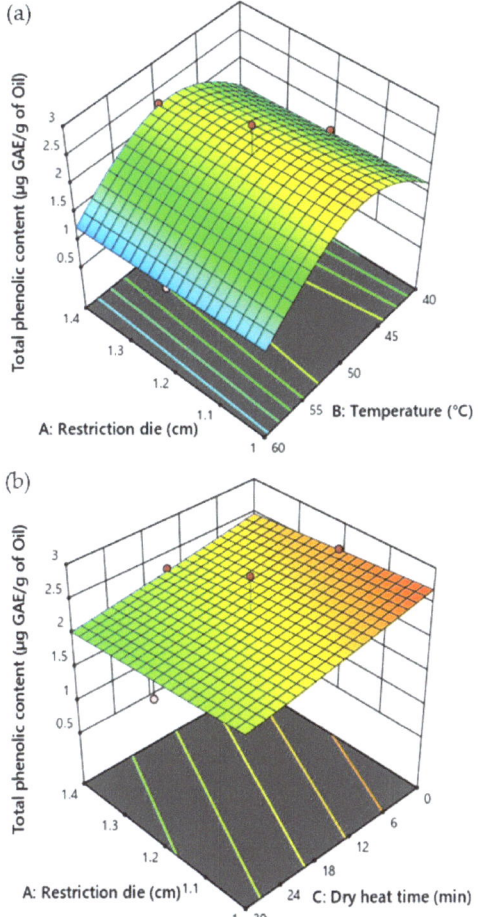

Figure 3. *Cont.*

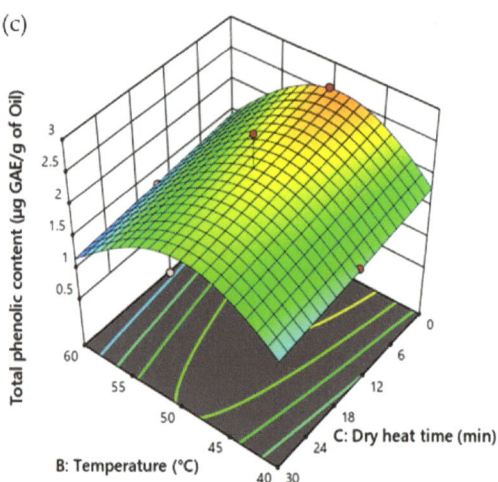

Figure 3. Response surface plot illustrating the influence of restriction die size, temperature, and duration of dry heat time on the total phenolic content of CSO. (**a**) Influence of restriction die and temperature (**b**) restriction die and dry heat time, and (**c**) temperature and dry heat time on the total phenolic content of the CSO.

3.4. Acidity of CSO

The predicted acid value of CSO in the three center-point STDs was 0.92 mg KOH/g of oil. The observed acid values of CSO center-point STDs were 1.18, 0.93, and 0.95 mg KOH/g of oil (Table 2). The smallest acid values of CSO were predicted (0.50 mg KOH/g of oil) and observed (0.85 mg KOH/g of oil) in STD 5 (size of the restriction die: 1 cm; temperature: 40 °C; and duration of dry heat exposure at 100 °C: 30 min), STD 8 (size of the restriction die: 1.4 cm; temperature: 60 °C; and duration of dry heat exposure at 100 °C: 30 min), and STD 9 (size of the restriction die: 1 cm; temperature: 50 °C; and duration of dry heat exposure at 100 °C: 15 min) (Table 2). The 2FI model was employed to evaluate variances in the acid value of CSO (Table 4).

The 2FI model for the acid value of CSO was significant ($p = 0.0056$) with adjusted R^2 and predicted R^2 values of 0.8279 and 0.6874, respectively, and a nonsignificant lack of fit ($p = 0.2351$). The results showed that the 2FI model equation was suitable for predicting the acid value of CSO.

The CCD-generated 2FI model equation for the acid value of CSO (mg KOH/g of oil) was as follows (Equation (6)):

$$\text{Acid value} = 2.9194 - 1.5386A - 0.0377B - 0.0984C + 0.0286AB + 0.0467AC + 0.0009BC \quad (6)$$

where A is the size of the restriction die, B is the pressing temperature, and C is the duration of dry heat exposure.

The size of the restriction die ($p = 0.0007$), temperature ($p = 0.0032$), size of the restriction die and temperature ($p = 0.0058$), size of the restriction die and dry heat exposure time ($p = 0.0008$), and temperature and dry heat exposure time ($p = 0.0012$) significantly affected the acid value of CSO (Table 5). RSM prediction indicated that a lower temperature and reduced size of the restriction die could produce CSO with a lower acid value. All factors greatly influenced the acid value, while dry heat exposure time was not an influencing factor for the acid value (Figure 4).

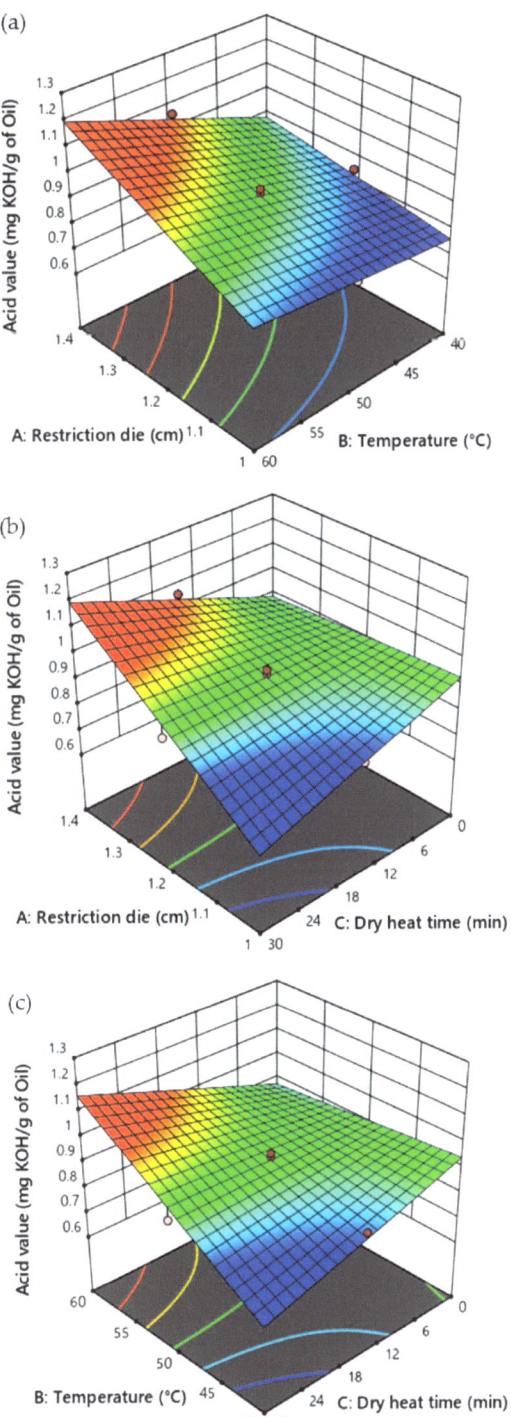

Figure 4. Response surface plot illustrating the influence of restriction die size, temperature, and duration of dry heat time on the acid value of CSO. (**a**) Influence of restriction die and temperature (**b**) restriction die and dry heat time, and (**c**) temperature and dry heat time on the acid value of the CSO.

3.5. Peroxide Value of CSO

The predicted peroxide value of CSO in the three center-point STDs was 6.15 mEq/Kg of oil. The observed peroxide values of CSO center-point STDs were 6.37, 6.64, and 7.39 mEq/Kg of oil (Table 2). The smallest peroxide value of CSO was predicted (3.35 mEq/Kg of oil) and observed (3.53 mEq/Kg of oil) in STD 3 (size of the restriction die: 1 cm; temperature: 60 °C; and duration of dry heat exposure at 100 °C: 0 min) and STD 13 (size of the restriction die: 1.2 cm; temperature: 50 °C; and duration of dry heat exposure at 100 °C: 0 min) (Table 2). The reduced cubic model was employed to evaluate variances in the peroxide value of CSO (Table 4).

The reduced cubic model for the peroxide value of CSO was significant ($p = 0.0005$) with adjusted R^2 and predicted R^2 values of 0.8389 and 0.7110, respectively, and a nonsignificant lack of fit ($p = 0.2604$). The results revealed that the reduced cubic model equation was suitable for predicting the peroxide value of CSO.

The CCD-generated reduced cubic model equation for the acid value of CSO (mEq/Kg of oil) was as follows (Equation (7)):

$$\text{Peroxide value} = \frac{1}{\sqrt[0.25]{(0.4650 + 0.2438A + 0.0063B - 0.0530C - 0.0059AB + 0.0008BC + 0.0018C^2 - 3.0026 \times 10 - 5BC^2)}} \quad (7)$$

where A is the size of the restriction die, B is the pressing temperature, and C is the duration of dry heat exposure.

The temperature ($p = 0.0039$), dry heat exposure time ($p = 0.0161$), dry heat exposure time^2 ($p < 0.0001$), and temperature and dry heat exposure time^2 ($p < 0.0017$) significantly influenced the peroxide value of CSO (Table 5). RSM prediction indicated that a high temperature could produce CSO with a lower peroxide value. Additionally, the high (30 min) and low (0 min) durations of dry heat exposure presented lower peroxide values. However, the size of the restriction die did not significantly affect the peroxide values (Figure 5).

3.6. Pressing Time

The pressing time for CSO extraction in the three center-point STDs was 66.03 min. The observed pressing times for CSO extraction in center-point STDs were 80.00, 51.00, and 68.00 min (Table 2). The lowest pressing times for CSO extraction were predicted (46 min) and observed (80.00 min) in STD 7 (size of the restriction die: 1 cm; temperature: 60 °C; and duration of dry heat exposure at 100 °C: 30 min) and STD 1 (size of the restriction die: 1 cm; temperature: 40 °C; and duration of dry heat exposure at 100 °C: 0 min) (Table 2). The 2FI model was utilized to evaluate variances in the pressing time for CSO extraction (Table 4).

The 2FI model for the pressing time of CSO extraction was significant ($p = 0.0233$) with adjusted R^2 and predicted R^2 values of 0.6175 and 0.5119, respectively, and a nonsignificant lack of fit ($p = 0.9798$). The results revealed that the 2FI model equation was suitable for predicting the pressing time of CSO extraction.

The CCD-generated 2FI model equation for the pressing time of CSO extraction (min) was as follows (Equation (8)):

$$\text{Time} = -79.6031 + 116.5072A + 2.1827B + 1.7049C - 1.8126AB + 1.2082AC - 0.0558BC \quad (8)$$

where A is the size of the restriction die, B is the pressing temperature, and C is the duration of dry heat exposure.

The size of the restriction die ($p = 0.0258$) and temperature ($p = 0.0328$) influenced the extraction time of CSO significantly (Table 5). RSM prediction showed that the reduced size of the restriction die and high temperature could reduce the pressing time needed for CSO extraction. However, the duration of dry heat exposure time did not affect the pressing time (Figure 6).

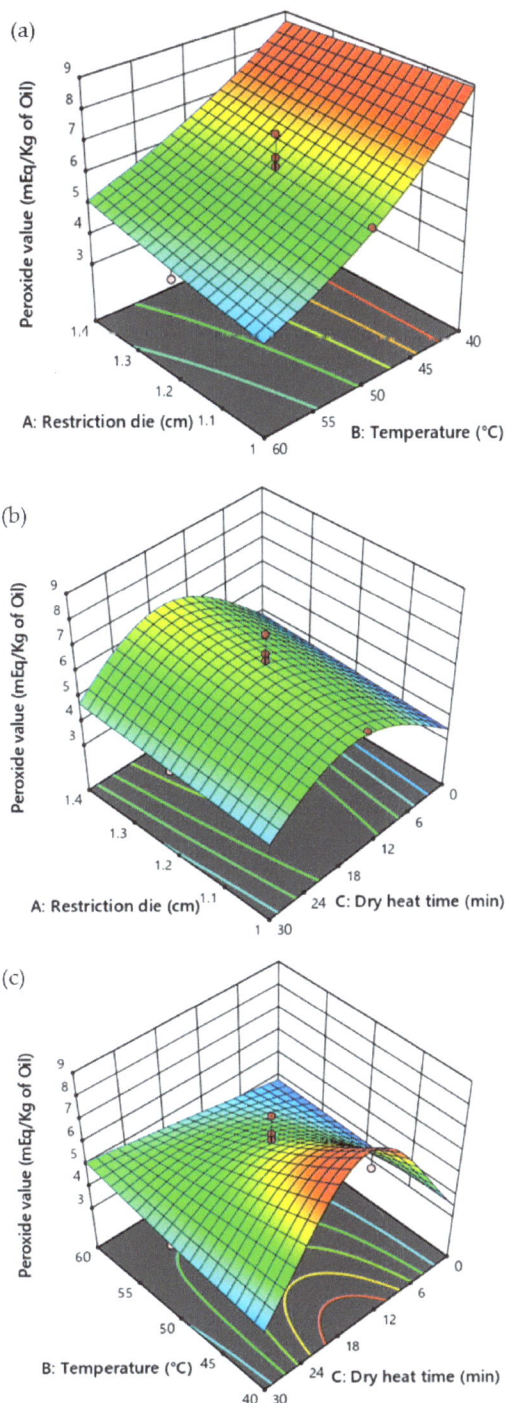

Figure 5. Response surface plot illustrating the influence of restriction die size, temperature, and duration of dry heat time on the peroxide value of CSO. (**a**) Influence of restriction die and temperature (**b**) restriction die and dry heat time, and (**c**) temperature and dry heat time on the peroxide value of the CSO.

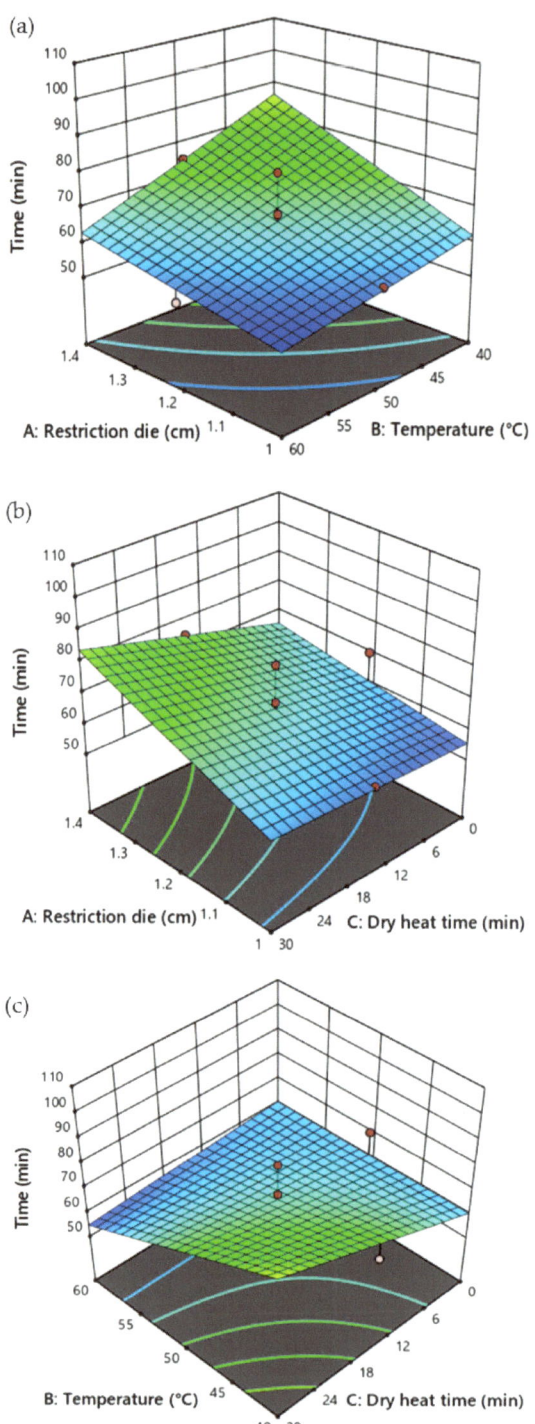

Figure 6. Response surface plot illustrating the influence of restriction die size, temperature, and duration of dry heat time on the pressing time for CSO extraction. (**a**) Influence of restriction die and temperature (**b**) restriction die and dry heat time, and (**c**) temperature and dry heat time on the pressing time for CSO extraction.

3.7. Fatty Acid Composition

CSO extracted under optimal conditions (based on the design-expert-recommended conditions for the desirable traits) and STD 13 (based on the observed desirable traits) were selected for the fatty acid analysis. The fatty acid composition of the representative CSO samples was reported (Table 6). The order of fatty acid abundance in CSO observed in the study was as follows: α-linolenic acid > linoleic acid > palmitic acid > oleic acid > stearic acid > eicosenoic acid > palmitoleic acid > myristic acid. About 1.49% and 1.54% of unidentified peaks were observed in the CSO extracted under optimal conditions and STD 13, respectively. Other fatty acids were not detected.

Table 6. The fatty acid content of representative CSO samples.

Fatty Acid	Fatty Acid Content (%)	
	Oil Extracted under the Optimal Condition *	Oil Extracted under the Condition of STD 13 **
α-Linolenic acid (C18:3 n − 3)	58.62	58.46
Linoleic acid (C18:2 n − 6 cis)	20.81	20.93
Palmitic acid (C16:0)	7.84	7.73
Oleic acid (C18:1 n − 9 cis)	7.49	7.52
Stearic acid (C18:0)	3.55	3.64
Eicosenoic acid (C20:1 n − 9)	0.13	0.10
Palmitoleic acid (C16:1)	0.04	0.04
Myristic acid (C14:0)	0.03	0.03

* Optimal condition: size of the restriction die = 1 cm, temperature = 53 °C, and no dry heat. ** STD 13: restriction die size = 1.2 cm, temperature = 50 °C, and no dry heat.

3.8. Antioxidant Capacity of CSO

The treatments that had an impact on the TPC were selected to evaluate their antioxidant capacity. STD 13 (size of the restriction die: 1.2 cm; temperature: 50 °C; and duration of dry heat time: 0 min) presented the greatest ABTS and DPPH scavenging activities. The maximum value of 19.21 µg TE/g of oil was observed in the ABTS assay, and the values were significantly different from those of other STDs. The smallest value of 8.18 µg TE/g of oil was observed in STD 5 (size of the restriction die: 1.0 cm; temperature: 40 °C; and duration of dry heat time: 30 min) (Figure 7A). In the DPPH assay, the highest (STD 13) and lowest (STD 4) values of 5.69 ± 0.21 and 2.66 µg TE/g of oil, respectively, were detected (Figure 7B). The maximum NO scavenging activity of 186.68 µg CE/g of oil was observed in STD 7 (size of the restriction die: 1 cm; temperature: 60 °C; and duration of dry heat time: 30 min). The NO scavenging activity of the CSO samples of STD 7 did not differ significantly from that of the other STDs (Figure 7C).

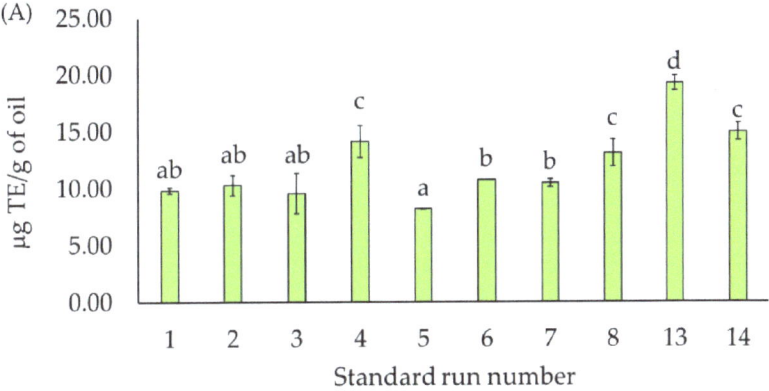

Figure 7. Cont.

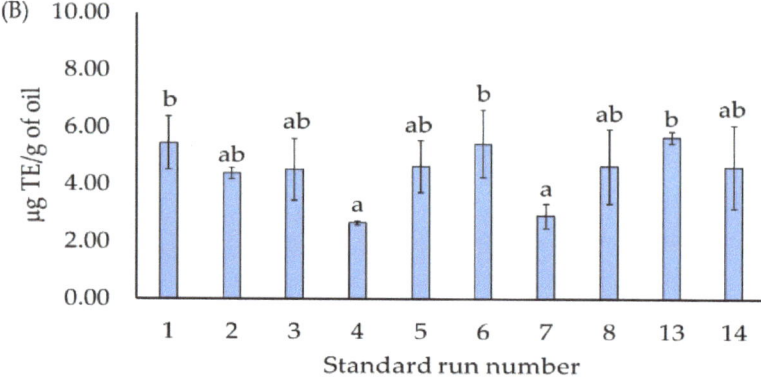

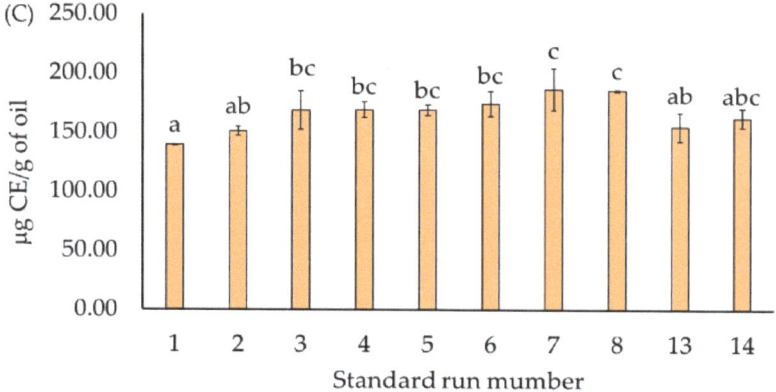

Figure 7. Antioxidant capacity of CSO: (**A**) ABTS, (**B**) DPPH, and (**C**) NO radical scavenging assays. a–d: Significant differences ($p < 0.05$) between the standard runs.

4. Discussion

Previous studies reported the influence of several factors in CSO extraction and its quality. Fernandes et al. (2012) reported the influence of different methods on the yield of CSO. The cold solvent, pressing, and supercritical CO_2 methods yielded 19.28 ± 0.74, 20.01 ± 0.95, and 24.64% of oil, respectively [20]. Similarly, the effects of different extraction methods (solvent extraction, Soxhlet extraction, and screw-pressing methods) were reported. The results showed that solvent extraction was the optimal method for high-quality oil extraction [21]. Maximum (92.8%) oil yield was obtained from Mexican chia seed using the supercritical CO_2 method (with a pressure of 450 bar and extraction time of 300 min) [22]. Uribe et al. (2011) reported that high oil yield was obtained from chia seeds via the supercritical CO_2 method (at a pressure of 408 bar and temperature of 80 °C). The study stated that pressure had a greater influence on oil yield [23]. The type of solvents and solvent and seed ratio affected the yield of CSO [24].

The optimal conditions to extract the maximum residual oil from partially defatted chia flour were 10.2% moisture content and a 58.5 °C pressing temperature [25]. Ghafoor et al. (2020) reported that roasting temperature affects the yield of CSO. Maximum oil was extracted from chia seeds roasted at a high temperature (180 °C) [2]. Martínez et al. (2012) [11] discovered that seed moisture and pressing speed affected oil yield when using a Komet screw press but that the pressing temperature and restriction die had no effect. Chia seed by-products (chia seed meal and fibrous fractions) subjected to the pressing extraction method produced a high yield of residual oil [26]. The maximum yield (30.15%) of CSO was observed with the following conditions: size of the restriction die: 1.2 cm;

temperature: 50 °C; and duration of dry heat exposure at 100 °C: 15 min (Table 2). The size of the restriction die, temperature, and duration of dry heat exposure had no effect on the yield, while the interaction and quadratic model influenced the yield (Figure 2). Additionally, the moisture content of the chia seed did not affect the CSO yield.

Martínez-Cruz and Paredes-López (2014) [7] reported on chia seed extract (from chia powder) containing 1.63 mg GAE/g of chia seed and revealed that the primary phenolic compounds were rosmarinic, protocatechuic, caffeic, and gallic acids and daidzin. Chia seeds roasted at different temperatures were subjected to oil extraction. The TPC of CSO varied depending on the roasting temperature. TPC was found to be decreased in samples undergoing high-temperature roasting [2]. The TPC of roasted CSO was lower than that of non-roasted oil, although cold-press extraction was not different from Soxhlet extraction in terms of TPC [27]. However, Ixtaina et al. (2011) reported that total polyphenolic compounds were higher in CSO extracted using the pressing method than those obtained via the solvent extraction method [3]. The size of the restriction did not affect the TPC of CSO. The increase in the duration of dry heat exposure reduced the TPC of CSO, indicating that prolonged heat exposure might degrade the phytocompounds in the oil. Pressing temperature also affected the TPC (Figure 3). The experimental data revealed that maximum TPC was observed in one of the center-point STDs. In detail, the restriction die size (1.2 cm), temperature (50), and duration of dry heat treatment (15 min) aided in producing CSO with a TPC of 2.86 µg GAE/g of oil (Table 2).

The breakdown of triacylglycerol could contribute to an increase in the total acidity of CSO [2]. CSO had higher acid values when extracted with a solvent (using a Soxhlet apparatus and thermal cycles at 80 °C for 8 h) than when extracted via pressing (using a Komet screw press at 25 to 30 °C) [3]. This result could be explained by the fact that heating causes oil oxidation, which raises the acid value. The results demonstrated that high temperatures influenced the acid values. Acid values of CSO extracted by the cold solvent, pressing, and supercritical methods were 1.13 ± 0.21, 1.18 ± 0.06, and 1.41 ± 0.46 mg KOH/g oil, respectively [20]. However, in the present study, none of the samples exceeded the upper limit of the acid value (4 mg KOH/g of oil) established by the Thai community product standards (Thailand Ministry of Public Health (No. 421), B.E. 2564 Issued under the Food Act B.E. 2522 on oils and fats) (Table 2). RSM prediction showed that a low temperature, lowered restriction die size, and high dry heat exposure time (i.e., low moisture content) aided in the extraction of CSO with a lower acid value (Figure 4). The experimental data indicated that die size (1 or 1.4 cm), temperature (50 or 60), and dry heat time (15 or 30 min) produced CSO with 0.80 mg KOH/g of oil (Table 2).

According to the study on oil quality, the peroxide value trends for canola, corn, and sunflower oils fluctuate. This fluctuation is caused by the primary oxidation products of lipid oxidation and hydroperoxides, which are formed from unsaturated fatty acids. The initial increase in the peroxide value demonstrated a larger concentration of hydroperoxides. However, the peroxide value was reduced in the secondary oxidation products [28,29]. The peroxide values of oil decreased with an increase in temperature under microwave treatments [30]. According to Ghafoor et al. (2020) [2], increasing the roasting temperature led to a rise in the peroxide values of CSO from 3.21 meqO$_2$/kg (control) to 18.42 meqO2/kg (180 °C). Peroxide content was used to estimate the degree of oxidation of edible oils. Applying heat during chia seed roasting may influence the oxidation reaction, thus increasing the peroxide index of CSO [2]. CSO extracted from pretreated (water boiling, microwave roasting, oven drying, and autoclaving) chia seeds with pressing methods presented peroxide values ranging from 0.69 to 2.67 mEq/Kg oil [12]. The peroxide index values of CSO ranged from 3.12 to 8.08 mEq/Kg of oil (Table 2), which were higher than the values previously reported [31]. These changes may be associated with the cultivar of chia and processing conditions. The peroxide values of CSO extracted using the cold solvent, pressing, and supercritical methods were 0, 10.98 ± 0.61, and 0.37 meq O$_2$/Kg oil, respectively [20], indicating that CSO extracted via pressing had a higher peroxide value. These results indicated that the extraction methods greatly influenced the quality of CSO.

However, further comparative studies are required to confirm the above statement. In the present study, none of the samples exceeded the upper limit of the peroxide index value (10.0 mEq active oxygen/kg oil) established by the Thai community product standards (Thailand Ministry of Public Health (No. 421), B.E. 2564, issued under the Food Act B.E. 2522 on oils and fats) (Table 2). Additionally, the RSM prediction showed that a high temperature would aid in the extraction of CSO with a lower peroxide value. The moisture content of the seed also affected the peroxide value, while the size of the restriction die had no effect on the peroxide value (Figure 5). The experimental data indicated that a die size of 1.2 cm, a temperature of 50 °C, and no dry heat treatment helped extract CSO with 3.12 mEq/Kg of oil (Table 2).

The restriction die and temperature significantly affected the pressing time (Figure 5). The pressing time increased as the size of the restriction die was raised, while the pressing time decreased as the temperature was raised. Because of insufficient friction during pressing, high seed moisture content resulted in poor oil recovery. This effect could be attributed to the development of an external gelatinous structure with water-retaining properties [11]. According to Santoso and Inggrid (2014) [32], the average oil yield increases as pressing time increases. In the present study, the RSM prediction demonstrated that a reduced size of the restriction die (1 cm), no dry heat exposure, and high temperature might reduce the pressing time needed for CSO extraction (Figure 6). The experimental data also demonstrated the above statement (Table 2).

Several studies have indicated that Mexican CSO extracted via solvent and supercritical CO_2 methods is rich in α-linolenic acid, linoleic acid, palmitic acid, oleic acid, and stearic acid [22,23,33]. Fernandes et al. (2019) revealed that Brazilian CSO extracted via different extraction methods (cold solvent, pressing, and supercritical CO_2) also showed high α-linolenic acid content, followed by linoleic and oleic acids and saturated fatty acids [20]. Chilean CSO contains about 62.8% α-linolenic acid; other fatty acids include linoleic, oleic, palmitic, and stearic acids [31]. The present study also revealed that Thai CSO had a high content of α-linolenic acid, followed by linoleic acid, palmitic acid, oleic acid, stearic acid, eicosenoic acid, palmitoleic acid, and myristic acid (Table 6), indicating that the fatty acid profile of Thai CSO was also like that in the previous report.

Compared to the other oils (sunflower, safflower, canola, and soybean oil), CSO presented the lowest DPPH and ABTS radical scavenging capabilities [33,34]. The composition and proportion of fatty acids, including myristic, palmitic, palmitoleic, stearic, oleic, linoleic, linolenic, behenic, arachidonic, lignoceric, trianoic, and arachidic acids [2,3,11,27,33,35], as well as tocopherols and phenolic compounds, in CSO may affect its NO scavenging activities. Roasting harmed the TPC and antioxidant activity of CSO, as these properties were significantly lower in oil extracted from roasted chia seeds using the Soxhlet and cold-press extraction methods compared to non-roasted seeds, which could be attributed to the degradation or polymerization of phytocompounds during thermal treatments [27]. CSO extracted through solvent extraction (n-hexane) had an antioxidant activity of 33.94 IC_{50} mg/mL (DPPH assay) and 28.51 IC_{25} mg/mL (ABTS assay) [33]. Our data also revealed that avoiding dry heat treatment produced high-quality CSO in terms of ABTS, DPPH, and NO radical scavenging activities. Additionally, pressing temperature affected the ABTS and DPPH values of CSO (Figure 7).

The literature revealed that the seed cultivar; extraction methods; extraction conditions such as pretreatments, temperature, and pressure; and seed physical conditions such as moisture influenced the yield and quality of CSO. The quality and quantity of oil extracted by the solvent or through physical methods varied. Thus, the results are inconsistent, and a single optimal condition to achieve high-quality CSO has yet to be determined. However, the results suggest that edible oil could be stored with natural antioxidants to prevent oxidation and reduce acid and peroxide values.

5. Conclusions

This study reported the influence of the restriction die size, temperature, and dry heat exposure time on the quantity and quality of Thai CSO. The restriction die size of 1 cm, pressing temperature of 53 °C, and absence of dry heat were the optimal conditions for extracting the desired quality of CSO, according to the predicted RSM model. The result showed that the average yield, TPC, acid value, peroxide value, and pressing time were 29.47%, 2.20 µg GAE/g of oil, 0.96 mg KOH/g of oil, 2.87 mEq/Kg of oil, and 45.10 min, respectively. The size of the restriction die influenced the extraction time and acid value of CSO. The pressing temperature affected the extraction time, acid and peroxide values, and TPC of CSO. The duration of dry heat exposure affected the TPC and peroxide value of CSO. The fatty acid composition did not change significantly between the predicted and real-time experimental setup. The fatty acid composition of Thai CSO was comparable to that of previously reported studies. The results obtained from the present study demonstrated that chia seeds contain several phytochemicals, some of which are sensitive to heat treatment and thus require proper attention when selecting a processing technique. Further studies are required to confirm these findings and evaluate other parameters influencing CSO quality and quantity.

Author Contributions: Conceptualization, C.C. and P.S.; methodology, M.S.; software, M.S., C.T. and B.S.S.; validation, C.C., C.T. and B.S.S.; formal analysis, M.S. and C.T.; investigation, P.S.; resources, C.C.; data curation, M.S. and C.T.; writing—original draft preparation, M.S., B.S.S., P.K., P.S. and C.C.; writing—review and editing, M.S., B.S.S., P.K. and C.C.; visualization, P.S.; supervision, C.C. and P.S.; project administration, S.P.; funding acquisition, C.C., M.S. and P.S. All authors have read and agreed to the published version of the manuscript.

Funding: The study was partially funded by Chiang Mai University, Chiang Mai, Thailand. M.S. was funded by the Postdoctoral Fellowship (No. 03/2022) from Mae Fah Luang University, Thailand.

Institutional Review Board Statement: Not applicable.

Informed Consent Statement: Not applicable.

Data Availability Statement: The data presented in this study are available within the article.

Acknowledgments: The authors gratefully acknowledge the Chiang Mai University, Chiang Mai, Thailand, and Mae Fah Luang University, Chiang Rai, Thailand.

Conflicts of Interest: The authors declare no conflict of interest.

References

1. Baginsky, C.; Arenas, J.; Escobar, H.; Garrido, M.; Valero, N.; Tello, D.; Pizarro, L.; Valenzuela, A.; Morales, L.; Silva, H. Growth and yield of chia (*Salvia hispanica* L.) in the Mediterranean and desert climates of Chile. *Chil. J. Agric. Res.* **2016**, *76*, 255–264. [CrossRef]
2. Ghafoor, K.; Ahmed, I.A.M.; Özcan, M.M.; Al-Juhaimi, F.Y.; Babiker, E.E.; Azmi, I.U. An evaluation of bioactive compounds, fatty acid composition and oil quality of chia (*Salvia hispanica* L.) seed roasted at different temperatures. *Food Chem.* **2020**, *333*, 127531. [CrossRef] [PubMed]
3. Ixtaina, V.Y.; Martínez, M.L.; Spotorno, V.; Mateo, C.M.; Maestri, D.M.; Diehl, B.W.; Nolasco, S.M.; Tomás, M.C. Characterization of chia seed oils obtained by pressing and solvent extraction. *J. Food Compos. Anal.* **2011**, *24*, 166–174. [CrossRef]
4. Orona-Tamayo, D.; Valverde, M.E.; Paredes-López, O. Chia—The new golden seed for the 21st century: Nutraceutical properties and technological uses. In *Sustainable Protein Sources*; Elsevier: Amsterdam, The Netherlands, 2017; pp. 265–281.
5. Weber, C.W.; Gentry, H.S.; Kohlhepp, E.A.; McCrohan, P.R. The nutritional and chemical evaluation of chia seeds. *Ecol. Food Nutr.* **1991**, *26*, 119–125. [CrossRef]
6. Coates, W. Protein content, oil content and fatty acid profiles as potential criteria to determine the origin of commercially grown chia (*Salvia hispanica* L.). *Ind. Crops Prod.* **2011**, *34*, 1366–1371.
7. Martínez-Cruz, O.; Paredes-López, O. Phytochemical profile and nutraceutical potential of chia seeds (*Salvia hispanica* L.) by ultra high performance liquid chromatography. *J. Chromatogr. A* **2014**, *1346*, 43–48. [CrossRef]
8. Enes, B.N.; Moreira, L.P.; Silva, B.P.; Grancieri, M.; Lúcio, H.G.; Venâncio, V.P.; Mertens-Talcott, S.U.; Rosa, C.O.; Martino, H.S. Chia seed (*Salvia hispanica* L.) effects and their molecular mechanisms on unbalanced diet experimental studies: A systematic review. *J. Food Sci.* **2020**, *85*, 226–239. [CrossRef]

9. Reyes-Caudillo, E.; Tecante, A.; Valdivia-Lopez, M.A. Dietary fibre content and antioxidant activity of phenolic compounds present in Mexican chia (*Salvia hispanica* L.) seeds. *Food Chem.* **2008**, *107*, 656–663. [CrossRef]
10. Akinfenwa, A.O.; Cheikhyoussef, A.; Cheikhyoussef, N.; Hussein, A.A. Cold pressed chia (*Salvia hispanica* L.) seed oil. In *Cold Pressed Oils*; Elsevier: Amsterdam, The Netherlands, 2020; pp. 181–190.
11. Martínez, M.L.; Marín, M.A.; Faller, C.M.S.; Revol, J.; Penci, M.C.; Ribotta, P.D. Chia (*Salvia hispanica* L.) oil extraction: Study of processing parameters. *LWT-Food Sci. Technol.* **2012**, *47*, 78–82. [CrossRef]
12. Muhammad, I.; Muhammad, N.; Manzoor, M.; Amna, J.; Zafar, A.; Akhtar, M.; Muhammad, A.; Yasir, H. Fatty acids characterization, oxidative perspectives and consumer acceptability of oil extracted from pre-treated chia (*Salvia hispanica* L.) seeds. *Lipids Health Dis.* **2016**, *15*. [CrossRef]
13. Sosa, A.; Ruiz, G.; Rana, J.; Gordillo, G.; West, H.; Sharma, M.; Liu, X.; Torre, R. Chia crop (*Salvia hispanica* L.): Its history and importance as a source of polyunsaturated fatty acids omega-3 around the world: A review. *J Crop Res Fert* **2016**, *1*, 1–9. [CrossRef]
14. Sirilun, S.; Sivamaruthi, B.S.; Pengkumsri, N.; Saelee, M.; Chaiyasut, K.; Tuntisuwanno, N.; Suttajit, M.; Peerajan, S.; Chaiyasut, C. Impact of different pre-treatment strategies on the quality of fatty acid composition, tocols content & metabolic syndrome related activities of Perilla frutescens seed oil. *J. Appl. Pharm. Sci.* **2016**, *6*, 001–008.
15. Saelee, M.; Sivamaruthi, B.S.; Tansrisook, C.; Duangsri, S.; Chaiyasut, K.; Kesika, P.; Peerajan, S.; Chaiyasut, C. Response Surface Methodological Approach for Optimizing Theobroma cacao L. Oil Extraction. *Appl. Sci.* **2022**, *12*, 5482. [CrossRef]
16. Fuentes, E.; Báez, M.E.; Bravo, M.; Cid, C.; Labra, F. Determination of total phenolic content in olive oil samples by UV–visible spectrometry and multivariate calibration. *Food Anal. Methods* **2012**, *5*, 1311–1319. [CrossRef]
17. Thitipramote, N.; Pradmeeteekul, P.; Nimkamnerd, J.; Chaiwut, P.; Pintathong, P.; Thitilerdecha, N. Bioactive compounds and antioxidant activities of red (Brown Red Jasmine) and black (Kam Leum Pua) native pigmented rice. *Int. Food Res. J.* **2016**, *23*, 410.
18. Sivamaruthi, B.S.; Pengkumsri, N.; Saelee, M.; Kesika, P.; Sirilun, S.; Peerajan, S.; Chaiyasut, C. Impact of physical treatments on stability and radical scavenging capacity of anthocyanidins. *Health* **2016**, *1*, 2.
19. Pengkumsri, N.; Chaiyasut, C.; Saenjum, C.; Sirilun, S.; Peerajan, S.; Suwannalert, P.; Sirisattha, S.; Sivamaruthi, B.S. Physicochemical and antioxidative properties of black, brown and red rice varieties of northern Thailand. *Food Sci. Technol* **2015**, *35*, 331–338. [CrossRef]
20. Fernandes, S.S.; Tonato, D.; Mazutti, M.A.; de Abreu, B.R.; da Costa Cabrera, D.; D'Oca, C.D.R.M.; Prentice-Hernández, C.; de las Mercedes Salas-Mellado, M. Yield and quality of chia oil extracted via different methods. *J. Food Eng.* **2019**, *262*, 200–208. [CrossRef]
21. Noshe, A.S.; Al-Bayyar, A.H. Effect of extraction method of Chia seeds Oil on its content of fatty acids and antioxidants. *Int. Res. J. Eng. Technol.* **2017**, *234*, 1–9.
22. Ixtaina, V.Y.; Vega, A.; Nolasco, S.M.; Tomás, M.C.; Gimeno, M.; Bárzana, E.; Tecante, A. Supercritical carbon dioxide extraction of oil from Mexican chia seed (*Salvia hispanica* L.): Characterization and process optimization. *J. Supercrit. Fluids* **2010**, *55*, 192–199. [CrossRef]
23. Uribe, J.A.R.; Perez, J.I.N.; Kauil, H.C.; Rubio, G.R.; Alcocer, C.G. Extraction of oil from chia seeds with supercritical CO_2. *J. Supercrit. Fluids* **2011**, *56*, 174–178. [CrossRef]
24. Silva, C.; Garcia, V.; Zanette, C. Chia (*Salvia hispanica* L.) oil extraction using different organic solvents: Oil yield, fatty acids profile and technological analysis of defatted meal. *Int. Food Res. J.* **2016**, *23*, 998–1004.
25. Aranibar, C.; Pigni, N.B.; Martínez, M.L.; Aguirre, A.; Ribotta, P.D.; Wunderlin, D.A.; Borneo, R. Influence of the extraction conditions on chia oil quality and partially defatted flour antioxidant properties. *J. Food Sci. Technol.* **2021**, *59*, 1982–1993. [CrossRef] [PubMed]
26. Capitani, M.; Spotorno, V.; Nolasco, S.M.; Tomás, M.C. Physicochemical and functional characterization of by-products from chia (*Salvia hispanica* L.) seeds of Argentina. *LWT-Food Sci. Technol.* **2012**, *45*, 94–102. [CrossRef]
27. Özcan, M.M.; Al-Juhaimi, F.Y.; Ahmed, I.A.M.; Osman, M.A.; Gassem, M.A. Effect of soxhlet and cold press extractions on the physico-chemical characteristics of roasted and non-roasted chia seed oils. *J. Food Meas. Charact.* **2019**, *13*, 648–655. [CrossRef]
28. Kaleem, A.; Aziz, S.; Iqtedar, M. Investigating changes and effect of peroxide values in cooking oils subject to light and heat. *FUUAST J. Biol.* **2015**, *5*, 191–196.
29. Guillén, M.A.D.; Cabo, N. Fourier transform infrared spectra data versus peroxide and anisidine values to determine oxidative stability of edible oils. *Food Chem.* **2002**, *77*, 503–510. [CrossRef]
30. Cerretani, L.; Bendini, A.; Rodriguez-Estrada, M.T.; Vittadini, E.; Chiavaro, E. Microwave heating of different commercial categories of olive oil: Part I. Effect on chemical oxidative stability indices and phenolic compounds. *Food Chem.* **2009**, *115*, 1381–1388. [CrossRef]
31. da Silva Marineli, R.; Moraes, É.A.; Lenquiste, S.A.; Godoy, A.T.; Eberlin, M.N.; Maróstica Jr, M.R. Chemical characterization and antioxidant potential of Chilean chia seeds and oil (*Salvia hispanica* L.). *LWT-Food Sci. Technol.* **2014**, *59*, 1304–1310. [CrossRef]
32. Santoso, H.; Inggrid, M. Effects of temperature, pressure, preheating time and pressing time on rubber seed oil extraction using hydraulic press. *Procedia Chem.* **2014**, *9*, 248–256. [CrossRef]
33. Shen, Y.; Zheng, L.; Jin, J.; Li, X.; Fu, J.; Wang, M.; Guan, Y.; Song, X. Phytochemical and biological characteristics of mexican chia seed oil. *Molecules* **2018**, *23*, 3219. [CrossRef] [PubMed]

34. Xuan, T.D.; Gangqiang, G.; Minh, T.N.; Quy, T.N.; Khanh, T.D. An overview of chemical profiles, antioxidant and antimicrobial activities of commercial vegetable edible oils marketed in Japan. *Foods* **2018**, *7*, 21. [CrossRef] [PubMed]
35. Ayerza, R.; Coates, W. Ground chia seed and chia oil effects on plasma lipids and fatty acids in the rat. *Nutr. Res.* **2005**, *25*, 995–1003. [CrossRef]

Disclaimer/Publisher's Note: The statements, opinions and data contained in all publications are solely those of the individual author(s) and contributor(s) and not of MDPI and/or the editor(s). MDPI and/or the editor(s) disclaim responsibility for any injury to people or property resulting from any ideas, methods, instructions or products referred to in the content.

Article

Nutritional Profiling and Cytotoxicity Assessment of Protein Rich Ingredients Used as Dietary Supplements

Bianca-Maria Tihăuan [1,2], Ioana-Cristina Marinaș [1,*], Marian Adascălului [2], Alina Dobre [2], Grațiela Grădișteanu Pîrcălăbioru [1], Mădălina Axinie [3], Laura Mihaela Ștefan [4] and Denisa Eglantina Duță [2]

1. Research Institute of the University of Bucharest—ICUB, 91-95 Spl. Independentei, 50567 Bucharest, Romania
2. National Institute of Research & Development for Food Bioresources—IBA Bucharest, 5 Băneasa Ancuța, 020323 Bucharest, Romania
3. Department of Science and Engineering of Oxide Materials and Nanomaterials, Faculty of Applied Chemistry and Materials Science, University Politehnica of Bucharest, 011061 Bucharest, Romania
4. National Institute of Research and Development for Biological Sciences, 296 Splaiul Independenței, 060031 Bucharest, Romania
* Correspondence: ioana.cristina.marinas@gmail.com

Featured Application: Nutritional Profiling for subsequent dietary supplements/nutraceuticals development and formulation as well as by-products valorification.

Abstract: In recent years, the scientific community has made significant progress in understanding nutrition, leading consumers to shift their preferences away from animal-based protein products and towards natural, plant-based protein sources. This study aimed to determine the nutritional value, in vitro cytotoxicity and antioxidant activity for different sources of high protein content products (pea, yeast, almond, spirulina and *Pleurotus* spp.) with potential usage as raw materials for dietary supplements, especially since these products do not benefit from stricter regulation requirements regarding their actual health benefits. The characterization of raw materials consisted in evaluation of their nutritional profile (by addressing moisture content, crude protein content, extractable fat, ash, carbohydrates) and microbial contamination (TAMC, TYMC, *Enterobacteriaceae* and β-glucuronidase positive *Escherichia coli*), total content of free amino acids, soluble proteins, phenols and flavonoids, as well as antioxidant activity through chemical assays. We used 3-(4,5-dimethylthiazol-2-yl)-2,5-diphenyl-2H-tetrazolium bromide (MTT) assay and lactate dehydrogenase (LDH) release to evaluate the potential cytotoxicity of selected raw materials. Results obtained indicate high percentages of proteins for the pea powder (77.96%) and *Spirulina* powder (64.79%), *Pleurotus* spp. flour had strong antioxidant activity, while the highest contamination values were registered for *Pleurotus* spp. powder (4.6×10^5 CFU/g or 5.66 log CFU/g). Cytotoxicity results demonstrate that tested ingredients have an impact on the metabolic activity of cells, affecting cellular integrity and provoking leakage of DNA at several concentrations. While plant-based protein supplementation may appear to be a promising solution to balance our busy lives, there are several advantages and disadvantages associated with them, including issues related to their absorption rate, bioavailability, cytotoxicity and actual nutritional benefits.

Keywords: high protein ingredients; cytotoxicity; antioxidant activity; nutritional value; dietary supplements

Citation: Tihăuan, B.-M.; Marinaș, I.-C.; Adascălului, M.; Dobre, A.; Pîrcălăbioru, G.G.; Axinie, M.; Ștefan, L.M.; Duță, D.E. Nutritional Profiling and Cytotoxicity Assessment of Protein Rich Ingredients Used as Dietary Supplements. *Appl. Sci.* **2023**, *13*, 6829. https://doi.org/10.3390/app13116829

Academic Editor: Monica Gallo

Received: 15 April 2023
Revised: 26 May 2023
Accepted: 28 May 2023
Published: 5 June 2023

Copyright: © 2023 by the authors. Licensee MDPI, Basel, Switzerland. This article is an open access article distributed under the terms and conditions of the Creative Commons Attribution (CC BY) license (https://creativecommons.org/licenses/by/4.0/).

1. Introduction

In the past, protein supplements and dietary supplements, in general, were mostly associated with bodybuilders and individuals who engaged in heavy exercise. However, in recent years, there has been a significant shift in consumer behavior, with a large segment of the general population seeking out protein supplements for meal replacement, weight loss and purported health benefits [1].

This shift can be attributed to several factors, including a growing interest in healthy living and wellness, as well as an increased awareness of the importance of proper nutrition for overall health and well-being. Additionally, advancements in nutritional science have led to a better understanding of the role that proteins play in the body and the potential benefits of protein supplementation for various health goals. Overall, this shift towards protein supplementation and dietary supplementation, in general, is indicative of a larger trend towards a greater focus on nutrition and wellness in modern society [2].

Therefore, proteins, being essential macronutrients, play a crucial role in the growth and maintenance of the human body as they provide energy, amino acids, and possess various physiochemical and sensory properties, making them useful functional ingredients for promoting health [3]. Despite their importance, protein deficiency remains a major concern. Furthermore, increased protein intake has been associated with several health benefits due to its fundamental role in all body cells and maintenance of the immune system. Notably, higher protein consumption has been linked to increased muscle mass and endurance [4–7]. The origin of protein can have a significant impact on the health benefits it provides. Proteins from animal sources, such as meat, dairy and eggs tend to be high in saturated fat and cholesterol, which can increase the risk of heart disease and other chronic health conditions when consumed in excess [4,5]. On the other hand, proteins from plant sources, such as legumes, nuts, seeds, and whole grains tend to be lower in saturated fat and cholesterol, and higher in fiber, vitamins, minerals, and antioxidants, which have been linked to a reduced risk of chronic diseases [6,7]. Additionally, plant-based proteins can provide several health benefits that animal-based proteins may not. For example, the use of protein isolates (hydrolysates) from peas have been shown to lower blood pressure and reduce the risk of kidney disease [8,9], the body metabolizing this protein similar to whey-based ones [10]. Almond protein powders are also metabolized similar to well-known whey-based proteins [11], being as well associated with beneficial effects against cardiometabolic diseases [12,13]. Spirulina powder's beneficial effects are identified in particular reference to obesity, hypertension and cardiovascular diseases [14,15]. Additionally, *Pleurotus* spp. protein powders have been associated with beneficial aids in various chronic diseases [16–18], as well as yeast protein powders promoting muscle protein synthesis, rapid recovery and biodiversity of gut-microbiota [19,20]. In summary, the origin of protein is an important consideration for optimizing health benefits. Consuming a variety of plant-based proteins can provide a range of health benefits, while minimizing the consumption of animal-based proteins may reduce the risk of chronic diseases associated with a high intake of saturated fat and cholesterol.

In addition, it is well known that compounds with antioxidant activity contribute to the cells' amelioration and/or protection of oxidative stress [21]. Synthetic antioxidants have been widely used in different industries being known for their health hazards such as liver damage and carcinogenesis [22]. Natural antioxidant peptides and proteins are an alternative in terms of safety and availability [23]. In the case of antioxidant peptides, the potential to reduce the risk of aging, inflammation and cardiovascular disease have been demonstrated, thus improving human health [24,25].

Up to date, numerous antioxidant peptides from various animal, plant and microbial proteins have been identified [26–29] with different mechanisms of action: inhibition of lipid oxidation, scavengers of free radicals and/or chelate pro oxidative metals [25,28,30,31]. Various studies have shown that the antioxidant properties of proteins are related to their composition, sequence, structure and hydrophobicity [11,32–34].

Keeping this in mind, we can highlight several advantages of consuming proteins of plant origin, such as lower saturated fat and cholesterol content, higher values of fiber, vitamins, minerals, and antioxidants compared to animal-based proteins, which are associated with a reduced risk of chronic diseases [35], environmental sustainability [36], cost-effectiveness, versatility and suitability for various dietary restrictions [37–40]. Overall, consuming proteins of plant origin can be a beneficial choice for both personal health and the health of the planet.

Although dietary supplementation seems like a viable solution that can balance our busy and demanding life and enhance or help maintain normal physiological functions of the human body, varying pros and cons have been associated with them regarding the absorption rate, bioavailability, cytotoxicity and actual nutritional benefits. Especially since according to the 2020 Agri-Food Fraud Network (FFN) report based on Regulation (EU) No. 1169/2011 on the food information to consumers, the majority of non-compliances regarded mislabeling, support documentation and product dilution/addition [41].

Therefore, our study analyzed the nutritional profile, physico-chemical properties, antioxidant activity and cytotoxicity degree of five high protein ingredients represented by: pea powder, almond powder, yeast powder, spirulina powder and *Pleurotus* spp. flour. These ingredients are frequently used in various diets, workout regimes or as meal replacements; as we know, the supplements are not as restricted by legislation, therefore mistakes and misconceptions are more likely to occur. Therefore, our study focuses on analyzing three key domains capable of outlining the safety consumption limits of ingredients such as these.

2. Materials and Methods

2.1. Reagents and Standards

Bovine albumin serum, Bradford regent, ninhydrin, 2,2-Diphenyl-1-picrylhydrazyl (DPPH), ferric reducing antioxidant potential (FRAP), 2,4,6-tripyridyl-s-triazine (TPTZ), gallic acid and Trolox (6-hydroxy-2,5,7,8-tetramethylchroman-2-carboxylic acid) were purchased from Sigma-Aldrich Chemie GmbH (Buchs, Switzerland). Folin–Ciocalteu's phenol reagent was purchased from Merck (Darmstadt, Germany). All chemicals used were of analytical grade.

2.2. Raw Materials

The raw materials/ingredients tested were procured from Romanian manufacturers. Pea powder from SC Paradisul Verde SRL, Brasov, Romania, spirulina powder from SC Fytovital SRL, Oradea, Romania, yeast powder from SC SANOVITA SRL, Romania, almond powder from SC PRONAT SRL, Romania and *Pleurotus* spp. flour was obtained by IBA, Bucharest, from fresh *Pleurotus* through drying at 50 °C and grinding.

2.3. Assessment of the Nutritional Profile of Dry Raw Materials

2.3.1. The Moisture Content

The moisture content was achieved through a gravimetric method that involved drying the samples at 150 °C until a constant mass was reached, method described by ISO 6673:2003 [42]. Approximately 5 g of the sample were weighed into a vial, which was dried to a constant mass in an oven (Memmert UF 110, Memmert, Schwabach, Germany). The analysis was performed in triplicate. The moisture content (M, %) was determined using the following formula:

$$M\,(\%) = \frac{w - d}{d} \times 100, \quad (1)$$

where w—wet weight, d—dry weight

2.3.2. Crude Protein Content

Total protein content was determined according to ISO 20483:2013 [43]. The protein content was analysed by the Kjeldahl method with a FOSS Kjeltec 2300 analyzer (FOSS Group, Hillerød, Denmark) after acid hydrolysis in an auto-digester (Behrotest InKjel, 450 P, Germany). According to the classical Kjeldahl method, samples were digested using concentrated sulfuric acid. The ammonium sulfate salt and alkali generated ammonia, which was trapped in boric acid via steam distillation. Titration was performed using a hydrochloric acid 0.2 N solution.

2.3.3. The Total Extractable Fat Content

Total fat content was measured using the Soxhlet method described in ISO 6492:1999 [44]. Briefly, the samples were transferred into the extraction cartridge. About 100 mL of petroleum ether was added to the flask attached to the Soxhlet apparatus (SoxtecTM 2050, Foss Analytical, Hilleroed, Denmark). The extraction lasted about 6 h. The total amount of lipids was determined by a gravimetric method as follows: the solvent of the extract was distilled, the residue was dried in an oven at 105 °C for 2 h and the flasks were weighed until constant mass. The fat content (FC) was calculated as follows:

$$FC\ (\%) = \frac{m2 - m1}{m} \times \frac{100}{100 - M(\%)} \times 100, \qquad (2)$$

where $m1$—empty flask weight, $m2$—flask with fat weight, m—sample weight, $M(\%)$—moisture content

2.3.4. Determination of Ash

The ash content was determined according to ISO 2171:2010 [45] with slight modifications. The total ash content is based on the evaluation of the residue amount obtained after sample calcination at 650 ± 25 °C and the ashes were dried to constant mass at 105 °C.

2.3.5. The Total Carbohydrates Content (CC) and Energy Value

The equation used to calculate the total carbohydrate content was [46]: total carbohydrates (% FW) = 100 − moisture (%) − protein content (% FW) − crude fat (% FW) − ash (% FW). The results show total carbohydrates as g/100 g FW. The energy value was calculated using the following factors: protein 4 kcal/g, fat 9 kcal/g and carbohydrates 4 kcal/g, according to the following formula:

$$EV\ (kcal) = 4 \times PC\ (\%) + 9 \times FC\ (\%) + 4 \times CC\ (\%) \qquad (3)$$

2.3.6. Fourier Transform Infrared Spectroscopy (ATR-FTIR)

The functional groups within extracts were evaluated by Fourier transform infrared spectroscopy (FTIR, Agilent Cary 630 FTR spectrophotometer, Agilent, CA, USA) equipped with an ATR device. The spectra were recorded from 4000–650 cm^{-1} with an average of 400 scans and a resolution of 4 cm^{-1} resolution, at 22 °C.

2.4. Assessment of the Physico-Chemical Properties for the Extracts Obtained from Raw Materials

To determine the content of soluble proteins, free amino acids, phenolics and flavonoids, an aqueous extraction corresponding to a concentration of 10 mg dry material/mL distillated water was performed. The extracts were vortexed for 15 min and centrifuged at 8000 rpm for 15 min at 10 °C using a centrifuge (Thermo Scientific, Waltham, MA, USA). All assays were performed in triplicate. The results were expressed as mean ± SD.

2.4.1. Determination of Total Soluble Protein

The protein concentration from the pea powder, almond powder, yeast powder, spirulina powder and *Pleurotus* spp. flour was determined by the Bradford assay [47]. The calibration curve was made using bovine albumin serum (BSA) for concentrations that ranged between 1.4–0.1 mg/mL ($R^2 = 0.9928$). The extracts obtained in distilled water were used for spectrophotometric determination of protein at λ = 595 nm, using FlexStation 3 UV-Vis (Molecular Devices, San Jose, CA, SUA) spectrophotometer.

2.4.2. Determination of Total Free Amino Acids

Total free amino acids were quantified after protein precipitation with trichloroacetic acid (TCA). The samples were centrifuged (15 min, 10,000 rpm, Thermo Scientific, Waltham, MA, USA) and 10 µL of the supernatant was inserted into 300 cL of Cd-ninhydrin reagent. The samples were kept for 10 minutes at 84 °C and the absorbance was read at λ = 507 nm.

The calibration curve (glycine—$R^2 = 0.9940$) was measured for concentrations that ranged between 10–0.625 mg/mL. The Cd-ninhydrin reagent was prepared as follows: 1 mL of 1 g/mL $CdCl_2$ was added to a solution containing 0.8 g ninhydrin dissolved in 90 mL absolute ethanol: glacial acetic acid 8:1 (*v*:*v*). The solution formed was diluted with distilled water in a volume ratio of 1:1.5 [20].

2.4.3. Total Phenolic Content

Total phenolic content was measured by the Folin–Ciocalteu assay [48]. Over 50 µL of sample was transferred along with 50 µL Folin–Ciocalteu reagent, 0.45 mL H_2O and 0.50 mL of 7% Na_2CO_3. The samples/standard solutions were incubated in the dark for 60 min. The samples were centrifuged (5 min, 10,000 rpm, Thermo Scientific, Waltham, MA, USA) and the absorbance was read at 765 nm with FlexStation 3 UV-Vis (Molecular Devices, GA, USA) Spectrophotometer. The calibration curve was measured using gallic acid in the same condition with samples ($R^2 = 0.9955$). The linear domain was between 250–12.5 µg/mL. Total phenolic content was rendered as mg gallic acid equivalent (GAE)/g sample.

2.4.4. Total Flavonoid Content

The flavonoid content was determined through the $AlCl_3$ assay [49]. Briefly, over 100 µL sample/standard solution was added along with 100 µL sodium acetate 10%, 12 µL $AlCl_3$ 2.5% and 680 µL ethanol 70%. The sample/standard solutions were incubated at dark for 45 min. After that, the samples were subjected to centrifugation (5 min, 10,000 rpm). The supernatant absorbance was recorded at $\lambda = 430$ nm. A standard curve was created with quercetin ($R^2 = 0.9935$). The linear domain was between 200–12.5 µg/mL. Total flavonoid content was rendered as mg quercetin equivalent (QE)/g sample.

2.5. Antioxidant Activity

The DPPH assay was performed according to the method described by Madhu (2013) [50] with some minor changes. The linearity range of Trolox standard through this method was between 0 and 100 mM ($R^2 = 0.9943$). The absorbance was determined using a FlexStation 3 UV–VIS spectrophotometer (Molecular Devices Company, Sunnyvale, CA, USA) at $\lambda = 515$ nm.

FRAP assay. 285 µL of the FRAP reagent, prepared according to Benzie and Strain (1999) [51] were added over the 15 µL of sample. The reaction mixture was incubated at 37 °C for 30 min in the dark. The samples were centrifuged (Thermo Scientific, Waltham, MA, USA) and the absorbance was recorded at 593 nm using FlexStation 3 UV–VIS spectrophotometer (Molecular Devices Company, Sunnyvale, CA, USA). Trolox (50 to 250 µM) was used for the calibration curve ($R^2 = 0.997$).

2.6. Evaluation of the Microbiological Status

The vegetal high protein ingredients were tested for bacterial and fungal contamination. The monitoring of microbiological status and the quality assessment of yeast, pea, almond, spirulina and *Pleurotus* spp. powder were performed using national and international food microbiology standards. The purpose was to assess the microbiological status of these powders because the bacterial and fungal content are an important parameter for quality assurance and safe consumption.

Each sample was prepared according to specific standard requirements. An amount of 10 g of each sample was dispersed in 90 mL of either peptone water or buffered peptone water and homogenized using a Stomacher SEWARD 400 (Seward, NY, USA). Incubation steps were performed using Panasonic and Memmert incubators; the water activity was determined using an Aquaspector AQS-2-TC instrument (Nagy GmbH, Gäufelden, Germany), and additional required equipment such as Dilumat Start instrument (BioMérieux, Marcy-l'Étoile, France) and Laminar flow Faster vs.—4 (Faster SRL, Cornaredo, Italy) cabinet were used for performing microbiological assays.

2.6.1. Determination of the Total Aerobic Microbial Count (TAMC) I

Determination of contamination with both mesophilic and psychrotrophic microorganisms was performed according to standard ISO 4833-1:2014 [52] by plating 0.1 mL of each sample on Plate Count Agar (PCA—Oxoid, UK) using a Drigalski spatula, followed by incubation at 30 °C for three days. After the incubation period, all the colonies from the plates were numbered. Interpretation of results was performed using the following Formula (4):

$$CFU = (\sum C)/((n_1 + 0.1 n_2) \times d) \quad (4)$$

where: CFU = average no. of colony forming units from two serial dilutions; $\sum C$ = sum of colonies counted in all retained plates; n_1 = number of plates retained at first dilution; n_2 = number of plates retained at the second dilution; d = dilution from which the first counts were made.

2.6.2. Determination of the Total Yeast and Mold Count (TYMC)

Determination of TYMC was performed according to standard ISO 21527-2:2009 [53] by dispersing 0.1 mL of samples inoculum using a Drigalski spatula on the surface of Dichloran Glicerol Agar (DG 18) followed by incubation at 25 °C for seven days. Various molds and yeasts grow on this medium, DG-18 (Oxoid, UK) being specifically recommended for selective isolation of xerophilic molds from food samples. After the incubation period, the colonies were counted and analyzed according to Formula (4).

2.6.3. Determination of *Enterobacteriaceae*

The presence of *Enterobacteriaceae*, especially in food grade products, is in general considered an indicator of inadequate factory hygiene, prominently conditioned by the incipient microbial load of raw materials. According to the ISO 21528:2 [54] method, 1 mL of each sample was inoculated in liquefied Violet Red Bile Glucose (VRBG) media (Oxoid, UK), followed by a second layer of the same media after the solidification of the first layer. The plates were incubated at 37 °C for 24 ± 2 h. For the typical colonies (red–pink colonies), with or without a precipitation zone, biochemical tests for oxidase and glucose fermentation were performed for confirmation.

2.6.4. Determination of β-Glucuronidase Positive *Escherichia coli*

The presence of *E. coli* in food grade products is an indicator of contamination with organisms of fecal origin, thus corroborating this analysis with results from Section 2.4.3 indicates significant issues and concerns regarding compliance with hazard analysis and critical control points regulations applied to food grade products. The method described in ISO 16649-2:2007 [45] is based on the enumeration of *E. coli* using colony count technique. According to standard requirements, molten selective culture medium containing X-β-D-glucuronide, named Tryptone Bile X-glucuronide Agar (Oxoid, UK) was used for determination of β-glucuronidase positive *Escherichia coli*. Samples were incubated at 44 °C for 24 h in order to allow the selective growth of *E. coli*. On this selective chromogenic media, presence of blue colonies indicates detection of *E. coli*. The number of colony forming units (CFU) of β-glucuronidase-positive *E. coli* per gram (g) of sample was calculated using Formula (4).

2.7. Cytotoxicity Assessment by MTT and LDH

Evaluation of cytotoxicity requires sensitive, quantitative, reliable and automated methods for the precise determination of cell viability and death, therefore, we selected the MTT and LDH assays for this task.

The mouse fibroblasts L929 cell line (ECACC—European Collection of Authenticated Cell Cultures) was selected for cytotoxicity assessment of pea powder, almond powder, yeast powder, spirulina powder and *Pleurotus* spp. flour. All samples were analyzed from a starting concentration of 500 mg/mL and serially diluted in order to obtain the following concentrations: 250 mg/mL, 62.5 mg/mL, 7.813 mg/mL, and 0.651 mg/mL.

Cultivation of L929 cells was performed according to ECACC specifications in DMEM (Dulbecco's Modified Eagle Medium, Sigma-Aldrich, St. Louis, MO, USA) media supplemented with 10% FBS (Fetal Bovine Serum—Sigma-Aldrich) and 1% Pen/Strep (penicillin/streptomycin solution, 50 µg/mL—Sigma-Aldrich) for 24 h at 37 °C, at 95% humidity with 5% CO_2. After 24 h cells were washed using PBS (Phosphate Buffered Solution—Sigma-Aldrich), harvested using trypsin (Sigma-Aldrich) and counted using Trypan Blue (Sigma-Aldrich) and a hemocytometer. The seeding density for the MTT and LDH assays was optimized at 5×10^4.

Using the MTT we measured the cellular metabolic activity, which is an indicator of cell viability, proliferation and cytotoxicity. This colorimetric assay is based on the reduction of a yellow tetrazolium salt (3-(4,5-dimethylthiazol-2-yl)-2,5-diphenyltetrazolium bromide or MTT) to purple formazan crystals by metabolically active cells [55]. The mechanism of function relies on viable cells containing NAD(P)H-dependent oxidoreductase enzymes that reduce the MTT to formazan. Moreover, the insoluble formazan crystals formed are dissolved using a cell grade solvent and the resulting colored solution is quantified by measuring its absorbance at 500–600 nm, depending on various kits specifications. This non-radioactive, colorimetric assay system using MTT was first described by Mosmann T. (1983) [56] and improved in subsequent years by several other investigators. Since disturbances in cellular metabolism cannot fully confirm the actual toxicity of a certain compound, elucidation of the process associated with reduction of cellular population is important, therefore we selected LDH assay for quantification of plasma membrane damage. Upon cellular damage, NAD^+ is reduced to $NADH/H^+$ by the LDH-catalyzed conversion of lactate to pyruvate. Subsequently, the catalyst transfers H/H+ from NADH/H+ to the tetrazolium salt INT which is reduced to formazan. An increase in the amount of dead or plasma membrane-damaged cells results in an increase in LDH enzyme activity in the culture supernatant [56].

Cells seeded at 5×10^4 density in a clear 96 well cell culture plate were treated with pea powder, almond powder, yeast powder, spirulina powder and *Pleurotus* spp. flour (samples concentrations: 500 mg/mL, 250 mg/mL, 62.5 mg/mL, 7.813 mg/mL, and 0.651 mg/mL) and incubated for 24 h at 37 °C, 95% humidity with 5% CO_2. After 24 h of exposure to tested compounds, cells were incubated for 4 h with MTT reagent (Roche) at 37 °C, 95% humidity with 5% CO_2. After incubation, cells were treated with MTT solvent (Roche) for 15 min at room temperature. Absorbance was measured using a spectrophotometric microplate reader (ELISA reader) at OD = 570 nm. For the LDH assay we used the LDH Cytotoxicity Detection Kit (Roche). Cells exposed to tested compounds were treated with LDH reagent mixture for 15 min at 37 °C. Afterwards, the LDH activity was measured using a spectrophotometric microplate reader (ELISA reader) at 492 nm with a 600 nm wavelength reference.

2.8. Statistical Analysis

Statistical evaluations were performed using GraphPad Prism 9 (San Diego, CA, USA). Biological data were analyzed using the two-way ANOVA with Dunnett's multiple comparison test. For physico-chemical assessments, data were expressed as means ± SD determined by triplicate analysis. The statistical analysis was conducted using a one-way ANOVA with Tukey test. The level of significance was set to $p < 0.05$.

3. Results

3.1. Assessment of the Nutritional Profile and Physico-Chemical Properties

Determining key physico-chemical properties is an important step in analyzing any kind of dietary supplement.

Determination of total proteins for our samples put out high percentages of protein for the pea powder (77.96 ± 0.70%) and Spirulina powder (64.79 ± 2.20%). Determination of total extractable fat revealed that almond powder has the highest lipid content out of the tested samples (6.91 ± 0.10%). Results for all tested samples are presented in Table 1.

Table 1. Nutritional profile for the raw materials with high protein content.

Samples	Moisture (%)	Protein (%)	Fat (%)	Ash (%)	Carbohydrates (%)	Energy Values (kcal)
Pea powder	7.43 ± 0.06 [b]	77.96 ± 0.70 [a]	0.29 ± 0.03 [d]	3.77 ± 0.19 [e]	10.55 ± 0.73 [e]	356.65 ± 0.56 [c]
Yeast powder	4.43 ± 0.18 [d]	46.23 ± 1.44 [d]	0.30 ± 0.03 [d]	5.80 ± 0.30 [c]	43.24 ± 1.27 [b]	360.58 ± 1.26 [b]
Spirulina powder	7.31 ± 0.35 [b]	64.79 ± 2.20 [b]	1.10 ± 0.05 [bc]	7.54 ± 0.26 [a]	19.26 ± 2.35 [d]	346.10 ± 2.22 [e]
Almond powder	6.04 ± 0.10 [c]	52.11 ± 1.55 [c]	6.91 ± 0.10 [a]	6.60 ± 0.07 [b]	28.34 ± 1.77 [c]	383.99 ± 0.23 [a]
Pleurotus spp. flour	8.20 ± 0.15 [a]	16.75 ± 0.26 [e]	1.36 ± 0.24 [b]	5.66 ± 0.20 [cd]	68.03 ± 0.70 [a]	351.36 ± 0.97 [d]

a–e: different superscript letters in the same column indicate significantly different values for $p < 0.05$ by Tukey's test; same superscript letters in the same column indicate not significantly different values for $p > 0.05$ by Tukey's test

Figure 1 shows the average spectra of the five high protein ingredients analyzed. Similar bands can be observed for all five ingredients but of different intensity, which indicates the same components but of different concentrations. The FTIR spectra was divided into two distinct parts. The first one, between 750 cm^{-1} and 1900 cm^{-1} is associated with stretching vibration of C=C, C=O, C-C, C-N and C-O (Figure 1, light blue zone). The range between 2750 cm^{-1} and 3700 cm^{-1} (Figure 1, light pink zone), corresponds to the vibration of bands that contain hydrogen atoms, such as C-H, O-H and N-H. The assignment of representative bands (Table 2) was performed by analyzing lipids and protein samples and taking into account the literature data [57–59].

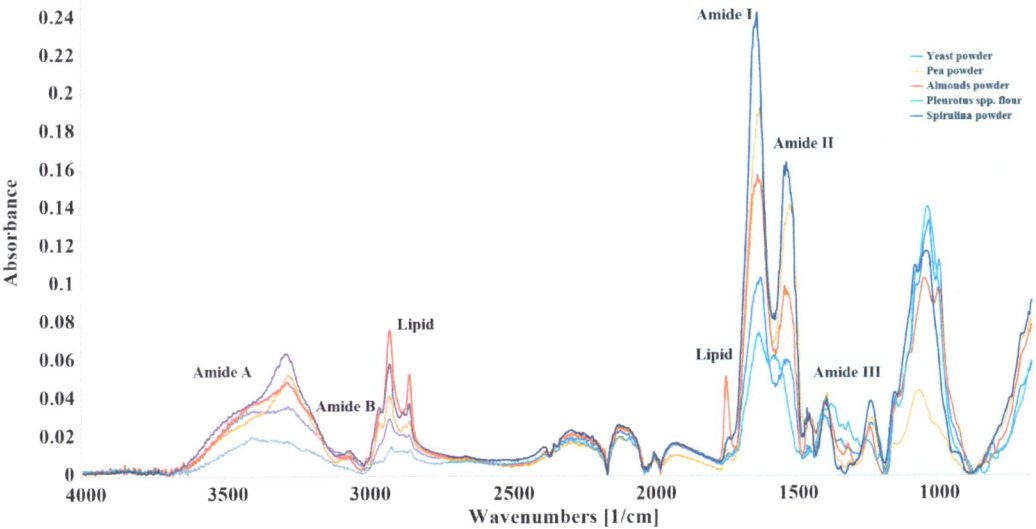

Figure 1. ATR-FTIR spectra of yeast powder (orange), pea powder (blue), almond powder (green), *Pleurotus* spp. flour (light blue), spirulina powder (red).

In order to facilitate the comparison of the five analyzed FTIR spectra, they were all normalized to the maximum. The peptide group, the structural repeat unit of proteins, shows characteristic bands named amide A, B, I, II, III. The two prominent features, namely, the amide I (1600–1650 cm^{-1}) and amide II (approximately at 1530 cm^{-1}) bands are the two major bands of the protein infrared spectrum. The amide I band is mainly associated with the C=O stretching vibration with minor contribution from the out-of-phase C-N stretching vibration, the CCN deformation and is directly related to the backbone conformation. Amide II results from the N-H bending vibration and from the C-N stretching vibration with smaller contributions from C=O in plane bend and the C-C and N-C stretching vibrations. Like for the amide I vibration, the amide II vibration of proteins is affected by side chain vibration. Amide III (1200–1400 cm^{-1}) is a very complex band resulting from a mixture of several coordinated displacements, such as the in-phase contribution from the N-H bending and the C-N stretching vibration. The bands in the region of 1740 cm^{-1} and 2830–3010 cm^{-1} are associated with the lipids from the sample composition. The peak intensity from 1744 cm^{-1} for the almond powder sample, compared to the other samples, is in correlation with the results obtained for the total fats analysis, which reveals that almond powder has the highest lipid content out of the tested samples (6.91%) (Table 1).

Table 2. Wavenumbers assignment to the characteristic bands.

Sample	Amide A (cm^{-1})	Amide B (cm^{-1})	Amide I (cm^{-1})	Amide II (cm^{-1})	Amide III (cm^{-1})	Lipid (cm^{-1})
Yeast powder	3277	3010–3060	1625	1525	1180–1430	1733 2835–2990
Pleurotus spp. flour	3207	3020–3230	1631	1571	1185–1430	1710 2800–2945
Pea powder	3265	3010–3115	1626	1522	1185–1425	1740 2830–3000
Almond powder	3279	3020–3100	1636	1533	1200–1475	1744 2800–3020
Spirulina powder	3281	3010–3112	1638	1533	1185–1480	1734 2830–3000

The N-H stretching vibration gives rise to the amide A band between 3130 cm^{-1} and 3400 cm^{-1}. Its frequency depends on the strength of the hydrogen bond. The amide A band is usually part of a Fermi resonance doublet with the second component absorbing weakly between 3010 cm^{-1} and 3100 cm^{-1} (amide B).

The protein content obtained by the Bradford assay was comparatively lower than that obtained using the Kjeldahl method for all the samples (Tables 1 and 3), the main cause being that only the total nitrogen content was measured through Kjeldahl, whereas other biomolecules containing nitrogen, such as amino acids, nucleic acids and other nitrogenous compounds [60] or insoluble proteins may have been present. Among the products tested, the highest soluble protein content was obtained for almond powder while yeast powder had the lowest content, probably due to its nucleic acid content.

Table 3. Evaluation of total soluble proteins, free amino acids, phenol and flavonoid content and their antioxidant activity.

Sample	Total Soluble Proteins (mg/g)	Total Free Amino Acids (mg/g)	Total Phenols (mg GAE/g)	Total Flavonoids (mg QE/g)	DPPH (mM TEAC/g)	FRAP (mM TEAC/g)
Yeast powder	2.91 ± 0.16 [d]	13.00 ± 0.24 [b]	0.88 ± 0.04 [a]	0.60 ± 0.12 [e]	2.16 ± 0.69 [d]	0.82 ± 0.04 [d]
Pleurotus spp. flour	6.15 ± 0.55 [cd]	42.25 ± 1.61 [a]	6.51 ± 0.09 [b]	4.68 ± 0.1 [a]	5.48 ± 0.29 [c]	13.15 ± 1.19 [a]
Pea powder	12.73 ± 0.51 [b]	6.31 ± 0.05 [cd]	1.25 ± 0.02 [c]	0.98 ± 0.07 [d]	1.04 ± 0.17 [e]	0.52 ± 0.04 [d]
Almonds powder	95.25 ± 4.69 [a]	8.04 ± 0.5 [c]	3.90 ± 0.15 [d]	3.26 ± 0.21 [b]	6.56 ± 0.37 [b]	8.11 ± 0.63 [b]
Spirulina powder	6.23 ± 0.86 [c]	6.55 ± 0.43 [cd]	5.83 ± 0.43 [d]	2.68 ± 0.1 [c]	10.2 ± 0.13 [a]	7.52 ± 0.47 [bc]

a–e: different superscript letters in the same column indicate significantly different values for $p < 0.05$ by Tukey's test; same superscript letters in the same column indicate not significantly different values for $p > 0.05$ by Tukey's test.

In this study, the antioxidant activity was measured through two independents assays: FRAP and DPPH. Table 3 shows that the *Pleurotus* spp. flour had strong antioxidant activity through the FRAP method and was correlated with the capacity of total free amino acids (chelation metals) and in a low manner with the phenol content and flavonoids. For almond powder, where a higher content of soluble proteins was obtained, the antioxidant activity was low through both methods. For spirulina powder, the antioxidant activity could be given by c-phycocyanins [61].

3.2. Evaluation of the Microbiological Status

The results obtained from the microbiological analysis of the five products reveal the lack of contamination with pathogenic bacteria, in this case, *Escherichia coli*. The total number of aerobic bacteria, expressed in cfu/g, recorded the highest values in the case of *Pleurotus* spp. powder (4.6×10^5 CFU/g or 5.66 ± 0.1 log CFU/g), and in the case of spirulina and pea powder, the degree of contamination was lower, 10^3 cfu/g (Table 4). *Pleurotus* spp. powder has a high degree of contamination with TAMC, TYMC and *Enterobacteriaceae*, so the sample presents a microbiological risk regarding the quality of the finished products that may include this powder.

Table 4. Results for microbiological and stability indicators.

Samples	Total Aerobic Count CFU/g	E. coli CFU/g	Enterobacteriaceae CFU/g	Yeast and Molds CFU/g	Water Activity Value
Yeast powder	<10	<10	<10	<10	0.355 ± 0.1
Pea powder	7.0×10^3 $3.85 \log \pm 0.07$	<10	<10	<10	0.408 ± 0.17
Almond powder	<10	<10	<10	<10	0.424 ± 0.20
Spirulina powder	3.1×10^3 $3.49 \log \pm 0.31$	<10	<10	<10	0.363 ± 0.18
Pleurotus powder	4.6×10^5 $5.66 \log \pm 0.1$	<10	5.2×10^3 $3.72 \log \pm 0.27$	2.7×10^3 $3.43 \log \pm 0.18$	0.550 ± 0.17

Almond and yeast powder do not show microbiological contamination, thus presenting a satisfactory quality. Water activity is a critical parameter when it comes to food safety. It is a crucial factor in the microbiological control of food products. Water activity (aw) is defined as the equilibrium state achieved when a hygroscopic material is placed in a sealed container, and a balance is established between the material and the air above it. The relative humidity, which occurs at a constant air temperature, corresponds to the value of water activity multiplied by 100 (aw = relative humidity (%)/100). The majority of food items have a water activity level above 0.95, providing enough moisture to support the growth of bacteria, yeasts and mold.

The samples tested showed a low value of water activity, but the samples with aw value close to 0.600 presented microbial growth; in this case, *Pleurotus* spp. powder was the most contaminated sample tested (aw was 0.550).

3.3. Cytotoxicity Assessment by MTT and LDH

We selected colorimetric assays for cytotoxicity assessment. The MTT assay is a very popular and widely used colorimetric assay for the evaluation of cytotoxicity [62] as it provides beneficial aspects such as rapidity, reliability and significant knowledge regarding the metabolic activity of cells. The LDH assay provides information about cellular damage (lysis of cell's membranes) after cell treatment, and complements the MTT assay in drawing conclusions about the potential mechanism of action.

Our samples were analyzed from a starting concentration of 500 mg/mL, as the scientific literature indicates [63–68]. The results obtained show that this concentration provokes a significant cell viability reduction in the case of *Pleurotus* spp. flour (up to 77% viability loss), followed by yeast powder (up to 69% viability loss). The subsequent concentrations (250 mg/mL, 62.5 mg/mL, 7.813 mg/mL and 0.651 mg/mL) tend to follow the same pattern (as seen in Figure 2A). Pea powder and almond powder present similar results, with viability reduction rates of up to 30%. Spirulina powder has great biocompatibility results (except for a concentration of 500 mg/mL—which is higher than the recommended concentration for most similar supplements). In the case of yeast powder, concentrations lower that 7 mg/mL have beneficial effects and act as nutritive substrates for cells.

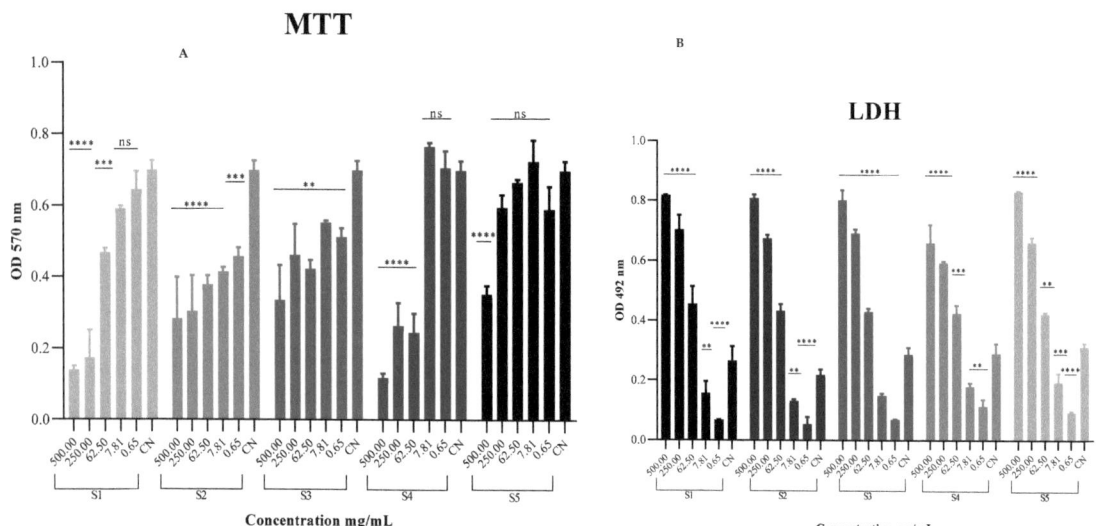

Figure 2. (**A**) Assessment of cytotoxicity using MTT; (**B**). Assessment of cytotoxicity using LDH assay. S1—*Pleurotus* spp. flour; S2—pea powder; S3—almond powder; S4—yeast powder; S5—spirulina powder; CN = Control—untreated NCTC cells with culture media; ****, ***, ** = p value < 0.05; ns = statistically non-significant.

Results obtained by the LDH assay (Figure 2B) conferee great evidence regarding cellular membrane integrity. All samples analyzed showcased membrane damage over 7.81 mg/mL. This result demonstrates that tested ingredients have an impact on the metabolic activity of cells—as MTT assay results indicate, and affect the cells integrity, provoking leakage of DNA.

4. Discussion

The human body does not store amino acids in the same way that it does fatty acids or carbohydrates. As a result, it is essential to ensure that the daily intake of amino acids needed for protein synthesis and other specific metabolic functions is sufficient. The concentration of amino acids in the blood remains relatively stable, which means that inadequate dietary protein intake can increase muscle protein breakdown, while excessive protein intake can lead to the breakdown of proteins for energy [69].

In the United States, the population's average recommended dietary allowance for protein is 0.8 grams of protein per kilogram of body weight per day [49]. In the European Union, based on nitrogen balance data, the European Food Safety Authority (EFSA) has established an average requirement of 0.66 grams of protein per kilogram of body weight per day for healthy adults, regardless of gender. Therefore, the recommendation for daily protein intake has been set at 0.83 grams of protein per kilogram of body weight per day [70], which corresponds to around 10–12% of total energy intake (E%). However, these recommendations may vary at the national level. For example, the Spanish recommendation is 0.93–1.2 g of protein per kilogram of body weight per day [71] and the Finnish recommendation is 1.1–1.3 grams of protein per kilogram of body weight per day [72].

The protein content of raw materials is strongly influenced by the environment (pedo-climatic conditions, soil chemistry, pollution), variety and processing grade, which justifies the variability between the studies published so far [17,73–76]. Results obtained by our study indicate a high percentage of protein for the pea powder (77.96 ± 0.70%) and Spirulina powder (64.79 ± 2.20%); out of the tested ingredients, these are the most balanced in terms of fat and carbohydrate percentages, and their correlation with energy values. *Pleurotus* spp. has a low fat content, which is in accordance with the literature data. The lipid fraction of mushrooms generally comprises representative compounds from all lipid classes (free fatty acids, mono-, di- and triglycerides, sterols, sterol esters and phospholipids). Pea and yeast powders had a low fat content but with a high energy value due to a high protein content in the case of pea powder, and carbohydrates for yeast powder.

According to Soenen et al. (2012) [77], it was found that the relatively high protein intake highlights the success of the so-called "low-carbohydrate" diet, which is usually high in protein. Reduced carbohydrate intake had no effect on decreasing body weight and fat mass during energy restriction, while increasing daily absolute protein promoted body weight loss while reducing fat mass during the weight loss phase. Even though the highest energy value was obtained for almond powder due to the higher fat content and yeast powder for the high carbohydrate content, the preferred raw materials are pea and spirulina powder due to the total protein content.

According to Friedman et. al., 2004 [78], it was shown that the ninhydrin test for the quantification of amino groups in protein structure was more difficult than the analysis of free amino acids. The method of quantifying amino acids by Cd-ninhydrin has proven to be more effective in highlighting α-amino acids. In this context, a higher content of free amino acids for *Pleurotus* spp. flour was shown, while the soluble protein content was 6.15 mg/g. The administration of nutritional sources with a higher content of free amino acids are more indicated due to the absorption rate and assimilation efficiency higher than proteins [79].

Flavonoids are phenolic compounds with strong antioxidant activity by scavenging free radicals based on ET (electron transfer) mechanisms (DPPH) or by chelating transition metals (FRAP). Results obtained show that *Pleurotus* spp. flour had strong antioxidant activity through the FRAP method and was correlated with the capacity of total free amino acids (chelation metals) and in a low manner with phenol content and flavonoids. Almond powder has a high content of soluble proteins but with low antioxidant activity, an effect demonstrated by both antioxidant methods. For the spirulina powder, the antioxidant effect may be due to c-phycocyanins according to Zhou et. al., 2005 [61]. These compounds are known as the main light-harvesting protein pigment with nutritional and therapeutical properties.

Mesophilic aerobic bacteria and fungi are two of the most common types of microorganisms responsible for food spoilage. To prevent the contamination of finished food products with spoilage microorganisms, it is important to maintain a high level of hygiene throughout the entire food processing and handling chain. Preventing the contamination of food products with spoilage microorganisms requires a combination of good hygiene practices, effective sanitation protocols and ongoing monitoring and quality control. Results obtained for the TAMC and TYMC assessments indicate <10 CFU/g values for the yeast and almond powders. Pea, spirulina and *Pleurotus* spp. powders had $> 3 \times 10^3$ microbial contamination levels. If for pea and spirulina powders the microbial load is considered acceptable according to specific standards [52,53], for *Pleurotus* spp. powder, TAMC value of 4.6×10^5 CFU/g – 5.66 log ± 0.1 CFU/g and TYMC of 2.7×10^3 – 3.43 log ± 0.18 CFU/g are exceeding standard limits and cannot be considered suitable for integration for consumption or for composition of other products. As for determinations regarding β-glucuronidase positive *Escherichia coli*, neither of the samples tested positive, results being in accordance with standardized requirements [80].

Another category of microorganisms that are present in food in poor hygiene cases are those of the family *Enterobacteriaceae*. Bacteria from the *Enterobacteriaceae* family are commonly found in either dry or humid environments, are sensitive to heat treatment and sanitation and are desirable to be monitored in food environments. The presence of *Enterobacteriaceae* in post-hygienic processes may draw attention to the inefficiency of hygiene procedures. Our results indicate values of <10 CFU/g for yeast, pea, almond and spirulina powders. According to standardized requirements [54], these results comply. *Pleurotus* spp. powder showed a contamination of $5.2 \times 10^3 - 3.72 \log \pm 0.27$ CFU/g that exceeds standard limitations. These results indicate a high microbial contamination risk for the *Pleurotus* spp. powder. In order to exploit its beneficial effects, decontamination pre-treatments are necessary.

The assessment of toxicity of any given substance that is intended to be exposed to living organisms relies on in vitro evaluation of the dose–response. The dose–response relationship, or exposure–response relationship describes the magnitude of the response of an organism as a function of exposure (or doses) to a stimulus or stressor (usually a chemical) after a certain exposure time.

Results obtained demonstrate that tested ingredients have an impact on the metabolic activity of cells—as MTT assay results indicate, and affect the cells' integrity, provoking leakage of DNA. All in all, slow or delayed metabolic cellular activity indicates loss of balance and survival cues from the microenvironment. Such modifications do not support proper development and healthy homeostasis of mature tissues [81]. Here, we struggle with a debacle, as for some ingredients the recommended dosage for optimal effects is cytotoxic. Therefore, it should be reduced and supplemented with other products that help build the desired effect.

According to the EU Food Safety outlines, foodstuffs of animal and plant origin may present a microbiological risk. Microbiological criteria regarding the acceptability of foodstuffs and their manufacturing processes are obligated to comply with Good Hygiene and Manufacturing Practices (GHP, GMP) and the Hazard Analysis Critical Control Point (HACCP) principles that contribute to achieving food safety. Microbiological testing alone cannot guarantee the safety of a foodstuff tested, but these criteria provide objectives and reference points to assist food manufacturers and competent authorities in their activities to manage and monitor the safety of foodstuffs, respectively. Moreover, Commission Regulation (EC) EN No 2073/2005 on microbiological criteria for foods lays down food safety criteria for relevant foodborne bacteria, their toxins, and metabolites. Besides the high protein content that supports certain dietary needs (e.g., peas), the high-energy yield and notable antioxidant activity, increased levels of contamination in products tested through this study can pose several risks to human health (e.g., allergies and improper interactions with other drugs), besides false claims and regulatory issues.

5. Conclusions

Dietary supplementation does not benefit from stricter regulations regarding dosage (as pharmaceuticals do), health claims, contaminants or fraudulent practices, therefore consumers should be advised. At this moment, it falls into their responsibility to better examine what they consume and to genuinely evaluate safety. As our study presented, beyond the desirable physico-chemical and antioxidant properties, the wary microbiological status and cytotoxicity degree of such products are important parameters to consider for safe consumption. As for future perspectives, further studies are envisioned. We seek to demonstrate and confirm, on suitable final formulations, the potential beneficial effects of obtained products in order to safely place them into the current market.

Our view is that the regulatory body responsible for overseeing the placement of dietary supplements on the Romanian market should encourage manufacturers to contribute more to the research community by conducting additional testing on their products. This would be beneficial for both the manufacturers and the research community, as it could help improve existing regulations and promote the placement of safer dietary supplements in the therapeutic market.

Author Contributions: Conceptualization, B.-M.T. and I.-C.M.; data curation, B.-M.T., I.-C.M. and L.M.Ș.; formal analysis, I.-C.M.; funding acquisition, D.E.D.; methodology, B.-M.T., I.-C.M., M.A. (Marian Adascălului), A.D., G.G.P. and M.A. (Mădălina Axinie); project administration, D.E.D.; resources, M.A. (Marian Adascălului), M.A. (Mădălina Axinie), L.M.Ș. and D.E.D.; writing—original draft, B.-M.T., I.-C.M., M.A. (Marian Adascălului), A.D., G.G.P. and M.A. (Mădălina Axinie); writing—review and editing, B.-M.T., I.-C.M., L.M.Ș. and D.E.D. All authors have read and agreed to the published version of the manuscript.

Funding: This work was supported by a grant of the Ministry of Research and Innovation through Program 1—Development of the National R&D System, Subprogram 1.2—Institutional Performance—Projects for Excellence Financing in RDI, project no. 26 PFE/17.10.2018. Also, this study was achieved through Core Programme (PN 19 02 01 01) and Program 1–Development of the National R&D System, Subprogram 1.2-Institutional Performance-Projects for Excellence Financing in RDI, project no. 17 PFE/2021 with the support of the Ministry of Research, Innovation and Digitization.

Institutional Review Board Statement: Not applicable.

Informed Consent Statement: Not applicable.

Data Availability Statement: The data presented in this study is available at links associated with each reference and within manuscript.

Acknowledgments: We acknowledge the support of ICUB—Earth Environment and Life Science Department.

Conflicts of Interest: The authors declare no conflict of interest. The funders had no role in the design of the study, in the collection, analyses, or interpretation of data, in the writing of the manuscript, or in the decision to publish the results.

References

1. Bartels, C.L.; Miller, S.J. Dietary Supplements Marketed for Weight Loss. *Nutr. Clin. Pract.* **2003**, *18*, 156–169. [CrossRef] [PubMed]
2. Samal, J.R.K.; Samal, I.R. Protein Supplements: Pros and Cons. *J. Diet. Suppl.* **2017**, *15*, 365–371. [CrossRef]
3. Europe Dietary Supplements Market (2021–2027): Market Forecast by Types, Function Type, Form, End User, Country and Competitive Landscape. Available online: https://www.researchandmarkets.com/reports/5360076/europe-dietary-supplements-market-2021-2027 (accessed on 26 April 2023).
4. Micha, R.; Michas, G.; Mozaffarian, D. Unprocessed Red and Processed Meats and Risk of Coronary Artery Disease and Type 2 Diabetes – An Updated Review of the Evidence. *Curr. Atheroscler. Rep.* **2012**, *14*, 515–524. [CrossRef] [PubMed]
5. Schwingshackl, L.; Hoffmann, G. Dietary fatty acids in the secondary prevention of coronary heart disease: A systematic review, meta-analysis and meta-regression. *BMJ Open* **2014**, *4*, e004487. [CrossRef] [PubMed]
6. Pan, A.; Chen, M.; Chowdhury, R.; Wu, J.H.; Sun, Q.; Campos, H.; Mozaffarian, D.; Hu, F.B. α-Linolenic acid and risk of cardiovascular disease: A systematic review and meta-analysis. *Am. J. Clin. Nutr.* **2012**, *96*, 1262–1273. [CrossRef]
7. Yokoyama, Y.; Nishimura, K.; Barnard, N.D.; Takegami, M.; Watanabe, M.; Sekikawa, A.; Okamura, T.; Miyamoto, Y. Vegetarian Diets and Blood Pressure: A Meta-analysis. *JAMA Intern. Med.* **2014**, *174*, 577–587. [CrossRef]
8. Shanthakumar, P.; Klepacka, J.; Bains, A.; Chawla, P.; Dhull, S.B.; Najda, A. The Current Situation of Pea Protein and Its Application in the Food Industry. *Molecules* **2022**, *27*, 5354. [CrossRef]
9. Ge, J.; Sun, C.; Corke, H.; Gul, K.; Gan, R.; Fang, Y. The health benefits, functional properties, modifications, and applications of pea (*Pisum sativum* L.) protein: Current status, challenges, and perspectives. *Compr. Rev. Food Sci. Food Saf.* **2020**, *19*, 1835–1876. [CrossRef]
10. Salles, J.; Guillet, C.; Le Bacquer, O.; Malnero-Fernandez, C.; Giraudet, C.; Patrac, V.; Berry, A.; Denis, P.; Pouyet, C.; Gueugneau, M.; et al. Pea Proteins Have Anabolic Effects Comparable to Milk Proteins on Whole Body Protein Retention and Muscle Protein Metabolism in Old Rats. *Nutrients* **2021**, *13*, 4234. [CrossRef]
11. Maykish, A.; Nishisaka, M.M.; Talbott, C.K.; Reaves, S.K.; Kristo, A.S.; Sikalidis, A.K. Comparison of Whey Versus Almond Protein Powder on Nitrogen Balance in Female College Students; The California Almond Protein Powder Project (CAlmond-P³). *Int. J. Environ. Res. Public Health* **2021**, *18*, 11939. [CrossRef]
12. Becerra-Tomás, N.; Paz-Graniel, I.; Kendall, C.W.C.; Kahleova, H.; Rahelić, D.; Sievenpiper, J.L.; Salas-Salvadó, J. Nut consumption and incidence of cardiovascular diseases and cardiovascular disease mortality: A meta-analysis of prospective cohort studies. *Nutr. Rev.* **2019**, *77*, 691–709. [CrossRef]

13. Barreca, D.; Nabavi, S.M.; Sureda, A.; Rasekhian, M.; Raciti, R.; Silva, A.S.; Annunziata, G.; Arnone, A.; Tenore, G.C.; Süntar, I.; et al. Almonds (*Prunus dulcis* Mill. D. A. Webb): A Source of Nutrients and Health-Promoting Compounds. *Nutrients* **2020**, *12*, 672. [CrossRef] [PubMed]
14. Alfadhly, N.K.Z.; Alhelfi, N.; Altemimi, A.B.; Verma, D.K.; Cacciola, F.; Narayanankutty, A.; Rohn, S.; Alfadhly, N.K.Z.; Alhelfi, N.; Altemimi, A.B.; et al. Trends and Technological Advancements in the Possible Food Applications of Spirulina and Their Health Benefits: A Review. *Molecules* **2022**, *27*, 5584. [CrossRef] [PubMed]
15. Gogna, S.; Kaur, J.; Sharma, K.; Prasad, R.; Singh, J.; Bhadariya, V.; Kumar, P.; Jarial, S. Spirulina—An Edible Cyanobacterium with Potential Therapeutic Health Benefits and Toxicological Consequences. *J. Am. Nutr. Assoc.* **2022**, 1–14. [CrossRef] [PubMed]
16. Kumar, K. Nutraceutical Potential and Processing Aspects of Oyster Mushrooms (*Pleurotus* Species). *Curr. Nutr. Food Sci.* **2020**, *16*, 3–14. [CrossRef]
17. Lavelli, V.; Proserpio, C.; Gallotti, F.; Laureati, M.; Pagliarini, E. Circular reuse of bio-resources: The role of *Pleurotus* spp. in the development of functional foods. *Food Funct.* **2018**, *9*, 1353–1372. [CrossRef]
18. Raman, J.; Jang, K.Y.; Oh, Y.L.; Oh, M.; Im, J.H.; Lakshmanan, H.; Sabaratnam, V. Cultivation and Nutritional Value of Prominent *Pleurotus* spp.: An Overview. *Mycobiology* **2020**, *49*, 1–14. [CrossRef]
19. Jach, M.E.; Serefko, A.; Ziaja, M.; Kieliszek, M. Yeast Protein as an Easily Accessible Food Source. *Metabolites* **2022**, *12*, 63. [CrossRef]
20. Liao, Y.; Zhou, X.; Peng, Z.; Li, D.; Meng, S.; Xu, S.; Yang, X.; Liu, L.; Yang, W. Muscle aging amelioration by yeast protein supplementation was associated with gut microbiota. *J. Funct. Foods* **2022**, *89*, 104948. [CrossRef]
21. Rizzello, C.G.; Tagliazucchi, D.; Babini, E.; Rutella, G.S.; Saa, D.L.T.; Gianotti, A. Bioactive peptides from vegetable food matrices: Research trends and novel biotechnologies for synthesis and recovery. *J. Funct. Foods* **2016**, *27*, 549–569. [CrossRef]
22. Agrawal, H.; Joshi, R.; Gupta, M. Isolation, purification and characterization of antioxidative peptide of pearl millet (*Pennisetum glaucum*) protein hydrolysate. *Food Chem.* **2016**, *204*, 365–372. [CrossRef] [PubMed]
23. Sarmadi, B.H.; Ismail, A. Antioxidative peptides from food proteins: A review. *Peptides* **2010**, *31*, 1949–1956. [CrossRef] [PubMed]
24. Zarei, M.; Ebrahimpour, A.; Abdul-Hamid, A.; Anwar, F.; Abu Bakar, F.; Philip, R.; Saari, N. Identification and characterization of papain-generated antioxidant peptides from palm kernel cake proteins. *Food Res. Int.* **2014**, *62*, 726–734. [CrossRef]
25. Lu, X.; Zhang, L.; Sun, Q.; Song, G.; Huang, J. Extraction, identification and structure-activity relationship of antioxidant peptides from sesame (*Sesamum indicum* L.) protein hydrolysate. *Food Res. Int.* **2018**, *116*, 707–716. [CrossRef] [PubMed]
26. Coda, R.; Rizzello, C.G.; Pinto, D.; Gobbetti, M. Selected Lactic Acid Bacteria Synthesize Antioxidant Peptides during Sourdough Fermentation of Cereal Flours. *Appl. Environ. Microbiol.* **2012**, *78*, 1087–1096. [CrossRef] [PubMed]
27. Maestri, E.; Marmiroli, M.; Marmiroli, N. Bioactive peptides in plant-derived foodstuffs. *J. Proteom.* **2016**, *147*, 140–155. [CrossRef] [PubMed]
28. Elias, R.J.; Kellerby, S.S.; Decker, E.A. Antioxidant Activity of Proteins and Peptides. *Crit. Rev. Food Sci. Nutr.* **2008**, *48*, 430–441. [CrossRef]
29. Wen, C.; Zhang, J.; Zhang, H.; Duan, Y.; Ma, H. Plant protein-derived antioxidant peptides: Isolation, identification, mechanism of action and application in food systems: A review. *Trends Food Sci. Technol.* **2020**, *105*, 308–322. [CrossRef]
30. Qian, Z.-J.; Jung, W.-K.; Kim, S.-K. Free radical scavenging activity of a novel antioxidative peptide purified from hydrolysate of bullfrog skin, Rana catesbeiana Shaw. *Bioresour. Technol.* **2008**, *99*, 1690–1698. [CrossRef]
31. Rajapakse, N.; Mendis, E.; Jung, W.-K.; Je, J.-Y.; Kim, S.-K. Purification of a radical scavenging peptide from fermented mussel sauce and its antioxidant properties. *Food Res. Int.* **2005**, *38*, 175–182. [CrossRef]
32. Nwachukwu, I.D.; Aluko, R.E. Structural and functional properties of food protein-derived antioxidant peptides. *J. Food Biochem.* **2019**, *43*, e12761. [CrossRef] [PubMed]
33. Du, Z.; Comer, J.; Li, Y. Bioinformatics approaches to discovering food-derived bioactive peptides: Reviews and perspectives. *TrAC Trends Anal. Chem.* **2023**, *162*, 117051. [CrossRef]
34. Zhu, Y.; Lao, F.; Pan, X.; Wu, J. Food Protein-Derived Antioxidant Peptides: Molecular Mechanism, Stability and Bioavailability. *Biomolecules* **2022**, *12*, 1622. [CrossRef]
35. Satija, A.; Bhupathiraju, S.N.; Spiegelman, D.; Chiuve, S.E.; Manson, J.E.; Willett, W.; Rexrode, K.M.; Rimm, E.B.; Hu, F.B. Healthful and Unhealthful Plant-Based Diets and the Risk of Coronary Heart Disease in U.S. Adults. *J. Am. Coll. Cardiol.* **2017**, *70*, 411–422. [CrossRef] [PubMed]
36. Poore, J.; Nemecek, T. Reducing food's environmental impacts through producers and consumers. *Science* **2018**, *360*, 987–992. [CrossRef] [PubMed]
37. Mistry, K.; Sardar, S.D.; Alim, H.; Patel, N.; Thakur, M.; Jabbarova, D.; Ali, A. Plant Based Proteins: Sustainable Alternatives. *Plant Sci. Today* **2022**, *9*, 820–828. [CrossRef]
38. Langyan, S.; Yadava, P.; Khan, F.N.; Dar, Z.A.; Singh, R.; Kumar, A. Sustaining Protein Nutrition through Plant-Based Foods. *Front. Nutr.* **2022**, *8*, 1237. [CrossRef]
39. Alcorta, A.; Porta, A.; Tárrega, A.; Alvarez, M.D.; Vaquero, M.P. Foods for Plant-Based Diets: Challenges and Innovations. *Foods* **2021**, *10*, 293. [CrossRef]
40. Hertzler, S.R.; Lieblein-Boff, J.C.; Weiler, M.; Allgeier, C. Plant Proteins: Assessing Their Nutritional Quality and Effects on Health and Physical Function. *Nutrients* **2020**, *12*, 3704. [CrossRef]

41. The EU Agri-Food Fraud Network and the Administrative Assistance and Cooperation System 2020 Annual Report Health and Food Safety. Available online: https://food.ec.europa.eu/safety/eu-agri-food-fraud-network_en (accessed on 14 April 2023).
42. DIN ISO 6673; Green Coffee—Determination of Loss in Mass at 105 °C (ISO 6673:2003). Available online: https://www.en-standard.eu/din-iso-6673-green-coffee-determination-of-loss-in-mass-at-105-c-iso-6673-2003/ (accessed on 14 April 2023).
43. ISO 20483:2013; Cereals and Pulses—Determination of the Nitrogen Content and Calculation of the Crude Protein Content—Kjeldahl Method. Available online: https://www.iso.org/standard/59162.html (accessed on 31 May 2023).
44. ISO 6492; Animal Feeding Stuffs—Determination of Fat Content. 1999. Available online: https://www.iso.org/standard/12865.html (accessed on 14 April 2023).
45. BS EN ISO 2171:2010; Cereals, Pulses and By-Products. Determination of Ash Yield by Incineration. 2010. Available online: https://landingpage.bsigroup.com/LandingPage/Standard?UPI=000000000030213066 (accessed on 14 April 2023).
46. Mattila, P.; Mäkinen, S.; Eurola, M.; Jalava, T.; Pihlava, J.-M.; Hellström, J.; Pihlanto, A. Nutritional Value of Commercial Protein-Rich Plant Products. *Plant Foods Hum. Nutr.* **2018**, *73*, 108–115. [CrossRef]
47. Bradford, M.M. A rapid and sensitive method for the quantitation of microgram quantities of protein utilizing the principle of protein-dye binding. *Anal. Biochem.* **1976**, *72*, 248–254. [CrossRef]
48. Singleton, V.L.; Orthofer, R.; Lamuela-Raventós, R.M. Analysis of total phenols and other oxidation substrates and antioxidants by means of folin-ciocalteu reagent. *Methods Enzymol.* **1999**, *299*, 152–178. [CrossRef]
49. Woisky, R.G.; Salatino, A. Analysis of propolis: Some parameters and procedures for chemical quality control. *J. Apic. Res.* **1998**, *37*, 99–105. [CrossRef]
50. Madhu, G.; Bose, V.C.; Aiswaryaraj, A.; Maniammal, K.; Biju, V. Defect dependent antioxidant activity of nanostructured nickel oxide synthesized through a novel chemical method. *Colloids Surfaces: A Physicochem. Eng. Asp.* **2013**, *429*, 44–50. [CrossRef]
51. Benzie, I.F.F.; Strain, J.J. Ferric reducing/antioxidant power assay: Direct measure of total antioxidant activity of biological fluids and modified version for simultaneous measurement of total antioxidant power and ascorbic acid concentration. *Methods Enzymol.* **1999**, *299*, 15–27. [CrossRef] [PubMed]
52. European Standards. UNE EN ISO 4833-1:2014; Microbiology of the Food Chain—Horizontal Method for the Enumeration of Microorganisms—Part 1: Colony Count at 30 Degrees C by the Pour Plate Technique (ISO 4833-1:2013). Available online: https://www.en-standard.eu/une-en-iso-4833-1-2014-microbiology-of-the-food-chain-horizontal-method-for-the-enumeration-of-microorganisms-part-1-colony-count-at-30-degrees-c-by-the-pour-plate-technique-iso-4833-1-2013/ (accessed on 14 April 2023).
53. EVS-ISO 21527-2:2009; Microbiology of Food and Animal Feeding Stuffs — Horizontal Method for the Enumeration of Yeasts and Moulds — Part 2: Colony Count Technique in Products with Water Activity Less Than or Equal to 0,95 (ISO 21527-2:2008). Available online: https://www.evs.ee/en/evs-iso-21527-2-2009 (accessed on 31 May 2023).
54. ISO 21528-2:2017; Microbiology of the Food Chain—Horizontal Method for the Detection and Enumeration of Enterobacteriaceae—Part 2: Colony-Count Technique. International Organization for Standardization: Geneva, Switzerland, 2017.
55. Tihăuan, B.M.; Berca, L.M.; Adascalului, M.; Sanmartin, A.M.; Nica, S.; Cimponeriu, D.; Duță, D. Experimental in vitro cytotoxicity evaluation of plant bioactive compounds and phytoagents: A review. *Rom. Biotechnol. Lett.* **2020**, *25*, 1832–1842.
56. TMosmann, T. Rapid colorimetric assay for cellular growth and survival: Application to proliferation and cytotoxicity assays. *J. Immunol. Methods* **1983**, *65*, 55–63. [CrossRef]
57. Naumann, A.; Heine, G.; Rauber, R. Efficient discrimination of oat and pea roots by cluster analysis of Fourier transform infrared (FTIR) spectra. *Field Crop. Res.* **2010**, *119*, 78–84. [CrossRef]
58. Sharma, S.; Raj, N. FTIR Spectroscopic Characterisation of Almond Varieties (*Prunus dulcis*) from Himachal Pradesh (India). *Int. J. Curr. Microbiol. Appl. Sci.* **2018**, *7*, 887–898. [CrossRef]
59. Parmar, R.; Kumar, D. Study of chemical composition in wild edible mushroom *Pleurotus cornucopiae* (Paulet) from Himachal Pradesh, India by using Fourier transforms infrared spectrometry (FTIR), Gas chromatography-mass spectrometry (GCMS) and X-ray fluorescence (XRF). *Biol. Forum* **2015**, *7*, 1057–1066.
60. Mæhre, H.K.; Dalheim, L.; Edvinsen, G.K.; Elvevoll, E.O.; Jensen, I.-J. Protein Determination—Method Matters. *Foods* **2018**, *7*, 5. [CrossRef] [PubMed]
61. Zhou, Z.-P.; Liu, L.-N.; Chen, X.-L.; Wang, J.-X.; Chen, M.; Zhang, Y.-Z.; Zhou, B.-C. Factors that effect antioxidant activity of c-phycocyanins from *Spirulina platensis*. *J. Food Biochem.* **2005**, *29*, 313–322. [CrossRef]
62. Präbst, K.; Engelhardt, H.; Ringgeler, S.; Hübner, H. Basic colorimetric proliferation assays: MTT, WST, and resazurin. In *Methods in Molecular Biology*; Humana Press Inc.: Totowa, NJ, USA, 2017; Volume 1601.
63. Babault, N.; Païzis, C.; Deley, G.; Guérin-Deremaux, L.; Saniez, M.-H.; Lefranc-Millot, C.; Allaert, F.A. Pea proteins oral supplementation promotes muscle thickness gains during resistance training: A double-blind, randomized, Placebo-controlled clinical trial vs. Whey protein. *J. Int. Soc. Sports Nutr.* **2015**, *12*, 3–9. [CrossRef] [PubMed]
64. Guasch-Ferré, M.; Liu, X.; Malik, V.S.; Sun, Q.; Willett, W.C.; Manson, J.E.; Rexrode, K.M.; Li, Y.; Hu, F.B.; Bhupathiraju, S.N. Nut Consumption and Risk of Cardiovascular Disease. *J. Am. Coll. Cardiol.* **2017**, *70*, 2519–2532. [CrossRef] [PubMed]
65. Proserpio, C.; Lavelli, V.; Laureati, M.; Pagliarini, E. Effect of *Pleurotus ostreatus* powder addition in vegetable soup on ß-glucan content, sensory perception, and acceptability. *Food Sci. Nutr.* **2019**, *7*, 730–737. [CrossRef]

66. Takahashi, T.; Yu, F.; Zhu, S.-J.; Moriya, J.; Sumino, H.; Morimoto, S.; Yamaguchi, N.; Kanda, T. Beneficial Effect of Brewers' Yeast Extract on Daily Activity in a Murine Model of Chronic Fatigue Syndrome. *Evid. Based Complement. Altern. Med.* **2006**, *3*, 109–115. [CrossRef]
67. Torres-Duran, P.V.; Ferreira-Hermosillo, A.; Juarez-Oropeza, M.A. Antihyperlipemic and antihypertensive effects of Spirulina maxima in an open sample of mexican population: A preliminary report. *Lipids Health Dis.* **2007**, *6*, 33. [CrossRef]
68. Borges, A.; Ferreira, C.; Simões, M.; Wang, M.; Tao, L.; Xu, H.; Alexandri, M.; Papapostolou, H.; Vlysidis, A.; Gardeli, C.; et al. Antibacterial Activity and Mode of Action of Ferulic and Gallic Acids Against Pathogenic Bacteria. *Microb. Drug Resist.* **2013**, *19*, 256–265. [CrossRef]
69. Kårlund, A.; Gómez-Gallego, C.; Turpeinen, A.M.; Palo-Oja, O.-M.; El-Nezami, H.; Kolehmainen, M. Protein Supplements and Their Relation with Nutrition, Microbiota Composition and Health: Is More Protein Always Better for Sportspeople? *Nutrients* **2019**, *11*, 829. [CrossRef]
70. Agostoni, C.; Bresson, J.L.; Fairweather Tait, S.; Flynn, A.; Golly, I.; Korhonen, H.; Lagiou, P.; Løvik, M.; Marchelli, R.; Martin, A.; et al. Scientific Opinion on Dietary Reference Values for protein. *EFSA J.* **2012**, *10*, 2557. [CrossRef]
71. Sánchez-Muniz, F.J.; Bastida, S.L. En: *Libro Blanco de la Nutrición en España*; Fundación Española de la Nutrición (FEN): Madrid, Spain, 2013; pp. 113–124.
72. Fogelholm, M.; Hakala, P.; Kara, R. *Terveyttä Ruoasta-Suomalaiset Ravitsemussuositukset 2014 (Finnish Nutrition Recommendations 2014)*; State Nutrition Advisory Board: Helsinki, Finland, 2014.
73. PYen, P.P.; Pratap-Singh, A. Vacuum microwave dehydration decreases volatile concentration and soluble protein content of pea (*Pisum sativum* L.) protein. *J. Sci. Food Agric.* **2020**, *101*, 167–178. [CrossRef]
74. Wang, X.-X.; Zhao, Z.-H.; Chang, T.-S.; Liu, J.-G. Yeast screening from avermectins wastewater and investigation on the ability of its fermentation. *Bioprocess Biosyst. Eng.* **2011**, *34*, 1127–1132. [CrossRef] [PubMed]
75. Jin, S.-E.; Lee, S.J.; Kim, Y.; Park, C.-Y. Spirulina powder as a feed supplement to enhance abalone growth. *Aquac. Rep.* **2020**, *17*, 100318. [CrossRef]
76. Ahrens, S.; Venkatachalam, M.; Mistry, A.M.; Lapsley, K.; Sathe, S.K. Almond (*Prunus dulcis* L.) Protein Quality. *Plant Foods Hum. Nutr.* **2005**, *60*, 123–128. [CrossRef] [PubMed]
77. Soenen, S.; Bonomi, A.G.; Lemmens, S.G.; Scholte, J.; Thijssen, M.A.; van Berkum, F.; Westerterp-Plantenga, M.S. Relatively high-protein or 'low-carb' energy-restricted diets for body weight loss and body weight maintenance? *Physiol. Behav.* **2012**, *107*, 374–380. [CrossRef] [PubMed]
78. Friedman, M. Applications of the Ninhydrin Reaction for Analysis of Amino Acids, Peptides, and Proteins to Agricultural and Biomedical Sciences. *J. Agric. Food Chem.* **2004**, *52*, 385–406. [CrossRef]
79. Rønnestad, I.; Conceição, L.E.C.; Aragão, C.; Dinis, M.T. Free Amino Acids Are Absorbed Faster and Assimilated More Efficiently than Protein in Postlarval Senegal Sole (*Solea senegalensis*). *J. Nutr.* **2000**, *130*, 2809–2812. [CrossRef]
80. Microbiologia Alimentelor Şi Nutreţurilor. Metodă Orizontală Pentru Enumerarea Escherichia Coli Pozitive La Beta-Glucuronidază. Partea 2: Tehnica de Enumerare a Coloniilor La 44 Grade C Folosind 5-Bromo-4-Cloro-3-Indolil Beta-D-Glucuronat. Available online: https://magazin.asro.ro/ro/standard/142768 (accessed on 14 April 2023).
81. Mason, E.F.; Rathmell, J.C. Cell Metabolism: An Essential Link between Cell Growth and Apoptosis. *Biochim. Biophys. Acta Mol. Cell Res.* **2011**, *1813*, 645–654. [CrossRef]

Disclaimer/Publisher's Note: The statements, opinions and data contained in all publications are solely those of the individual author(s) and contributor(s) and not of MDPI and/or the editor(s). MDPI and/or the editor(s) disclaim responsibility for any injury to people or property resulting from any ideas, methods, instructions or products referred to in the content.

Article

The Influence of Texture Type and Grain Milling Degree on the Attenuation Limit, Protein Content, and Degradation in Wheat Wort

Vinko Krstanović [1], Kristina Habschied [1,*], Iztok Jože Košir [2] and Krešimir Mastanjević [1]

[1] Faculty of Food Technology Osijek, Josip Juraj Strossmayer University of Osijek, F. Kuhača 20, HR-31000 Osijek, Croatia; vkrstano@ptfos.hr (V.K.); kmastanj@gmail.com (K.M.)
[2] Slovenian Institute of Hop Research and Brewing, SI-3310 Žalec, Slovenia; iztok.kosir@ihps.si
* Correspondence: kristinahabschied@gmail.com; Tel.: +385-31-224-411

Featured Application: The results of the research should help in the application of wheat with a transitional (marbled) type of endosperm texture as a raw material for brewing purposes.

Abstract: Wheat of medium hardness with marbled endosperm (transitional form between hard and soft wheat) in which glassy and floury zones alternate, form almost exclusively available assortment for brewing needs in Southeastern Europe. The aim of this work is to establish the influence of the grain texture and the degree of milling on the attenuation limit of wheat wort obtained from this type of wheat. Wheat worts using hard, soft transitional, or marbled endosperm texture were produced. The indicators of proteolysis, cytolysis, and amylolysis were determined, with regard to the parameter attenuation limit (AL) or fermentability. From the results for the tested parameters, it was established that despite similar starting values for the most important quality parameters, transitional wheat produces significantly different wort, both among themselves and in comparison with hard and soft wheat, and also when looking at the results for different milling degree (fine or coarse). The obtained values for the attenuation limit for transition wheat are similar or even better when compared to soft wheat, with satisfactory values for almost all examined quality parameters of wort. It can be concluded that a transitional type of wheat can be used just as well as unmalted raw material for the production of wort, as well as a raw material for malting.

Keywords: hardness and vitreosity of endosperm; grist milling degree; fermentability of wheat wort

1. Introduction

Wheat (*Triticumm aestivum*) has been used for beer production for probably as long as barley. Wheat varieties selected for brewing are very rare and exist in only a few countries (mostly Germany). These varieties are commonly not favorable for growth in Southeast Europe [1]. Soft varieties are commonly suitable for brewing, particularly because they have lower protein content. Desirable characteristics for brewing wheat are given in a review paper by Faltermaier et al. [2], and the main remark is that a high protein content, which may be a benefit for farmers and bakers, is a fault for brewers, since it can prolong lautering times, cause filtration difficulties and fermentation problems in the brewery, as well as decreased flavor stability in the finished beer [3]. Ultimately, wheat's overall protein content is not much higher than barley's, but it contains a higher amount of high-molecular-weight (HMW) proteins which end up in wort [4,5] and contain a higher content of albumin and gluten [6,7].

European winter wheat varieties display a decreased content of proteins than spring varieties [8]. Red, hard wheat varieties commonly grown in the Pannonian Basin with typical continental climate which is often causing "forced maturation" phenomenon, are formally characterized as "hard". Regardless, they have many characteristics of soft wheat

varieties. These varieties are characterized by a pale reddish color, moderate hardness, almost completely transient vitreous grain, and the absence of awn. The forced maturation phenomenon is a result of a combination of increased air humidity and high temperatures during the grain-filling phase. This displays unsuitable effects on many cereal quality indicators, mostly affecting the protein content [9]. The hardness of wheat endosperm represents the main indicator of the wheat's suitability for its use for various end products and is closely related to its glassiness and protein content and structure, whereby glassy wheat can be considered to have an initial glassy grain content of $\geq 80\%$ [10]. According to the hardness of wheat grains, they are divided into hard and soft, and transitional or marbled. Transitional types of hardness occupy the largest share on the market and have determined hardness as hard, while in the industrial process, they behave like soft wheat, which is characterized by a high proportion of false or transient glassiness, i.e., that which is lost by soaking the wheat in water. These transitional forms of wheat are characterized by the so-called marbled endosperm, whose texture alternates between hard (glassy) and soft (floury) surfaces. Such wheat varieties typically belong to the II qualitative malting group. They are portrayed by increased total and soluble proteins and preferable wort viscosity) [11].

Unlike barley, there are no reference wheat varieties that would be used as standards. Narciss and Back [11], when assessing the quality of a certain variety for use in brewing, use a classification into 4 qualitative groups, of which 1 group is acceptable for brewing, and it is characterized by obtaining wort with a low proportion of soluble proteins (over 750 mg/L of soluble N in wort) and low viscosity (not over 1.6 mPa$\times$s 8.6%e), for the so-called soft wheat varieties. Such wheat is rarely present in Europe, so for the purposes of malt production or the use of wheat as unmalted raw material, suitable hybrids are used such as red hard wheat. They are characterized by their initial indicators (hardness, vitreosity) being more similar to true red hard wheat, while their behavior during milling is more similar to soft wheat (very high transitory vitreosity). As they have a significant proportion of glassy parts on the surface of the endosperm, the aim of this research was to establish how an increase in the degree of grain granulation can affect fermentability. On one hand, an increase in the degree of milling/granulation increases the ability of the enzyme complex (amylolysis, proteolysis, and cytolysis) to act on the cells of the endosperm, which would have a positive effect, and on the other hand, more non-fermentable ingredients (especially proteins) pass into the wort, which has a negative effect on the same.

Fermentability is usually affected by many agents such as variety, genetics, and phenotype. Process parameters during malting and mashing also contribute to the fermentability. All these factors complexly interlace and affect one another [12]. Thus, it is not easy to absolutely and doubtlessly connect fermentability with any of the quality indicators. The goal of the investigation was to establish how these transitional types of endosperm will behave during the milling process (fine and coarse granulation), that is, how the mentioned differences will affect the attenuation limit or fermentability of wheat wort.

2. Materials and Methods

2.1. Sample Preparation and Analysis

In order to avoid the influence of location and agrotechnical measures on the tested wheat, all samples were obtained from the same varietal experiment from the same location belonging to the Institute of Agriculture Osijek, and under the same agrotechnical growing conditions. The grain collected from these experiments was refined and untreated, separated, packaged, and stored in a dry and dark place for 3 months in order to overcome the so-called grain dormancy. For this experiment, two transitional or marbled varieties of medium hardness with similar hardness values, permanent and transient glassiness, and different proportions of starch and total proteins, were selected for testing. These are typical bread varieties (sample 2 = Tika-Taka; sample 3 = Bezostaja). Both varieties showed a very high proportion of transient vitreosity with the highest value for the glassy surface of the endosperm in the range between 20–60%. Along with them, one hard (sample 1 = Golubica) and one soft variety (sample 4 = Indira) were tested. Golubica is designated as a hard wheat variety, with high protein content and high hardness, while Indira is a variety with higher starch share and significantly lower hardness.

Wheat quality parameters were determined (for the vast majority of parameters) according to Analytica-EBC [13] methods: 3.4/4.4. thousand corn weight; 3.2/4.2 moisture; NIR-hardness; NIR-protein content (AACC Method 39–70A) [14] and starch (ICC method 169) using Infratec 1241 Grain Analyzer (Foss, Hilleroed, Denmark).

The vitreosity of wheat is determined by the ICC visual method Standard 129 [15], which is determined using Pohl's grain cutter (Farinotom, Sadkiewicz Instruments, Bydgoszcz, Poland). After the cut had been made, a visualization and designation of gray (hard, glassy) areas of the endosperm was performed by three trained analysts. The vitreosity of the cut grains was expressed as a percentage as described by [16].

Prior to wheat mash production, wheat malt was milled to two granulations, 1 mm and 0.2 mm using an IKA (Staufen, Germany) laboratory mill (Model MF10). Granulations were separated using a vibratory Sieve Shaker with a standard series of sieves (Retch, Haan, Germany).

Wheat mash was produced using a 50:50% wheat:barley malt ratio (2 L mash of each variety), by a standard mashing program (EBC® method 4.5.1.) and analyzed. The following EBC® methods were used for parameter determination: 4.1.4.5.1.1. total N; 49.1. soluble N; 4.1.4.11. Hartong number; 4.10 α-amino N (FAN); 4.5.1. the specific gravity of mash/wort, fine extract, saccharification time, appearance of mash/wort, and filtration time; 4.8. viscosity 4.11. attenuation limit of mash/wort (AL). High molecular N (HMW N) was determined according to MEBAK® [17] method 2.9.3.1., medium molecular N (MMW N) was determined as the difference between HMW N and LMW N and low molecular N (LMW N) MEBAK® method 2.9.3.2.

2.2. Statistical Analysis

Results were subjected to analysis of variance (ANOVA) and Fisher's least significant difference test (LSD). The *p*-value was set to be significant at <0.05. Statistica 13.1. (TIBCO Software Inc., Palo Alto, CA, USA) was the software of choice for this analysis.

3. Results and Discussion

The results of the initial analysis of the tested wheat are shown in Table 1. The hardness of the endosperm is an extremely important parameter for the use of wheat as a raw material in brewing because the harder endosperm delays the rate of hydration and enzyme modification during the malting process [18]. It is observed that only sample 1 can be considered as hard wheat with a borderline vitreosity (≥80%). All tested wheat have a very high transition vitreosity, even in the case of, as can be seen from the results for NIR-hardness, the hardest wheat (sample 1), which, as expected, also had the highest proportion of protein amounting to 13.3%.

Table 1. Basic raw material quality indicators used for the production of wheat wort.

ID No.	Mass 1000 Grains (g)	Hardness NIR-HD	Total Vitreosity (%)	Permanent Vitreosity (%)	Transient Vitreosity (%)	Protein Content (%)	Starch (%)
1	39.9 [c]	94.1 [a]	64 [a]	30 [a]	70 [d]	13.3 [a]	68.8 [c]
2	44.9 [a]	77.2 [b]	32 [c]	0 [c]	100 [a]	10.2 [c]	71.7 [b]
3	43.6 [b]	77.0 [b]	58 [b]	10 [b]	90 [c]	12.6 [b]	68.4 [c]
4	43.6 [b]	48.0 [c]	12 [d]	2 [c]	98 [b]	9.7 [d]	72.7 [a]

[a–d] Means within columns with different superscripts are significantly different ($p < 0.05$); sample 1 = Golubica (hard); sample 2 = Tika-Taka (medium hard); sample 3 = Bezostaja (medium hard); sample 4 = Indira (soft).

Similarly, as expected, soft wheat (sample 1) had the highest proportion of starch (72.7%) with the lowest proportion of protein (9.7%). In general, the results for the proportion of starch were very high for all tested varieties. A very high proportion of high transition vitreosity indicates what was already said in the introduction, that transitional wheats will behave more like soft wheats regardless of the relatively high initial (total) hardness, which is favorable for the production of mash [19].

This is confirmed by the results shown in Table 2 displaying the relationship between permanent and transition vitreuos vitreosity ity, which is particularly pronounced for the transitional type of endosperm hardness (samples 2 and 3). Vitreosity is a property easily tested by soaking the grains for 24 h period. Truly hard wheats show little changes in the results after 24 h of soaking, while transitional wheats show large changes in vitreosity, i.e., a decrease in it. From Table 2 it is visible that designated hard wheat (sample 1) had, before soaking, the highest share of 100–80% vitreous grains, 52%, which was lowered to 26% after 24 h soaking. Transitional wheats (samples 2 and 3) showed a significant shift towards lower vitreosity (below 60%) after soaking. This indicates that they are truly transient and have more soluble proteins which can cause problems during filtration of wort and beer. Similarly, this was the case with soft wheat (sample 4), where the vitreosity shifted towards the lower end after soaking for 24 h. However, further analysis is to show whether hard, soft, and transient wheats can be utilized for brewing with regards to grain milling degree. To test this thesis grain was milled into two granulations, coarse (1 mm) and fine (0.2 mm).

Table 2. Separated values for the degree and nature of vitreosity of wheat endosperm (before and after soaking of 24 h).

ID No.		Degree of Grain Vitreosity (%)						
		100	100–80	80–60	60–40	40–20	20–0	0
1	before soaking	12 [b]	52 [a]	20 [a]	12 [d]	4 [f]	0 [f]	0 [d]
	after soaking	4 [c]	26 [c]	22 [a]	20 [a]	20 [b]	8 [e]	0 [d]
2	before soaking	18 [a]	14 [d]	6 [cd]	26 [b]	12 [cd]	24 [c]	0 [d]
	after soaking	0 [d]	0 [f]	4 [d]	26 [b]	32	38 [b]	0 [d]
3	before soaking	18 [a]	40 [b]	20 [a]	8 [e]	10 [d]	2 [f]	2 [d]
	after soaking	2 [c]	8 [e]	8 [c]	30 [a]	32 [a]	14 [d]	6 [c]
4	before soaking	0 [d]	12 [d]	14 [b]	20 [c]	16 [c]	24 [c]	14 [b]
	after soaking	2 [c]	0 [f]	8 [c]	8 [e]	8 [e]	52 [a]	22 [a]

[a–f] Means within columns with different superscripts are significantly different ($p < 0.05$); sample 1 = Golubica (hard); sample 2 = Tika-Taka (medium hard); sample 3 = Bezostaja (medium hard); sample 4 = Indira (soft).

Reference or target values when it comes to wheat as a raw material do not actually exist, but general requirements are set for the weight of 1000 grains as high as possible, hardness as low as possible, glassiness as low as possible, and of a transient character as low as possible, proteins as low as possible. The low share of proteins simultaneously increases the proportion of starch because it is about the indicators that are in the so-called formal correlation (together they add up to 100) so decreasing one increases the other. Reference values in terms of target values or certain limits for the values of individual indicators exist when it comes to wheat malt and we often use them when presenting the results for the quality of wheat malt in our works; we have attached one such table with the specified values below. However, in this paper, we used unmalted wheat as a raw material for obtaining wheat wort and could not use those values for wheat malt for comparison with the values for wort obtained in this paper.

The results in Table 3 show the influence of the type of endosperm texture and differences in milling degree on the main quality indicators of wheat wort. When considering the influence of the type of endosperm texture as the main indicator in fine granulation samples, a very clear and significant difference was observed between hard and soft wheat, i.e., the corresponding wort. The main indicators of proteolysis are the Kolbach index, soluble proteins, and FAN values [19,20]. Hard wheat has significantly higher values for soluble protein and FAN, but also for HM N and MM N, with a significantly lower filtration time and wort viscosity. All tested kinds of wheat have very good values for soluble protein for both fine and coarse granulation.

Table 3. Results of the quality indicators of wheat wort for different types of hardness and granulations.

Sample No.	Granulations								Barley Malt
	1		2		3		4		
Parameters	Fine 0.2 mm	Coarse 1 mm	Fine 0.2 mm	Coarse 1 mm	Fine 0.2 mm	Coarse 1 mm	Fine 0.2 mm	Coarse 1 mm	
Moisture (grain) (%)	10.4 a	10.0 bc	9.9 c	10.2 abc	10.2 abc	10.1 bc	10.3 ab	10.2 abc	3.9
Extract (grain) (%)	71.3 e	71.9 cd	74.6 a	74.8 a	70.3 f	72.1 c	72.7 b	71.9 d	79.4
Total N (mg/100 mL)	59.5 b	55.0 d	53.5 f	56.0 c	59.4 b	61.6 a	53.1 g	54.2 e	75.4
HM N (mg/100 mL)	24.8 ab	24.1 c	20.8 e	22.4 d	24.5 bc	25.2 a	19.2 g	20.4 f	21.1
MM N (mg/100 mL)	5.2 d	2.1 g	2.6 f	4.8 e	5.7 b	6.9 a	5.6 bc	5.4 cd	6.7
LM N (mg/100 mL)	29.5 b	28.7 d	30.1 a	28.8 d	29.2 c	29.5 b	28.4 e	28.4 e	47.6
FAN α-amino N (mg/L)	76.8 d	78.3 b	66.2 g	84.3 a	68.7 e	77.5 c	66.6 f	77.4 c	133.8
Proteins (%)	12.7 a	12.5 b	10.1 d	10.0 d	11.9 c	12.0 c	9.8 d	9.6 e	9.8
Soluble N (mg/L)	595 b	550 d	535 f	560 c	594 b	616 a	531 g	542 e	754
Soluble N (%dm)	0.57 a	0.52 bc	0.51 c	0.54 b	0.57 a	0.59 a	0.51 c	0.52 c	0.67
Kolbach index (%)	25.2 f	23.7 g	28.4 c	30.4 a	26.8 e	27.6 d	29.4 b	30.5 a	41.0
Hartong 45° (%)	29.6 b	30.1 a	29.6 b	29.5 bc	29.2 c	28.7 d	28.2 e	28.8 d	40.8
Viscosity (mPa×s)	1.426 bc	1.430 b	1.451 a	1.451 a	1.401 d	1.417 c	1.423 bc	1.401 d	1.478
Specif. gravity (g/mL)	1.0321 cd	1.0323 bc	1.0329 a	1.0329 a	1.0319 d	1.0323 bc	1.0324 b	1.0322 bc	1.0340
Attenuation limit (%)	86.3 d	82.4 g	93.3 a	82.9 f	88.9 c	78.7 h	92.9 b	84.1 e	86.0
Extract (%)	75.67 e	75.92 d	77.27 b	77.40 a	75.17 f	76.03 d	76.34 c	75.94 d	79.41
Sacchar. time (min)	20	20	25	25	20	20	25	25	10
Filtration time (min)	35	20	30	25	25	20	45	25	70
Appearance of wort	clear	clear	clear	clear	clear	slightly opalescent	slightly opalescent	slightly opalescent	slightly opalescent

a–g Means within columns with different superscripts are significantly different ($p < 0.05$); dm—dry matter; N—nitrogen; HMW N—high molecular weight N; MMW N—medium molecular weight; LMW N—low molecular weight N; sample 1 = Golubica (hard); sample 2 = Tika-Taka (medium hard); sample 3 = Bezostaja (medium hard); sample 4 = Indira (soft).

For the attenuation limit (AL), a very clear difference between fine and coarse granulation was established, in such a way that fine granulation has a significantly higher AL compared to coarse in all types of wheat except soft, where this difference for AL is the smallest. The values for AL were very good for all tested types of wheat with the expected lowest value for hard wheat (coarse granulation, 82.4%), but it is interesting that both transition wheats have values for AL similar to soft wheat (both granulations). In the case of transitional types (samples 2 and 3), the only significant difference between these wheats is in the proportion of total proteins (Table 1) where sample 2 had 10.2%, and sample 3 had 12.6% of proteins.

The distribution of the glassiness of the endosperm surface after soaking (Table 2) is also very similar, with the final vitreosity being entirely transient. Compared to hard wheat, both worts from transitional types of wheat had a significantly lower proportion of protein and soluble N and FAN (Table 3). Sample 3 had the highest content of soluble N (616 mg/L) in coarse granulation, while FAN content was highest in sample 2 (84.3 mg/L for coarse granulation). In sample 2, which had the highest extract in the grain (74.8% in coarse granulation) and fine extract (77.27% in fine granulation) in the wort, a value obtained for AL was (higher than in soft wheat), together with values for total soluble proteins, HMW N, MMW N LMW N, and FAN, similar or even better in compared to soft wheat.

In this investigation, significant differences between fine and coarse granulation were found in all types of wheat. When it comes to the difference between fine and coarse milling granulation, Einsiedler et al. [21–23] found that an increase in milling degree results in an increase in the contact surface available for enzyme action, as a result of which, during isothermal mashing, the release of amino acids in mash occurs much faster and more complete, even in wheat with lower proteolytic power. He also established that when it comes to amylolysis, the increase in the milling degree has very little effect on the faster release of low molecular sugars, because the availability of places for the enzymatic reaction does not increase with the increase in the milling degree, while the activity of α- and β-amylase is halved in poorly modified endosperm.

When it comes to cytolysis, the authors concluded that the fineness of the granulation does not significantly affect starch degradation, but it does significantly affect the breakdown (transfer into mash) of β-glucan and protein degradation. Schneider [24] investigated the influence of the milling degree on the mashing process and found that the increase in the milling degree leads to an increase in the breakdown of high-molecular protein fractions, which consequently leads to an increase in the concentration of their degradation products in wort, with the largest increase of FAN concentration already at the very beginning of the mashing process. When using wheat as unmalted raw material (16% substitution for malt), it was found that the concentrations of soluble proteins, HMW, MWM, and LWM are similar to those of the control worts produced from barley malt, and the concentrations of formol N and FAN are even lower [25], while in the case of using wheat malt, the degree of grain modification (as a consequence of the applied malting procedure) has a very significant influence on their concentration in the wort [26].

Kühbeck et al. [24] examined the influence of different milling procedures (upward/isothermal) and preparation procedures (milling procedure, grits modification, grits: liquor ratio), from where indicators of cytolysis, proteolysis, and amylolysis were measured β-glucan, FAN and extract. He concluded that malt modification is the most important factor responsible for the breakdown of β-glucan and the release of FAN, while starch modification is responsible for extract yield, and for milling degree, he did not find that it significantly affects the mentioned indicators except in the case of poorly modified malt. From the results in (Table 3), it can be seen that in hard wheat (sample 1), the concentration of total N, medium molecular weight N (MM N), and soluble N increases significantly with the increase of milling degree, and with a significant increase of filtration time. For the attenuation limit (AL), a very clear difference between fine and coarse granulation was established, in such a way that fine granulation has a significantly higher AL compared to coarse in all types of wheat, except for soft, where this difference for AL is the smallest.

As for the main indicators of protein breakdown (Kolbach index, soluble proteins, and FAN), no significant difference was found between fine and coarse milling, nor for extract, viscosity, and Hartong 45°, so this significant increase in AL can be attributed to the increase in milling degree. The same trend of a significant increase in AL is also observed for transitional wheats (samples 2 and 3), although with them, with an increase in the milling degree, a significant decrease in MM N and FAN and a slightly less pronounced increase in the filtration time were observed. Compared to barley, wheat has a lower proportion of β-glucan and a higher arabinoxylan content, which should result in higher mash viscosity and reduced filtration volume [27,28]. As a reason for increasing the filtration time and in the case of very good values for the viscosity of wheat wort authors [25] state the possibility of forming a protein or protein-polysaccharide gel on the filter, which was also observed when unmalted wheat was used as a substitute for part of the malt in the infusion [25].

When the results for viscosity and Hartong number are observed, the already mentioned negative influence of the "forcing maturation" effect caused by the climate, characterized by high temperatures and high humidity towards the end of the growing season, results in a lower enzymatic strength compared to northern European wheats. This effect could be responsible for the lower proportion of soluble in total pentosans also compared to Northern European wheats [29], which also affects the viscosity of wort. In the case of soft wheat (sample 4), AL is also significantly higher for fine granulation with a very high, practically unacceptable, filtration time and a significant increase in viscosity for the same. The best values for AL, but also almost all tested parameters, were shown by sample no. 2 (Tika-Taka) with fine grit granulation (ø 0.2 mm). It can be concluded that the transitional type of wheat, which meets the criteria of classification into 2 qualitative groups, according to Narziss [11], can be just as good unmalted raw material for the production of wort, and raw material for malting, as soft wheat.

4. Conclusions

From the results for the tested transition wheats, it is evident that, despite similar starting values for the most important quality parameters, they behave very differently when it comes to mash obtained from them, both among themselves and in comparison with hard and soft wheat, and also when observing the results for different milling degree. The obtained values for the attenuation limit for transition wheats (fine grind—sample 2, 93.3% and sample 3, 88.9%) are similar or better when compared to soft wheats (92.9% for fine grist), with satisfactory values for almost all examined quality parameters of wort. It can be concluded that transitional type wheat can be utilized, as unmalted and even malted raw material for the production of wort, as soft wheat. Further research should be focused on obtaining hybrid, winter varieties that would display values for primarily proteolytic quality indicators as similar as possible to soft wheat. For breweries, this could mean an easier way to obtain wheat beer since the raw material would be suitable for brewing.

Author Contributions: Conceptualization, V.K.; methodology, I.J.K. and K.H.; software, K.M.; validation, I.J.K.; formal analysis, I.J.K.; investigation, V.K.; data curation, K.M.; writing—original draft preparation, V.K.; writing—review and editing, K.H.; visualization, K.H.; supervision, V.K. All authors have read and agreed to the published version of the manuscript.

Funding: This research received no external funding.

Institutional Review Board Statement: Not applicable.

Informed Consent Statement: Not applicable.

Data Availability Statement: Data are available upon request to the corresponding author.

Conflicts of Interest: The authors declare no conflict of interest.

References

1. Depraetere, S.; Delvaux, F.; Coghe, S.; Delvaux, F. Wheat variety and barley malt properties: Influence on haze intensity and foam stability of wheat beer. *J. Inst. Brew.* **2004**, *110*, 200–206. [CrossRef]
2. Faltermaier, A.; Waters, D.; Becker, T.; Arendt, E.; Gastl, M. Common wheat (*Triticum aestivum* L.) and its use as a brewing cereal—A review. *J. Inst. Brew.* **2014**, *120*, 1–15. [CrossRef]
3. Faltermaier, A.; Waters, D.; Becker, T.; Arendt, E.K; Gastl, M. Protein modifications and metabolic changes taking place during the malting of common wheat (*Triticum aestivum* L.). *J. Am. Soc. Brew. Chem.* **2013**, *71*, 153–160.
4. Delvaux, F.; Gys, W.; Michiels, J.; Delvaux, F.R.; Delcour, J.A. Contribution of wheat and wheat protein fractions to the col-loidal haze of wheat beers. *J. Am. Soc. Brew. Chem.* **2001**, *59*, 135–140.
5. Taylor, D.G. Brewing ales with malted cereals other than barley. *Ferment* **2000**, *1*, 18–20.
6. Brijs, K.; Delvaux, F.; Gilis, V.; Delcour, J.A. Solubilisationand degradation of wheat gluten proteins by barley malt proteolyticenzymes. *J. Inst. Brew.* **2002**, *108*, 348–354. [CrossRef]
7. Leach, A.A. Nitrogenous components of worts and beersbrewed from all-malt and malt plus wheat flour grists. *J. Inst. Brew.* **1968**, *74*, 183–192. [CrossRef]
8. Aiken, R.M.; O'Brien, D.M.; Olson, B.L.; Murray, L. Replacing fallow with continuous cropping reduces crop water productivity of semiarid wheat. *Agron. J.* **2013**, *105*, 199–207. [CrossRef]
9. Tabatabaei, S.A. The changes of germination characteristics and enzyme activity of Barley seeds under accelerated aging. *Cercet. Agron. Mold.* **2015**, *48*, 61–67. [CrossRef]
10. Baasandorj, T.; Ohm, J.B.; Simsek, S. Effect of dark, hard, and vitreous kernel content on protein molecular weight distribution and on milling and breadmaking quality characteristics for hard spring wheat samples from diverse growing regions. *Cereal. Chem.* **2015**, *92*, 570–577. [CrossRef]
11. Narziss, L.; Back, W. *Die Technologie Der Malzbereitung*, 7th ed.; Band, I., Ed.; Chapt. 10 Spezieler typ von malz; Wiley-VCH: Stuttgart, Germany, 1999.
12. Flavours of Wheat Beers. Available online: https://cdn.uclouvain.be/public/Exports%20reddot/inbr/documents/Zarnkow_Wheat_Flavour__Louvain_2008.pdf (accessed on 12 August 2019).
13. EBC Analysis Committee. *Analytica EBC*; Fachverlag Hans Carl: Nürnberg, Germany, 2010; ISBN 978-3-418-00759-5.

14. AACC. *Approved Methods of the AACC*, 10th ed.; AACC Method 39-70A; American Association of Cereal Chemists: St. Paul, MN, USA, 2000.
15. International Association for Cereal Science and Technology (ICC). Wheat. In *Quality Assurance and Safety of Crops & Foods*; Method 129; ICC: Vienna, Austria, 1980; ISBN 0913250791.
16. Hoseney, R.C. Structure of cereals. In *Principles of Cereal Sciences and Technology*; AACC: St. Paul, MN, USA, 1986; pp. 1–33.
17. Li, X.; Yang, J.; Xie, L.; Jin, Y.; Liu, J.; Xing, W. The effect of barley to wheat ratio in malt blends on protein composition and physicochemical characteristics of wort and beer. *Food Sci.* **2023**, *88*, 193–203. [CrossRef]
18. Sacher, B.; Narziß, L. Rechnerische Auswertungen von Kleinmälzeungsversuchen mit Winterweizen unter besonderer Berücksichtigung der Ernte 1991. *Mschr. Brauwiss.* **1992**, *45*, 404–412.
19. Guo, M.; Xu, K.; Wang, Z. Effect of kilning on the composition of protein and arabinoxylan in wheat malt. *J. Inst. Brew.* **2019**, *125*, 288–292. [CrossRef]
20. Psota, V.; Musilová, M. System for the evaluation of malting quality of wheat varieties. *Kvas. Prum* **2020**, *66*, 232–238. [CrossRef]
21. Einsiedler, F.; Schwill-Miedaner, A.; Sommer, K. Experimentelle Untersuchungen und Modellierung komplexer biochemischer und technologischer Prozesse am Beispiel des Maischens. Teil 1: Proteolyse. *Mschr. Brauwiss.* **1997**, *50*, 164–171.
22. Einsiedler, F.; Schwill-Miedaner, A.; Sommer, K.; Hämäläinen, J. Experimentelle Untersuchungen und Modellierung komplexer biochemischer und technologischer Prozesse am Beispiel des Maischens. Teil 2: Amylolyse. *Mschr. Brauwiss.* **1997**, *50*, 202–209.
23. Einsiedler, F.; Schwill-Miedaner, A.; Sommer, K.; Hämäläinen, J. Experimentelle Untersuchungen und Modellierung komplexer biochemischer und technologischer Prozesse am Beispiel des Maischens. Teil 3: Cytolyse. *Mschr. Brauwiss.* **1998**, *51*, 11–21.
24. Kühbeck, F.; Dickel, T.; Krottenthaler, M.; Back, W.; Mitzscherling, M.; Delgado, A.; Becker, T.J. Effects of Mashing Parameters on Mash β-Glucan, FAN and Soluble Extract Levels. *J. Inst. Brew.* **2005**, *111*, 316–327. [CrossRef]
25. Krstanović, V.; Habschied, K.; Lukinac, J.; Jukić, M.; Mastanjević, K. The Influence of partial substitution of malt with unmalted wheat in grist on quality parameters of Lager beer. *Beverages* **2020**, *6*, 7–20. [CrossRef]
26. Krstanović, V.; Habschied, K.; Mastanjević, K. Research of Malting Procedures for Winter Hard Wheat Varieties—Part II. *Foods* **2021**, *10*, 147–158. [CrossRef] [PubMed]
27. Lu, J.; Li, Y. Effects of arabinoxylan solubilization on wort viscosity and filtration when mashing with grist containing wheat and wheat malt. *Food Chem.* **2006**, *98*, 164–170. [CrossRef]
28. Dervilly, G.; Leclercq, C.; Zimmermann, D.; Roue, C.; Thibault, J.F.; Saulnier, L. Isolation and characterization of high molar mass water-soluble arabinoxylans from barley and barley malt. *Carbohydr. Polym.* **2002**, *47*, 143–149. [CrossRef]
29. Unbehend, L.; Unbehend, G.; Lindhauer, M.G. Comparison of the quality of some Croatian and German wheat varieties according to the German standard protocol. *Food/Nahrung* **2003**, *47*, 140–144. [CrossRef] [PubMed]

Disclaimer/Publisher's Note: The statements, opinions and data contained in all publications are solely those of the individual author(s) and contributor(s) and not of MDPI and/or the editor(s). MDPI and/or the editor(s) disclaim responsibility for any injury to people or property resulting from any ideas, methods, instructions or products referred to in the content.

Article

Pulses-Fortified Ketchup: Insight into Rheological, Textural and LF NMR-Measured Properties

Joanna Le Thanh-Blicharz [1,*], Jacek Lewandowicz [2], Patrycja Jankowska [1], Przemysław Łukasz Kowalczewski [3,*], Katarzyna Zając [1], Miroslava Kačániová [4,5] and Hanna Maria Baranowska [6]

1. Department of Food Concentrates and Starch Products, Institute of Agriculture and Food Biotechnology—State Research Institute, 61-361 Poznań, Poland
2. Department of Production Management and Logistics, Poznan University of Technology, 60-965 Poznań, Poland
3. Department of Food Technology of Plant Origin, Poznań University of Life Sciences, 60-624 Poznań, Poland
4. Institute of Horticulture, Faculty of Horticulture and Landscape Engineering, Slovak University of Agriculture, 94976 Nitra, Slovakia
5. School of Medical & Health Sciences, University of Economics and Human Sciences in Warsaw, 01-043 Warszawa, Poland
6. Department of Physics and Biophysics, Poznań University of Life Sciences, 60-637 Poznań, Poland
* Correspondence: joanna.lethanh-blicharz@ibprs.pl (J.L.T.-B.); przemyslaw.kowalczewski@up.poznan.pl (P.Ł.K.)

Citation: Le Thanh-Blicharz, J.; Lewandowicz, J.; Jankowska, P.; Kowalczewski, P.Ł.; Zając, K.; Kačániová, M.; Baranowska, H.M. Pulses-Fortified Ketchup: Insight into Rheological, Textural and LF NMR-Measured Properties. *Appl. Sci.* **2023**, *13*, 11270. https://doi.org/10.3390/app132011270

Academic Editor: Daniel Cozzolino

Received: 14 September 2023
Revised: 7 October 2023
Accepted: 12 October 2023
Published: 13 October 2023

Copyright: © 2023 by the authors. Licensee MDPI, Basel, Switzerland. This article is an open access article distributed under the terms and conditions of the Creative Commons Attribution (CC BY) license (https://creativecommons.org/licenses/by/4.0/).

Abstract: Tomato ketchup is one of the most popular foods eaten all over the world. To improve the texture of these sauces, modified starches are used most commonly. This may be negatively assessed by consumers. The solution to this problem could be the use of legume flours, as beyond thickening potential, they are recognized as plant foods of high nutritional value. The aim of the work was to estimate the applicability of pulse flour as a texture-forming agent for ketchup. A comprehensive assessment of the quality of ketchup was made, both in terms of sensory properties and instrumentally analyzed physicochemical features: acidity, color, texture, rheological properties, and dynamics of water molecules using the LF NMR method. It was stated that pulse flours are suitable for use as forming agents for ketchup, although they have slightly weaker thickening properties with a consistency index ranging from 5.06–6.82 $Pa \cdot s^n$, compared to acetylated distarch adipate (19.48 $Pa \cdot s^n$). Texture, which is the most important parameter for consumer acceptance of ketchup, can be successfully analyzed using instrumental methods. Firmness ranged from 0.51 N for lentil and pea-fortified ketchup to 0.55 N for the lupine variant. Ketchup thickened with different pulses and flours reveals slightly different individual sensory characteristics, so it makes it possible to create a new gamut of healthy tomato sauces. The highest overall sensory score was attributed accordingly to lentils (6.9), lupine (6.2), chickpeas (6.1), and peas (5.8).

Keywords: tomato sauces; pea; chickpea; lentil; lupin; sensory properties; instrumental analysis

1. Introduction

Tomato ketchup is one of the most popular foods eaten all over the world. It is a simple dish, just a sauce consisting of tomatoes, sucrose, vinegar, salt, and various species. It may also contain various types of texturizing agents such as pectin or xanthan and specially modified starches. Ketchup is often recognized as a condiment for snacks but just as often is eaten as a component of various foods such as pasta, pizza, etc. It has a large share of market sales. Moreover, further growth in its consumption, estimated to be 2.64% annually, is predicted in the period from 2021 to 2025 [1–4]. This popularity of ketchup is not only due to the fact that it improves the taste of popular dishes and snacks but also that consumers are aware of the pro-health activity of tomatoes [5]. Ketchup, as a tomato-based product, is nutritionally considered a source of valuable carotenoids, mainly lycopene, which reveals a potent antioxidative, hypolipidemic, and antidiabetic activity. The intake of this phytochemical is related to a lower threat of insulin resistance and metabolic syndrome. Moreover, it has been proven using in vivo studies that consumption of ketchup itself may improve blood glucose metabolism due to the beneficial regulatory effects in energy metabolism in hepatocytes [6,7].

Physicochemically, ketchup is a complex type of dispersed system. It may be considered as a suspension in which tomato pulp and spices are insoluble solids dispersed in the colloid aqueous phase containing water-soluble substances such as salt, sugar, and organic acids as well as polysaccharide macromolecules, e.g., starch. All these components contribute to the ketchup texture, the decisive factor for consumer acceptance of this food product. Both the degree of fragmentation of solid particles (mainly tomato pulp) and the broadly understood physicochemical properties of thickening hydrocolloids are important [3,8]. Rheological properties, that are strongly related to texture, of ketchup are mainly controlled by rheological properties of dispersing phase, i.e., colloidal system containing hydrocolloids, salt, sugar, and acids [9]. Hence, a proper choice of thickener is one of the most important challenges determining competitiveness in the food market. However, as modern consumers are more and more aware of the relationship between high quality nutrition and health there is a need for new thickeners that beyond its viscosity are also a source of health-promoting substances [10]. Application of flours obtained from legume seeds could be a good solution [11,12].

Legume seeds are a rich source of plant protein [13]. They also contain a lot of other nutritionally valuable components such as fat, slowly digestible carbohydrates, vitamins, minerals, and certain phytochemicals revealing pro-health activity. It should be, however, emphasized that legume seeds differ significantly in composition [14–16]. The most popular pulses, i.e., peas and beans are quite similar to each other as they rich in carbohydrates (above 60%) and low in fats (1–2%). The protein content is high (two to three times higher) compared to cereal grains, but the lowest among legumes and is close to 20%. Lupine is distinguished by a very high, similar to meat products, protein content. It is even higher than in soybean, used for oil extraction and production of vegetable protein preparations, and amounts of 44% [11,17–19]. Pulses phytochemicals include polyphenols, phytosterols, saponins, etc. reveal antioxidant activity and thus contribute in decreasing the risk of various non-communicable diseases such as diabetes, hypertension, cardiovascular diseases or cancer. Starch contained in pulses contains more amylose than popular commercial starches that results in slowing down of its digestion and lowering the glycemic index. Lupine is distinguished by a particularly high biological activity, however for a long time it was used mainly for animal nutrition due to the high content of anti-nutritional substances, mainly alkaloids. However, many years of breeders' efforts resulted in so called sweet varieties meeting high requirements regarding the quality of lupine seeds.

The high nutritional value of pulses induced interest in using their flours as ingredients in various foods, mainly cereal-based products such as bread, cakes, or snacks [20–23]. Much less attention has been paid to the use of pulses in other food categories [12]. Modern consumers more and more often are looking for food of high nutritional quality, especially so-called clean-label products, i.e., those that contain no food additives. Therefore, replacing the modified starch with pulse flour could not only contribute to increasing the nutritional value of ketchup but also meet the requirements of conscious consumers. Therefore, the purpose of the work was to estimate the applicability of pulse flour as a texture-forming agent for ketchup.

2. Materials and Methods

2.1. Ketchup

Tomato ketchup was prepared in 200 g batches according to the recipe in accordance with the procedure of Śmigielska et al. [24]. The ketchup formulation consisted of 60 g of concentrated (30%) tomato puree, 25 g of sucrose, 12 g of spirit vinegar (10%), 6.5 g of flour, and 3 g of table salt. The ketchup was manufactured using Thermomix (TM6, Vorwerk,

Wuppertal, Germany). Five different variants containing following thickeners were prepared: 1—acetylated distarch adipate (reference) (ZETPEZET, Piła, Poland), 2—lupine flour (Green Essence, Pyszyce, Poland), 3—pea flour (Crispy Natural, Kalisz, Poland), 4—chickpea flour (BIO PLANET, Leszno, Poland), 5—lentil flour (PRO-BIO, Staré Město, Czech Republic).

2.2. Titratable Acidity and pH

Titratable acidity and pH of ketchup were determined according to Pearson [25].

2.3. Color Measurements

Chroma Meter CR-310 (Konica Minolta Sensing Inc., Tokyo, Japan) was used for the color measurements of ketchup. Color parameters were expressed in CIE L*a*b* color space. Additionally, a/b ratio was calculated [26].

2.4. Universal Texture Profile

The universal texture profile was determined according to Lewandowicz et al. [27] with the assistance of TA.XT2 Texture Analyzer (Stable Micro Systems, Godalming, UK).

2.5. Rheological Properties

Rheological properties were determined using a RotoVisco1 rheometer (Haake Technik GmbH, Vreden, Germany) equipped with a Z20 DIN Ti coaxial measurement system. Flow curves were determined within 0.1–$600\ s^{-1}$ shear rate for increasing and decreasing shear speed, using the procedure described previously [27]. The obtained data were fitted to the Ostwald de Waele model using RheoWin 3.61 software.

2.6. Low Field NMR

NMR relaxation times were analyzed with a pulse NMR spectrometer PS15T operating at 15 MHz (Ellab, Poznań, Poland). The inversion–recovery (180-TI-90) pulse sequence was applied for measurements of the T_1 relaxation times. The 180 pulse was 4.8 μs, the distance between RF pulses (TI) changed from 2 to 1800 ms, and the repetition time TR between sequences was 18 s.

Measurements of the spin–spin (T_2) relaxation times were taken using the pulse train of the Carr–Purcell–Meiboom–Gill spin echoes (90-TE/2-(180)$_n$). The echo time TE was 3 ms, and the number of the spin-echoes signal (n) was 50. The five accumulations of the spin-echo trains were used. The repetition time between pulse trains was 15 s.

2.7. Sensory Analysis

Quantitative Descriptive Analysis (QDA) was performed by a panel consisting of six persons trained according to the PN-ISO 8586:2023-10 [28] standard. The following distinctive features were analyzed: color, texture (consistency, smoothness, grainy), smell (tomato, vinegary, foreign), taste (tomato, vinegary, sour, sweet, foreign), and overall assessment. The intensity of distinctive features was assessed using a 10 cm linear scale with specific boundary terms.

2.8. Statistical Analysis

The statistical analyses were performed using Statistica 13.3 (TIBCO Software Inc., Palo Alto, CA, USA). One-way analysis of variance and Tukey's post hoc test was performed to determine statistically homogenous subsets at $\alpha = 0.05$. Principal component analysis (PCA) was performed based on the correlation matrix.

3. Results and Discussion

All analyzed ketchup revealed basic physicochemical properties typical for this class of products (Table 1). Moreover, the differences between the values of the analyzed parameters were smaller than those found for ketchup offered commercially by the largest producers [29]. The lowest pH value was found for ketchup containing modified starch, whereas the highest was for lentil flour. The pH values of commercial products reach even lower values, up to 3.4. The titratable acidity of the analyzed products also differed slightly. It was the lowest for ketchup with lupine flour and the highest for products containing flours of other pulses. In the case of commercial products, the variation in the acidity value is significantly greater and ranges from 0.83 to 1.64 1/100 g [29].

Table 1. General characteristics of ketchup.

Thickener	Titratable Acidity (g/100 g)	pH	Lightness L*	Green/Magenta a*	Blue/Yellow b*	b/a
E-1422	1.10 ± 0.01 [b]	3.81	36.45 ± 0.03 [a]	16.35 ± 0.02 [a]	9.24 ± 0.04 [b]	1.8
Lupine	1.05 ± 0.01 [a]	3.88	39.57 ± 0.02 [c]	18.09 ± 0.05 [c]	13.00 ± 0.01 [c]	1.4
Pea	1.12 ± 0.02 [b]	3.89	38.06 ± 0.04 [b]	17.49 ± 0.06 [b]	2.36 ± 0.10 [a]	7.4
Chickpea	1.12 ± 0.01 [b]	3.85	40.17 ± 0.15 [d]	18.42 ± 0.08 [d]	13.83 ± 0.06 [d]	1.3
Lentil	1.12 ± 0.02 [b]	3.91	40.59 ± 0.04 [e]	18.61 ± 0.04 [e]	13.89 ± 0.05 [d]	1.3

Values marked with the same lowercase letter do not differ significantly $p > 0.05$.

Higher variability between the analyzed ketchup was observed in relation to their color parameters (Table 1). These differences are important as it is postulated that for the quality of tomato products (not just ketchup itself), lightness and a/b ratio are most indicative [26]. The darkest was the control sample, containing modified starch. The L* values increased in a series of ketchup containing sequentially pea, lupine, chickpea, and lentil flour. Nevertheless, these differences were not very large. Similar values of lightness have been reported by Ahouagi et al. [2] for ketchup containing small amounts of strawberry pulp. However, also significantly higher L* values of about 50 have been reported [26]. These differences may result both from the quality of tomatoes used to make the ketchup as well as from the difference in production technology. The changes in a* parameter relative to the green-red (magenta) color axis follow the pattern found for lightness. The highest shift towards a green hue for the reference sample was observed. Then, the series of increasing values of parameter a* follows that of the L* parameter. In contrast, many differences in b* parameter, relative to the blue-yellow opposite colors axis, were observed. The sample containing pea flour showed the highest shift towards blue. The control sample was similar. Much more yellowish, with b* parameter above 13, were samples containing other pulses flours. As a consequence, a huge difference in the values of the a/b ratio was observed. For tomato products, the a/b ratio should be at least 1.8 [26]. This requirement is met for the reference sample containing modified starch, significantly exceeded for the sample containing pea flour, and below the limit for the other ketchup. Similar observations were made for tomato pulp powder as a thickening agent. Too much addition of this thickener reduces the a/b ratio below the recommended value [26].

The texture is considered a crucial parameter in the consumer acceptability of ketchup [30]. However, the precise definition of texture is difficult, which implies multiplicity and diversity of methods for its assessment. Among the instrumental methods, the Texture Profile Analysis (TPA) should be mentioned primarily. It was originated by Surmacka-Szczesniak and is still being developed [31]. The parameters of the TPA of analyzed ketchup are presented in Table 2. Firmness, adhesiveness, and gumminess are parameters for which significant differences were observed. In comparison, cohesiveness and springiness were almost of the same value for all samples. Ketchup thickened with pulses flours showed similar firmness values as those reported by Ahouagi et. al, while the sample containing modified starch showed twice the value of this parameter [2].

Table 2. Universal texture profile analysis of analyzed ketchup.

Thickener	Firmness (N)	Adhesiveness (N × s)	Cohesiveness (-)	Springiness (-)	Gumminess (N)
E-1422	1.11 ± 0.06 [b]	−9.12 ± 0.18 [c]	0.80 ± 0.01 [a]	1.02 ± 0.01 [a]	0.89 ± 0.05 [b]
Lupine	0.55 ± 0.04 [a]	−0.60 ± 0.04 [b]	0.79 ± 0.00 [a]	1.00 ± 0.01 [a]	0.43 ± 0.03 [a]
Pea	0.51 ± 0.01 [a]	−0.34 ± 0.04 [a]	0.78 ± 0.01 [a]	1.02 ± 0.01 [a]	0.40 ± 0.01 [a]
Chickpea	0.54 ± 0.03 [a]	−0.59 ± 0.03 [b]	0.79 ± 0.01 [a]	1.00 ± 0.00 [a]	0.43 ± 0.02 [a]
Lentil	0.51 ± 0.02 [a]	−0.55 ± 0.05 [ab]	0.78 ± 0.01 [a]	1.02 ± 0.01 [a]	0.40 ± 0.01 [a]

Values marked with the same lowercase letter do not differ significantly $p > 0.05$.

Despite the popularity of the apparatus for analyzing the TPA profile, it is difficult to compare the data obtained. This phenomenon is due to the possibility of applying different probes and different parameters of texture analysis by different research groups. Therefore, to evaluate the texture of ketchup, rheological methods are used more often [4]. Such an approach provides better reproducibility of the results, especially if rheometers are used, i.e., apparatus with a defined and controlled shear rate [32]. The highest values of the consistency index were found for ketchup thickened with modified starch (Table 3). Samples containing pulse flours showed significantly lower K values, the smallest for peas and then in sequence for lupine, lentil, and chickpea. These data are consistent with those reported by Juszczak et al. for ketchup thickened with different modified starches [4]. Flow behavior index values showed much less variation. Its values below 1 mean that the apparent viscosity of all ketchup decreases with increasing shear rate. Moreover, all samples revealed rheopexy (antithixotropy). This phenomenon indicates that fluid reveals a time-dependent increase in viscosity by shearing. It means that ketchup, when shaken, would be fluid at first, becoming more viscous as the shaking continued. Rheopexy was considered a rather rare phenomenon, especially if not accompanied by thinning under an increasing shear rate. However, it is observed more and more often for different dispersed systems [33–35]. It is also believed that the relationship between thixotropy and rheopexy is complicated. Moreover, it has been stated that the thixotropic behavior of starch dispersion can change into rheopectic behavior by changing concentration or shear rate [36].

Table 3. Ostwald de Waele equation coefficients for analyzed ketchup.

Thickener	Consistency Index K (Pa × s^{-n})		Flow Behavior Index n (-)		Thixotropy (Pa × s^{-1})
	0–600 s^{-1}	600–0 s^{-1}	0–600 s^{-1}	600–0 s^{-1}	
E-1422	19.48 ± 1.26 [c]	20.22 ± 1.12 [d]	0.386 ± 0.007 [c]	0.384 ± 0.005 [c]	−2352 ± 128 [c]
Lupine	6.82 ± 0.08 [b]	9.02 ± 0.25 [c]	0.340 ± 0.007 [a]	0.299 ± 0.008 [a]	−1001 ± 170 [b]
Pea	5.06 ± 0.44 [a]	5.84 ± 0.01 [a]	0.388 ± 0.006 [c]	0.372 ± 0.007 [c]	−1262 ± 118 [b]
Chickpea	6.34 ± 0.17 [ab]	7.80 ± 0.01 [bc]	0.369 ± 0.002 [b]	0.341 ± 0.003 [b]	−1344 ± 169 [b]
Lentil	6.75 ± 0.01 [b]	6.89 ± 0.06 [ab]	0.355 ± 0.004 [b]	0.355 ± 0.002 [b]	−514 ± 71 [a]

Values marked with the same lowercase letter do not differ significantly $p > 0.05$.

Low-field nuclear magnetic resonance (LF NMR) is a method designed to study the dynamics of protons [37,38]. However, there has been a growing interest in applying this method to food studies in recent years [39–42]. This is possible as in food, the process of nuclear relaxation of both water and fat molecules is observed. As a result of measurements in the water system, the values of spin-lattice relaxation time T_1 and spin-spin relaxation time T_2 are recorded. T_1 is related to the ratio of free to entrapped water, whereas T_2 to water molecules dynamics. The spin-lattice relaxation time is affected by many factors, e.g., temperature, resonant frequency, or the presence of macromolecules, and it is always longer than T_2. This means that in the presence of large molecules (for example, starch), the rotation of water molecules is slowed down; thus, the relaxation time is shortened [43]. By incorporation of fats into the biopolymer solution and formation of a two-phase system, for example, in emulsions, each of the relaxation times separate into two components. This means that there are two fractions of protons relaxing at different rates in the system. Moreover, the chemical exchange between these fractions of protons is much slower than the relaxation time. The long components (T_{12} and T_{22}) reflect the relaxation processes of the fraction of protons mainly associated with starch paste [44]. By the analyses of ketchup employing the LF NMR method (Table 4), no splitting of relaxation times into two components was observed. This is despite the fact that ketchup is considered to be a complex two-phase system [4]. This observation proves that potato pulp particles suspended in a solution of low molecular mass components of ketchup did not form a phase in which proton relaxation proceeds differently than in the liquid phase. The spin-lattice relaxation time of ketchup containing acetylated distarch adipate is significantly lower than that of all others. Starch is a biopolymer of extremely high molecular mass, and its modification consists of cross-linking that additionally increases the size of its macromolecules [27]. Protein molecules have a much smaller molar mass, and starches of other plant species have a lower molar mass than the E 1422 starch used in this study [45]. The values of the spin-lattice relaxation times were less diversified, with the highest value observed for ketchup-containing pea flour (Table 4). Mean correlation time τ_c, which describes the time for a molecule to rotate by 1 radian, also showed a large variation, and its lowest value was observed for ketchup containing modified starch.

Table 4. LF NMR study results.

Thickener	Spin-Lattice Relaxation Time T_1 (ms)	Spin-Spin Relaxation Time T_2 (ms)	Mean Correlation Time τ_c (s)
E-1422	249.14 ± 0.58 [a]	101.60 ± 0.54 [a]	0.89×10^{-8}
Lupine	476.34 ± 1.41 [d]	110.38 ± 0.69 [d]	1.48×10^{-8}
Pea	419.07 ± 1.90 [c]	122.49 ± 0.72 [e]	1.21×10^{-8}
Chickpea	418.75 ± 2.04 [c]	107.81 ± 0.61 [c]	1.35×10^{-8}
Lentil	405.16 ± 0.99 [b]	104.92 ± 0.54 [b]	1.35×10^{-8}

Values marked with the same lowercase letter do not differ significantly $p > 0.05$.

The above-presented results of instrumental measured parameters of ketchup are an important indicator of their quality. Moreover, they are precisely repeatable and make it possible to find out the causes of observed phenomena. However, for consumer acceptability, sensory properties are of crucial importance. The most important sensory features of ketchup, apart from the appropriate flavor and texture, are intense red color, appropriate consistency, sweet tomato, and spicy taste [46]. The sensory profile of the analyzed samples is presented in Figure 1. It comprised, apart from the parameters analyzed with instrumental methods, several features that could be assessed only by a panel of specialists with appropriate sensory sensitivity. They include grainy, smoothness, and consistency, as well as different tastes (tomato, vinegar, sour, sweet, and foreign) and different smells (tomato, vinegar, sour, and foreign). The grainy of the evaluated products meant a structure similar to porridge, granularity. Ketchup with lupine and pea flour showed relatively high, but still moderate, grainy. The others, including the control sample, were characterized by small grainy. Smoothness is an attribute that gives the impression of delicacy, softness, and velvety on the tongue. The variability of ketchup smoothness was similar. The most smooth were ketchup with E1422 and lentil flour, whereas the least smooth with pea and lupine flour. With regard to the flavor, it is worth emphasizing that all ketchup did not reveal any foreign smell or taste. Moreover, all ketchup had the same slight sweetness (sweet taste). As for the most distinctive tomato flavor—the reference sample stood out from the others with the most intense tomato smell. The other ketchup revealed slightly less intensively of that feature. The tomato taste of all ketchup also differed only slightly, with the most intense for the reference sample and the least intense for that containing lentil flour. There were only slight differences in the sour and vinegar taste as well as vinegar smell assessment. It corresponds to the small differences in values of pH and titratable acidity of ketchup presented in Table 1. However, it should be mentioned that there was more significant diversity in flavor features stated by the sensory panel than in pH and titratable acidity values. Consistency was also subject to sensory evaluation. This parameter could be classified as characterizing the texture of the product. Ketchup containing modified starch stood out with the highest consistency. This observation corresponds both to the highest firmness (parameter of universal texture profile) and the highest consistency index derived from the Ostwald de Waele equation. This observation contradicts the theorem postulated in the literature about the impossibility of evaluating the texture of products with instrumental methods [30]. Color is also an important feature determining consumer acceptance. However, in contrast to texture, including rheological parameters, it is difficult to find a simple correlation between sensory and instrumental analysis as regards the color rated as the best was the ketchup containing modified starch (Figure 1). However, this sample was not distinguished by any of the parameters of CIE L*a*b* color space presented in Table 1. However, the a/b ratio is higher for the control sample than for ketchup with lupine, chickpea, and lentil flours. Nevertheless, its greatest value was recorded for ketchup with pea flour. Moreover, control ketchup only meets the requirement presented in the work by Farahanky et al. [26]. This phenomenon might be related to a large color shift towards blue for ketchup with pea flour, which resulted in a worse consumer acceptance of this product. This suggests that when evaluating the sensory qualities, it is important to balance individual product features, and the use of specific numerical parameters should be approached critically.

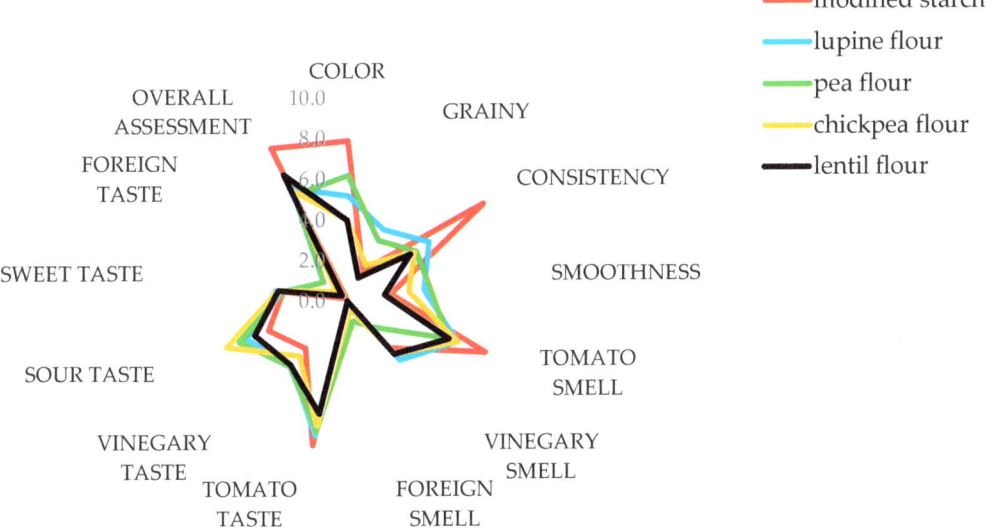

Figure 1. Characteristics of the sensory profile of analyzed ketchup.

The overall assessment of the products, defined as a general impression determining the comprehensive sensory quality of the product, taking into account all the features and their mutual harmonization, was found to be highest for the control sample. The other ketchup was rated slightly lower, with small differences between the lowest rated sample containing pea flour and the best sample, within pulses thickened ketchup, containing chickpea flour.

PCA analysis (Figure 2) was carried out to verify which factors determine the consumer acceptability of ketchup. The two principal components (factors 1 and 2) explain 81.27% of the total variance. The most noteworthy of the results presented in Figure 2A is the close correlation between consistency assessed during sensory study and firmness and gumminess examined instrumentally. Additionally, the consistency index analyzed using a rheometer with a defined and controlled shear rate reveals a pretty good correlation with the texture parameters. This means that contrary to the statement of Tauferova et al. [29] that the texture of products cannot be assessed instrumentally, such a procedure may be recommended. Moreover, the total assessment of ketchup reveals a high positive correlation with the above-mentioned texture parameters and also with cohesiveness and tomato smell. On the contrary, sour and vinegar taste, as well as adhesiveness, contribute to the negative assessment of ketchup. Vinegary smell, sweet taste, springiness, and flow behavior index seem to have little importance for consumer evaluation. Grainy, smoothness, titratable acidity, as well as foreign smell and taste, have a small but negative impact on the perception of ketchup. Regarding color, PCA analysis shows that sensory evaluation of this parameter is less important for the perception of the quality of ketchup than texture and tomato taste. Lightness, as well as a color parameter evaluated instrumentally, has a rather small but negative impact on consumer evaluation. Parameter b and a/b ratio seem to have even lesser effect. This may seem surprising in the context of a paper by Farahnaky et al. [26]. However, their work related to tomato products in general and not to ketchup specifically, which may explain this discrepancy. Spin-lattice relaxation time and mean correlation time revealed a negative correlation with the overall assessment of ketchup. This means that products in which water molecules reveal lower dynamics (stronger entrapped) are rated better. Spin-spin relaxation time is less important.

The PCA score plot (Figure 2B) highlights the similar properties of ketchup containing lupin, lentil, and chickpea flours. In contrast, products containing pea flour and modified starch were proven to have more unique characteristics. The differences between ketchup containing pulse flour and that with E 1422 appear to be mainly due to differences in texture. The uniqueness of ketchup containing pea flour lies in differences in other features, mainly in color. Summing up, ketchup thickened with chickpea, pea, lentil, soy, or lupine flour reveals good sensory properties; however, they differ significantly in texture and rheological properties. This may open the possibility of creating a whole branch of healthy tomato sauce.

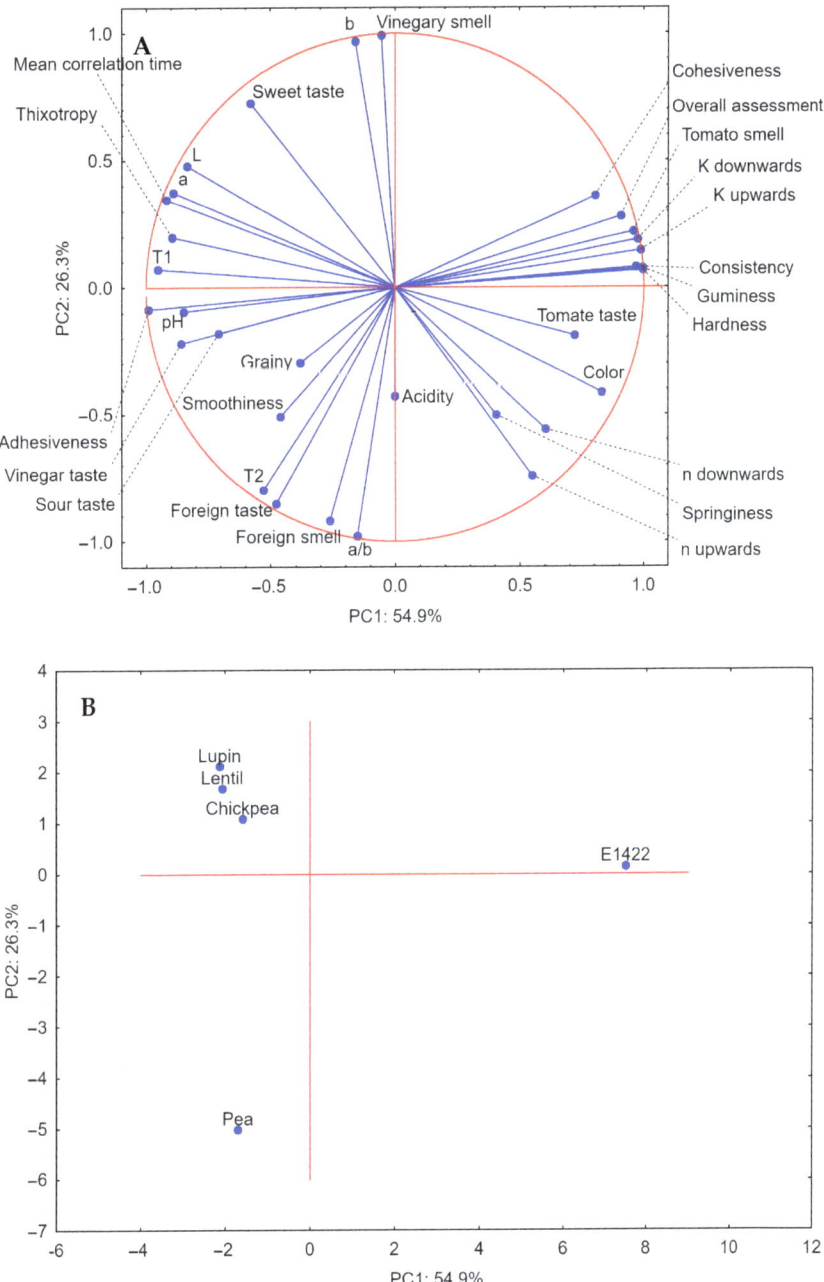

Figure 2. Principal Components Analysis of the results of sensory analysis (color, consistency, grainy, smoothness, tomato taste, tomato smell, vinegary taste, vinegary smell, foreign taste, foreign smell, sour taste, sweet taste, overall assessment) and instrumental analyses (pH, titratable acidity, L {lightness}, a {green/magenta ratio}, b {blue/yellow ratio}, Firmness, Adhesiveness, Cohesiveness, Springiness, Gumminess, K up {consistency index up}, K down {consistency index down}, n up {flow behavior index up}, n down {flow behavior index down}, thixotropy, T_1 {spin-lattice relaxation time}, T_2 {spin-spin relaxation time}, mean correlation time. (**A**) loadings plot; (**B**) score plot.

4. Conclusions

Pulse flours are suitable for use as texture-forming agents for ketchup. They have slightly weaker thickening properties than acetylated distarch adipate, but they do not significantly change the sensory properties of ketchup. The texture of ketchup is the most important parameter for consumer acceptance of this product. It can be successfully analyzed using instrumental methods. For this purpose, both rheological methods and the analysis of the universal texture profile are useful. The low-field nuclear resonance method also proved useful for assessing the quality of ketchup, as the dynamics of water protons strongly correlate with the texture of ketchup. Instrumental analysis of the color of ketchup is less useful for assessing the quality of ketchup than the previously mentioned methods. As ketchup containing different flours reveals slightly different individual characteristics, it opens the possibility of creating the whole gamut of healthy tomato sauces. Nevertheless, lentil flour should be recommended as the thickener of the first choice due to better overall sensory assessment of the final product.

Author Contributions: Conceptualization, J.L.T.-B.; Data curation, J.L.T.-B. and M.K.; Funding acquisition, P.Ł.K.; Investigation, J.L.T.-B., J.L., P.J., P.Ł.K., K.Z. and H.M.B.; Methodology, J.L.T.-B., J.L., K.Z. and H.M.B.; Project administration, J.L.T.-B.; Resources, P.Ł.K.; Supervision, J.L.T.-B. and P.Ł.K.; Visualization, J.L.; Writing—original draft, J.L.T.-B., J.L. and P.Ł.K.; Writing—review and editing, J.L.T.-B., J.L., P.Ł.K. and M.K. All authors have read and agreed to the published version of the manuscript.

Funding: This research received no external funding.

Institutional Review Board Statement: Not applicable.

Informed Consent Statement: Not applicable.

Data Availability Statement: The datasets generated for this study are available on request to the corresponding author.

Conflicts of Interest: The authors declare no conflict of interest.

References

1. Shokraneh, N.; Alimi, M.; Shahidi, S.-A.; Mizani, M.; Bameni Moghadam, M.; Rafe, A. Textural and Rheological Properties of Sliceable Ketchup. *Gels* **2023**, *9*, 222. [CrossRef] [PubMed]
2. Ahouagi, V.B.; Mequelino, D.B.; Tavano, O.L.; Garcia, J.A.D.; Nachtigall, A.M.; Vilas Boas, B.M. Physicochemical characteristics, antioxidant activity, and acceptability of strawberry-enriched ketchup sauces. *Food Chem.* **2021**, *340*, 127925. [CrossRef] [PubMed]
3. Baeghbali, S.; Shahriari, S.; Pazuki, G. Effect of pressure homogenization and modified starch on the viscosity of ketchup: Experimental and modeling. *J. Food Process Eng.* **2021**, *44*, e13683. [CrossRef]
4. Juszczak, L.; Oczadły, Z.; Gałkowska, D. Effect of Modified Starches on Rheological Properties of Ketchup. *Food Bioprocess Technol.* **2013**, *6*, 1251–1260. [CrossRef]
5. Collins, E.J.; Bowyer, C.; Tsouza, A.; Chopra, M. Tomatoes: An Extensive Review of the Associated Health Impacts of Tomatoes and Factors That Can Affect Their Cultivation. *Biology* **2022**, *11*, 239. [CrossRef] [PubMed]
6. Mirzaei, D.; Pedram Nia, A.; Jalali, M. Effect of inulin and date syrup from Kaluteh variety on the qualitative and microbial properties of prebiotic ketchup. *J. Food Sci. Technol.* **2021**, *58*, 4127–4138. [CrossRef]
7. Jeong, J.H.; Lee, H.L.; Park, H.J.; Yoon, Y.E.; Shin, J.; Jeong, M.-Y.; Park, S.H.; Kim, D.; Han, S.-W.; Kang, C.-G.; et al. Effects of tomato ketchup and tomato paste extract on hepatic lipid accumulation and adipogenesis. *Food Sci. Biotechnol.* **2023**, *32*, 1111–1122. [CrossRef]
8. Cai, X.; Du, X.; Zhu, G.; Cai, Z.; Cao, C. The use of potato starch/xanthan gum combinations as a thickening agent in the formulation of tomato ketchup. *CyTA J. Food* **2020**, *18*, 401–408. [CrossRef]
9. Le Thanh-Blicharz, J.; Lewandowicz, J. Functionality of Native Starches in Food Systems: Cluster Analysis Grouping of Rheological Properties in Different Product Matrices. *Foods* **2020**, *9*, 1073. [CrossRef]
10. Śmigielska, H.; Błaszczak, W.; Lewandowicz, G. Applicability of Food Grade Modified Starches as a Carrier of Microelements. *Processes* **2022**, *10*, 235. [CrossRef]
11. Kumar, S.; Pandey, G. Biofortification of pulses and legumes to enhance nutrition. *Heliyon* **2020**, *6*, e03682. [CrossRef] [PubMed]
12. Yazdanpanah, S.; Ansarifard, S.; Hasani, M. Development of Novel Gluten-Free Sausage Based on Chickpea, Corn Flour, and HPMC. *Int. J. Food Sci.* **2022**, *2022*, 3616887. [CrossRef] [PubMed]
13. Semba, R.D.; Ramsing, R.; Rahman, N.; Kraemer, K.; Bloem, M.W. Legumes as a sustainable source of protein in human diets. *Glob. Food Sec.* **2021**, *28*, 100520. [CrossRef]

14. Zielińska-Dawidziak, M.; Tomczak, A.; Burzyńska, M.; Rokosik, E.; Dwiecki, K.; Piasecka-Kwiatkowska, D. Comparison of Lupinus angustifolius protein digestibility in dependence on protein, amino acids, trypsin inhibitors and polyphenolic compounds content. *Int. J. Food Sci. Technol.* **2019**, *55*, 2029–2040. [CrossRef]
15. Tomczak, A.; Zielińska-Dawidziak, M.; Piasecka-Kwiatkowska, D.; Lampart-Szczapa, E. Blue lupine seeds protein content and amino acids composition. *Plant Soil Environ.* **2018**, *64*, 147–155. [CrossRef]
16. Azarpazhooh, E.; Ahmed, J. Composition of Raw and Processed Dry Beans and Other Pulses. In *Dry Beans and Pulses*; Wiley: Hoboken, NJ, USA, 2022; pp. 129–157.
17. Asif, M.; Rooney, L.W.; Ali, R.; Riaz, M.N. Application and Opportunities of Pulses in Food System: A Review. *Crit. Rev. Food Sci. Nutr.* **2013**, *53*, 1168–1179. [CrossRef] [PubMed]
18. Pereira, A.; Ramos, F.; Sanches Silva, A. Lupin (*Lupinus albus* L.) Seeds: Balancing the Good and the Bad and Addressing Future Challenges. *Molecules* **2022**, *27*, 8557. [CrossRef]
19. Bähr, M.; Fechner, A.; Hasenkopf, K.; Mittermaier, S.; Jahreis, G. Chemical composition of dehulled seeds of selected lupin cultivars in comparison to pea and soya bean. *LWT Food Sci. Technol.* **2014**, *59*, 587–590. [CrossRef]
20. Paladugula, M.P.; Smith, B.; Morris, C.F.; Kiszonas, A. Incorporation of yellow pea flour into white pan bread. *Cereal Chem.* **2021**, *98*, 1020–1026. [CrossRef]
21. Sopiwnyk, E.; Bourré, L.; Young, G.; Borsuk, Y.; Lagassé, S.; Boyd, L.; Sarkar, A.; Jones, S.; Dyck, A.; Malcolmson, L. Flour and bread making properties of whole and split yellow peas treated with dry and steam heat used as premilling treatment. *Cereal Chem.* **2020**, *97*, 1290–1302. [CrossRef]
22. Rachwa-Rosiak, D.; Nebesny, E.; Budryn, G. Chickpeas—Composition, Nutritional Value, Health Benefits, Application to Bread and Snacks: A Review. *Crit. Rev. Food Sci. Nutr.* **2015**, *55*, 1137–1145. [CrossRef] [PubMed]
23. Joshi, M.; Timilsena, Y.; Adhikari, B. Global production, processing and utilization of lentil: A review. *J. Integr. Agric.* **2017**, *16*, 2898–2913. [CrossRef]
24. Śmigielska, H.; Lewandowicz, J.; Le Thanh-Blicharz, J. Effect of type 4 resistant starch on colour and rheological properties of tomato ketchup. *Zywnosc Nauka Technol. Jakosc/Food Sci. Technol. Qual.* **2013**, *20*, 137–149. [CrossRef]
25. Pearson, D. *Chemical Analysis of Foods*, 7th ed.; Churchhill Livingstone: London, UK, 1976.
26. Farahnaky, A.; Abbasi, A.; Jamalian, J.; Mesbahi, G. The use of tomato pulp powder as a thickening agent in the formulation of tomato ketchup. *J. Texture Stud.* **2008**, *39*, 169–182. [CrossRef]
27. Lewandowicz, J.; Le Thanh-Blicharz, J.; Szwengiel, A. The Effect of Chemical Modification on the Rheological Properties and Structure of Food Grade Modified Starches. *Processes* **2022**, *10*, 938. [CrossRef]
28. PN-EN ISO 8586:2023-10; Sensory Analysis—Selection and Training of Sensory Assessors. The Polish Committee for Standardization: Warsaw, Poland, 2023.
29. Sharoba, A.M.; Senge, B.; El-Mansy, H.A.; Bahlol, H.E.; Blochwitz, R. Chemical, sensory and rheological properties of some commercial German and Egyptian tomato ketchups. *Eur. Food Res. Technol.* **2005**, *220*, 142–151. [CrossRef]
30. Tauferova, A.; Tremlova, B.; Bednar, J.; Golian, J.; Zidek, R.; Vietoris, V. Determination of Ketchup Sensory Texture Acceptability and Examination of Determining Factors as a Basis for Product Optimization. *Int. J. Food Prop.* **2015**, *18*, 660–669. [CrossRef]
31. Surmacka-Szczesniak, A. Classification of Textural Characteristics. *J. Food Sci.* **1963**, *28*, 385–389. [CrossRef]
32. Małyszek, Z.; Makowska, A.; Smentek, J.; Kubiak, P.; Le Thanh-Blicharz, J.; Lewandowicz, G. Assessment of factors determining accuracy in measuring rheological properties of modified starches. *Zywn. Nauk. Technol. Jakosc/Food. Sci. Technol. Qual.* **2015**, *22*, 160–175. [CrossRef]
33. Xu, X.; Zhang, D.; Tong, S.; Liu, F.; Wei, W.; Liu, Z. Experimental study on shear viscosity and rheopexy of *Escherichia coli* suspensions. *Rheol. Acta* **2022**, *61*, 271–280. [CrossRef]
34. Kaveh, Z.; Azadmard-Damirchi, S.; Yousefi, G.; Hosseini, S.M.H. Effect of different alcoholic-alkaline treatments on physical and mucoadhesive properties of tapioca starch. *Int. J. Biol. Macromol.* **2020**, *153*, 1005–1015. [CrossRef] [PubMed]
35. Zhao, H.; Zhang, K. The structure recovery capacity of highly concentrated emulsions under shear flow via studying their rheopexy. *J. Dispers. Sci. Technol.* **2018**, *39*, 970–976. [CrossRef]
36. Dewar, R.J.; Joyce, M.J. The thixotropic and rheopectic behaviour of maize starch and maltodextrin thickeners used in dysphagia therapy. *Carbohydr. Polym.* **2006**, *65*, 296–305. [CrossRef]
37. Ezeanaka, M.C.; Nsor-Atindana, J.; Zhang, M. Online Low-field Nuclear Magnetic Resonance (LF-NMR) and Magnetic Resonance Imaging (MRI) for Food Quality Optimization in Food Processing. *Food Bioprocess Technol.* **2019**, *12*, 1435–1451. [CrossRef]
38. Kamal, T.; Cheng, S.; Khan, I.A.; Nawab, K.; Zhang, T.; Song, Y.; Wang, S.; Nadeem, M.; Riaz, M.; Khan, M.A.U.; et al. Potential uses of LF-NMR and MRI in the study of water dynamics and quality measurement of fruits and vegetables. *J. Food Process. Preserv.* **2019**, *43*, e14202. [CrossRef]
39. Li, R.; Xia, Z.; Li, B.; Tian, Y.; Zhang, G.; Li, M.; Dong, J. Advances in Supercritical Carbon Dioxide Extraction of Bioactive Substances from Different Parts of *Ginkgo biloba* L. *Molecules* **2021**, *26*, 4011. [CrossRef]
40. Resende, M.T.; Osheter, T.; Linder, C.; Wiesman, Z. Proton Low Field NMR Relaxation Time Domain Sensor for Monitoring of Oxidation Stability of PUFA-Rich Oils and Emulsion Products. *Foods* **2021**, *10*, 1385. [CrossRef] [PubMed]
41. Brosio, E.; Gianferri, R.R. Low-resolution NMR—An analytical tool in foods characterization and traceability. In *Basic NMR in Foods Characterization*; Brosio, E., Ed.; Research Signpost: Kerala, India, 2009; pp. 9–37. ISBN 978-81-308-0303-6.

42. Abrami, M.; Chiarappa, G.; Farra, R.; Grassi, G.; Marizza, P.; Grassi, M. Use of low-field NMR for the characterization of gels and biological tissues. *ADMET DMPK* **2018**, *6*, 34. [CrossRef]
43. Makowska, A.; Dwiecki, K.; Kubiak, P.; Baranowska, H.M.; Lewandowicz, G. Polymer-Solvent Interactions in Modified Starches Pastes–Electrokinetic, Dynamic Light Scattering, Rheological and Low Field Nuclear Magnetic Resonance Approach. *Polymers* **2022**, *14*, 2977. [CrossRef]
44. Małyszek, Z.; Lewandowicz, J.; Le Thanh-Blicharz, J.; Walkowiak, K.; Kowalczewski, P.Ł.; Baranowska, H.M. Water Behavior of Emulsions Stabilized by Modified Potato Starch. *Polymers* **2021**, *13*, 2200. [CrossRef] [PubMed]
45. Hedley, C.L. *Carbohydrates in Grain Legume Seeds: Improving Nutritional Quality and Agronomic Characteristics*; CABI Publishing: Wallingrod, UK, 2000; ISBN 978-0-85199-467-3.
46. Bannwart, G.C.M.D.C.; Bolini, H.M.A.; Toledo, M.C.D.F.; Kohn, A.P.C.; Cantanhede, G.C. Evaluation of Brazilian light ketchups II: Quantitative descriptive and physicochemical analysis. *Ciência Tecnol. Aliment.* **2008**, *28*, 107–115. [CrossRef]

Disclaimer/Publisher's Note: The statements, opinions and data contained in all publications are solely those of the individual author(s) and contributor(s) and not of MDPI and/or the editor(s). MDPI and/or the editor(s) disclaim responsibility for any injury to people or property resulting from any ideas, methods, instructions or products referred to in the content.

Article

The Use of Thyme and Lemongrass Essential Oils in Cereal Technology—Effect on Wheat Dough Behavior and Bread Properties

Lucie Jurkaninová [1,*], Ivan Švec [2], Iva Kučerová [3], Michaela Havrlentová [4,5], Matěj Božik [1], Pavel Klouček [1] and Olga Leuner [3]

[1] Faculty of Agrobiology, Food and Natural Resources, Czech University of Life Sciences Prague, Kamýcká 129, 165 00 Prague, Czech Republic; bozik@af.czu.cz (M.B.); kloucek@af.czu.cz (P.K.)
[2] Faculty of Food and Biochemical Technology, University of Chemistry and Technology Prague, 166 28 Prague, Czech Republic; ivan.svec@vscht.cz
[3] Faculty of Tropical AgriSciences, Czech University of Life Sciences Prague, Kamýcká 129, 165 00 Prague, Czech Republic; kucerova@ftz.czu.cz (I.K.); leuner@ftz.czu.cz (O.L.)
[4] Faculty of Natural Sciences, University of St. Cyril and Methodius in Trnava, 917 01 Trnava, Slovakia; michaela.havrlentova@ucm.sk
[5] National Agricultural and Food Centre, Research Institute of Plant Production in Piešťany, 921 68 Piešťany, Slovakia
* Correspondence: jurkaninova@af.czu.cz

Citation: Jurkaninová, L.; Švec, I.; Kučerová, I.; Havrlentová, M.; Božik, M.; Klouček, P.; Leuner, O. The Use of Thyme and Lemongrass Essential Oils in Cereal Technology—Effect on Wheat Dough Behavior and Bread Properties. *Appl. Sci.* **2024**, *14*, 4831. https://doi.org/10.3390/app14114831

Academic Editors: Georgiana Gabriela Codină and Adriana Dabija

Received: 30 April 2024
Revised: 30 May 2024
Accepted: 31 May 2024
Published: 3 June 2024

Copyright: © 2024 by the authors. Licensee MDPI, Basel, Switzerland. This article is an open access article distributed under the terms and conditions of the Creative Commons Attribution (CC BY) license (https://creativecommons.org/licenses/by/4.0/).

Abstract: Consumers are more interested in replacing commonly used chemical preservatives with natural substances. The effect of 5, 10, 20, 40, and 80 mg of thyme and lemongrass essential oils (THY and LMG, respectively) per 100 g of wheat flour was studied from the viewpoints of dough rheology, dough leavening progress, and the results of laboratory baking trial. Changes in dough viscoelastic properties were evaluated by the Mixolab rheometer and the company software. The higher the thyme oil portion, the higher the dough structure destruction by kneading and heat input in torque point C2, and consecutively shorter stability of dough consistency (up to one-half of the values recorded for the control); reversely, the LMG did not affect both features verifiably. In the 90 min leavening test, a dough weight loss was decelerated by both essential oils similarly. During the baking test, the average volume of wheat small breads as control was evaluated on level 167 mL (bread yield 451 mL/100 g flour). Independently of the dose of the THY or LMG, small bread volumes oscillated between 148–168 and 135–161 mL (average bread yields 442 and 443 mL/100 g flour, respectively). The shelf life of the products with a higher portion of essential oil was extended by up to 7 days.

Keywords: wheat flour; thyme essential oil; lemongrass essential oil; dough rheology; Mixolab; leavening test; bread volume; bread yield; bread shelf life

1. Introduction

Bakery products have become an important part of the daily diet of people around the world due to rapid socio-economic development [1]. More than 9 billion kilograms of bakery goods are produced each year and approximately 70 kg of bakery products are consumed per person per year [2]. Bread comes in many types, sizes, shapes, and textures depending on national and regional traditions [3]. At the same time, the consumption of bakery products made from wholegrain and graham flour is increasing gradually, mainly due to a greater awareness of the need to reduce the consumption of simple carbohydrates and fats. The consumption of wholegrain and graham flour products elevates the intake of dietary fiber and vegetable protein [4].

To produce bakery products, the basic ingredients are flour, water, yeast (or other fluffing-up agent), and salt. Additional raw materials can be used to improve dough processing or to manufacture special products that often have a higher nutritional value [5]. The optimal processing depends on the type of baked goods and the requested properties of the final product [6]. For baked goods, important quality characteristics include large loaf volume, soft, uniform texture, and satisfactory shelf life [2]. Rolls and breads have a limited shelf life. They undergo physical, chemical, and microbial changes during cooling after baking and storage itself. Physical and chemical changes cause the loss of freshness, and deterioration in texture and taste. Microbiological spoilage is caused due to the growth of bacteria (mainly *Bacillus subtilis*), fungi (especially *Rhizopus nigricans*, *Penicillium expansum* or *P. stolonifer*, and *Aspergillus niger*), and yeasts (most often *Saccharomices cerevisiae*) [3]. It is responsible for the change in appearance, odor formation, and the production of toxic metabolites, monitored are mainly aflatoxin, ochratoxin A, deoxynivalenol, and zearalenone. These can threaten human health and generate significant economic losses for the baking industry and consumers [7]. According to the Food and Agriculture Organization of the United Nations (FAO), foodborne molds and their toxic metabolites cause about 25% of the global agricultural food losses [8].

The spoilage of bakery products is influenced by several factors. The importance could be considered for processing conditions including packaging and product characteristics. For a product, the ingredients used, nutritional composition, oxidative stability, moisture, water activity, and pH are decisive. In processing, the baking time, baking temperature, cooling, and hygiene of the production environment matter. For storage, temperature, access to light, relative humidity, and microorganism content are controlled. Packaging characteristics are also very important. For packaging, mechanical properties, thermal stability, gas permeability, UV barrier, gas composition, antioxidant, and antimicrobial activity are monitored [1].

A set of procedures is used to prevent the spoilage of bakery products. Conventional methods use preservatives. Preservatives are added to foods to increase or maintain nutritional value and quality. They prolong the shelf life of the products reasonably. They inhibit the growth of microorganisms and thus support consumer acceptability [9]. The amount and composition of each preservative is regulated by legislation. Preservatives are divided into two groups, chemical and natural ones. Chemical preservatives include synthetic substances such as benzoates, sorbates, nitrites, nitrates, sulfites, glutamates, glycerides, and others [10]. Natural preservatives include compounds that are obtained from natural sources such as salt, sugar, vinegar, honey, spices, and other substances [9].

Consumers have become more interested in healthy lifestyle forms and diets in recent years. There is a growing desire to reduce commonly used chemical preservatives (e.g., 'non-E' brand). In this context, essential oils are one of the important replacements for their antioxidant, antibacterial, antifungal, antiviral, antiparasitic, and insecticidal properties [11]. Moreover, the aroma of essential oils can positively enhance the sensory properties and increase the attractiveness of baked goods.

Essential oils are complex mixtures of the volatile compounds produced by plants. They can be synthesized in all plant organs, flower, bud, seed, leaf, stem, fruit, and root [12]. They are the secondary metabolites of aromatic plants [13]. Essential oils provide them protection against bacteria, viruses, fungi, and insects, and also enhance plant protection against herbivores [14].

Essential oil consists of approximately 20–60 different components. Usually, two or three components are dominant from 20–70%, while the remaining are present in trace amounts only [15]. The major components are mainly responsible for the biological function of the essential oil, but the remaining components also play a partial role [8].

Differences in the chemical composition and biological activity of essential oils depend on the climatic conditions of the growing area, the cultivation strategy, the plant variety, fertilization, and sufficient water intake during the growing season of the plant [16]. Variations in composition may also be due to the part of the plant from which they are obtained [15].

Spices and essential oils are used in food processing to enhance the flavor and extend the shelf life of foods [17]. The antimicrobials of plant origin led to an increase in overall quality and to the removal or reduction in pathogenic microorganisms. For example, eugenol in clove, cinnamaldehyde in cinnamon, carvacrol and thymol in oregano, sulfur compounds mainly diallyl trisulfide and diallyl disulfide in garlic, linalool in coriander, 1,8-cineol and camphor in rosemary, pulegone and limonene in parsley, citral in lemongrass, camphor in sage, and vanillin in vanilla have shown great antimicrobial effects [18].

However, the application of essential oils has some limitations despite its great potential. Essential oils can cause an allergic reaction and, from a certain amount, cause acute and chronic toxicity. Their use is also limited by the lack of scientific documentation for correct exposure dosages [14]. Another problem is their chemical variability. As the secondary metabolites of plants, essential oils are affected by external factors. This makes them extremely variable qualitatively and quantitatively over time. They have significant technological limitations for their application. The challenges are quality and standardization and strong sensory impact. Essential oils are volatile, unstable in air, and prone to oxidation. The extremely lipophilic nature of essential oils makes them difficult to use in polar solvents [19].

Thyme (*Thymus vulgaris* L.) is an aromatic, culinary, and medicinal herb of the *Lamiaceae* family. It is native to the Mediterranean region [11]. The main components of thyme essential oil are thymol and carvacrol, in levels of up to 70%, and p-Cymene in levels of up to 40%. The ratio of the components varies according to the species, chemotype, or geographical origin. The remainder consists of linalool, limonene, geraniol, borneol, terpinene, and camphene [20]. The high antioxidant activity is mainly due to flavonoids, but phenolic compounds such as rosmarinic and caffeic acids, thymol, and carvacrol also exhibit antioxidant activity [21]. Lemongrass (*Cymbopogon winterianus* Jowitt or *Cymbopogon nardus* (L.) W. Watson) is a grass of Southeast Asia. It contains a significant amount of essential oil consisting mainly of citronellal, geraniol, and citral, which is characterized by a lemony aroma due to the citral compound. The essential oil exhibits medicinal, antifungal, antimicrobial, and antioxidant effects [22].

The main reasons for using essential oils in bakery are their antimicrobial and antioxidant effects. To increase shelf-life, essential oils can be added as a direct ingredient in the formulation of bakery products or can be added to the packaging material or the internal atmosphere of the package [10]. Bakery products with medium and high water activity are sensitive to microbiological spoilage. These products include filled bakery products, soft biscuits, and laminated bakery products. Almost all bacteria, yeasts, and molds can occur in bakery products with a water activity above 0.94. Breads, fruit pies, cakes, and pizzas fall into this category [23]. The antibacterial effects of essential oils are manifested by limiting bacterial growth or directly destroying bacterial cells [8]. In general, Gram-positive bacterial species are more sensitive to natural compounds than Gram-negative bacteria due to the absence of an outer membrane [24]. The antibacterial activity of different essential oils has been investigated against both Gram-positive bacteria (*Bacillus subtilis*, *Staphylococcus aureus*, and *Listeria monocytogenes*) and Gram-negative bacteria (*Escherichia coli*, *Salmonella typhimurium*, *Pseudomonas aeruginosa*, and *Camplyobacter* spp.) [8]. Both cinnamon and clove essential oils can inhibit both Gram-positive and Gram-negative bacteria [24]. Mustard, cinnamon, and clove essential oils slow down the spoilage process of baked goods. Lemon grass, cinnamon essential oil, clove essential oil, and thyme essential oil have fungicidal activity. Turmeric, lemongrass, and cloves when added to the recipe retarded the growth of mold in butter cakes [25]. Bakery products are prone to rancidity [25]. Antioxidant activity may also be associated with non-phenolic compounds such as limonene, linalool, and citral. Cinnamon and clove essential oils consist of various monoterpenes and sesquiterpenes that show significant antioxidant activity; especially, cinnamaldehyde and eugenol are associated with antioxidant effects [8].

Rheological measurements can simulate and characterize the material properties during processing, thus allowing the quality control of products [26]. The rheological properties of dough are good indicators of behavior during kneading, fermentation, and baking [27]. Mixolab is a modern instrument for determining the quality control of cereals and flour. It is used to determine the thermomechanical properties of dough. The advantage of this method is that operating conditions such as kneader speed and temperature can be determined. This makes it possible to assess the effect of the addition of additives, kneading parameters, and the properties of the dough during heating [28]. For the analysis, 75 g of dough is used to determine the quality of protein and starch. The standard 'Chopin+' protocol analyzes dough kneading behavior, starch lubrication, amylase activity, and starch retrogradation in a single test [29]. Essential oils can influence the physicochemical and rheological properties of the dough. There are changes in the qualitative properties of baked goods such as volume, texture, and sensory characteristics [27]. The volume of the bread usually decreases, and the stiffness increases after the addition of essential oils. Essential oils are thought to bind to storage proteins or polysaccharides depending on their nature. This leads to a strengthening of the dough and a prolongation of the dough development time [30]. The use of thyme essential oil at concentrations equal to or greater than 0.15 mL/100 g of dough resulted in a significant reduction in bread volume [31].

The present study was aimed at the assessment of the effect of thyme and lemongrass essential oils on the bakery quality of common commercial wheat flour in several technological stages. Changes in behavior during kneading, i.e., in the viscoelastic properties of non-fermented dough were recorded on the Mixolab instrument. The course of fermentation was described by a row of the standard leavening microtests, and the final quality of bread in terms of laboratory baking trial was finalized by color evaluation.

2. Materials and Methods

2.1. Materials

Based on information from the literature and previous research, we determined the effective concentrations of essential oils (EOs) as follows: 0 mg/100 g for the control, and 5, 10, 20, 40, and 80 mg/100 g of flour. To ensure accurate EO concentration and to simplify dosing, the amounts were recalculated according to the actual density of both essential oils, and application was performed by a micropipette.

The wheat white, finely-milled flour (STD) was rendered by the producer Mlýny J. Voženílek, Private Limited Company, settled in town Předměřice nad Labem, Czech Republic. According to the Czech state standard (ČSN 56 0512 [32], corresponds to the ICC 104/1 [33]), the quality parameters specify an ash content of a maximum of 0.52% and a gluten content in dry matter of a minimum of 32.0%.

Thyme essential oil (THY) as well as lemongrass essential oil (LMG) were bought from Sigma-Aldrich Ltd. (Saint Louis, MO, USA), with the countries of origin being Spain and India, respectively, produced by the steam distillation method, thyme EO (44.34% thymol, 17.88% p-Cymene), lemongrass EO (40.06% geranial, 31.65% neral), determined by GC-MS and GC-FID.

The density of the THY was 0.917 g/mL, and of the LMG 0.896 g/mL. Both EOs were mixed with the STD immediately before each test, and selected doses increased stepwise from 0 mg/100 g of flour to 5, 10, 20, 40, and 80 mg/100 g of flour. The codes of all flour mixtures combined the actual dose and the type of EO: 5THY, 10THY, 20THY, 40THY, 80THY for the THY, and similarly 5LMG–80LMG for LMG.

For the leavening microtests, further raw materials were necessary: fresh pressed baker's yeast (*Saccharomyces cerevisiae*) 'EXTRA' was produced by the LLC 'BALEX COMPANY' (Kharkov, Ukraine), while salt and beet sugar were bought in the local retail market. In correspondence with the work's aim of EO effect testing, any other commonly used fat (rapeseed or sunflower oil, or margarine) was not used.

2.2. Methods

2.2.1. Mixolab Test

To describe the rheological behavior of wheat flour enhanced by essential oils, rheometer Mixolab (Chopin Technologies, Villeneuve-la-Garenne, France; today part of KMP Analytics, Westborough, MA, USA) with the company's software was employed. An advantage of this apparatus lies in low material consumption as well as in the description of the dough kneading phase together with the dough pasting one during one test (in principle, it represents a combination of the Farinograph rheometer and the Amylograph viscometer). The test settings included the 'Chopin+' protocol according to the ICC norm No. 173; details of the protocol settings were published earlier [34,35]. The description of the curves recorded was restricted to the primary parameters: torque points C1, C2, C3, C4, and C5 and times t_C1 as well as STA (time of the occurrence of the torque point C1 ≈ dough development time, and stability of dough consistency). The technological meaning of these five torque points was discussed in the same papers [34,35].

For each single test with the essential oil, a determination of the moisture content was carried out in advance by using an infrared moisture balance. Wheat flour mixture with THY or LMG was prepared manually—10 g of finely milled wheat flour was weighed into the beaker and the desired amount of essential oil was added with a pipette. The flour and essential oil were mixed carefully with a glass rod, and homogenization of the sample took 2 min. The mass was poured onto an aluminum plate and placed in an infrared balance. For the Mixolab test itself, flour–oil samples were prepared similarly; only a mixture homogenization was prolonged to 5 min.

2.2.2. Leavening Microtest—Dough Volume & Sample Weight Monitoring

A leavening (fermentation) test was carried out according to the internal methodology. It is the simplest method to monitor volume changes in yeasted dough during a technological phase of leavening. For more detailed and precise records, apparatuses called fermentograph or rheofermentometer could be employed after yeasted dough preparation on a laboratory kneader (typically the Farinograph or the AlveoConsistograph). Here, the recipe comprised 50.0 g flour, 1.5 g fresh compressed yeast, and 40 mL distilled water (30 °C). The correct amount of essential oil was pipetted into the flour, followed by the thorough homogenization of the sample by manual mixing for 5 min. A water suspension was prepared from the yeast first, followed by the preparation of the dough. Yeast suspension was combined with the flour–EO mixture and kneaded for 3 min to a uniform structure. The prepared dough mass was quantitatively transferred into 250 mL glass beakers or into 50 mL Falcon tubes for weight and volume monitoring, respectively. Each fulfilled laboratory glass was placed in a fermentation tank where the leavening test was carried out under pre-set fermentation conditions (34 °C, relative humidity (RH) 78%). The fermentation experiment took 90 min in total, and both the weight and volume of the samples were recorded at 10 min intervals. The weight was evaluated with an accuracy of ±0.001 g. Each result is the average of two measurements.

The leavening test was used also for testing lupin flour effect on semolina breadmaking quality—Spina et al. [36] prepared dough on Farinograph to consistency 500 Brabender units, weighted pieces of 25 g and measured in 50 mL gradual cylinders, previously oiled to avoid sticking. For dough leavening, they set the incubator to 30 ± 2 °C, and contrary to our own procedure, measurement in each 10 min was aimed at reading the volume level only.

2.2.3. Bread Preparation and Quality Evaluation

For the laboratory baking experiment, the internal modification of the standard Rapid Mix Test (RMT) method was used. The baking trial aimed at quality control and the evaluation of the effect of EO addition to the recipe on bread quality, bread volume, and sensory acceptability (ECC 2062/81). The used recipe is as follows: 200 g finely-milled white wheat flour (T530), 3.3 g salt, 114 mL water, 6.0 g baker's fresh yeast; THY or LMG in doses 5, 10, 20, 40, and 80 mg/100 g of flour. As in the case of the leavening test, the recipe did not involve any other fat.

In a Domino food processor (výrobce, město, stát) equipped with a kneading hook, the homogenization of the flour–EO mixtures took 2 min, followed by fast 2 min mixing with further ingredients and final slow 10 min kneading of dough. In an incubator with auto-controlled temperature and RH pre-set to 34 °C and 78% (výrobce, město, stát), dough mass and dough pieces spent 20 and 40 min for leavening and maturation. After the leavening stage, the dough mass was split into 60.0 ± 0.1 g pieces manually and rounded up on the Extensograph shaping unit. Baking was carried out in a professional oven (výrobce, město, stát)–(*nebo* spolupracující pekárna); the program 'BUN' with auto-steaming and an initial temperature of 260 °C, gradually reduced to 237 °C in 13 min was selected. Baked pieces were left on filtration paper for 2 h to cool down at laboratory conditions. Three representative samples were weighted, and their volumes were determined by the traditional rapeseed displacement method (AACCI Method 10-05.01) [37]. The collected data were transformed into bread yield on the basis of the known proportion of flour into 1 small-bread piece.

Bread yield was calculated by Equation (1):

$$Bread\ yield = \frac{100 \times volume\ of\ 3\ bulks}{weight\ of\ flour\ in\ 3\ bulks} \quad (mL/100\ gflour) \tag{1}$$

2.2.4. Bread Color Assessment

The evaluation of the color of the bread parts (crumb and crust separately) was made by using a colorimeter CM-600d (Minolta, Osaka, Japan) at a 10° angle and in daylight mode (D65). The CIE *Lab* colorimetric model was adopted, measuring the color coordinates L^* (luminosity, whiteness), a^* (greenness to redness), and b^* (blueness to yellowness). The measurements were carried out in five repetitions using two standards, crust standard L^* 67.5, a^* 11.47, and b^* 36.42, and crumb standard L^* 75.06, a^* 0.17, and b^* 19.03. Before each measurement, the instrument was calibrated using White Calibration Cap CM-A177.

In the CIE *Lab* color space, the difference between two colors (tints) is expressed by Equation (2):

$$\Delta E = \sqrt{(\Delta L^*)^2 + (\Delta a^*)^2 + (\Delta b^*)^2} \tag{2}$$

The final ΔE value is the difference of the measured pairs ($L^*_{SAMPLE} - L^*_{STD}$), ($a^*_{SAMPLE} - a^*_{STD}$), and ($b^*_{SAMPLE} - b^*_{STD}$) [38]. Simplified, it is a rectangular (Euclidean) distance between the tint of the enriched product and the tint of the control (e.g., bread, or crust and crumb separately). There is experimental knowledge about the so-called *just noticeable difference* between two independent samples—by human vision, they could be considered as clearly different (recognizable) at ΔE equal to 1.0–2.0 at least.

2.2.5. Statistical Analysis

Data collected from the rheological measurement, leavening microtest, and baking trial was described by Tukey's HSD test and correlation analysis (p = 95%) using the Statistica 13.0 software (TIBCO Software Inc., Palo Alto, CA, USA). The interrelations among the qualitative features were explored by multivariate Principal Component Analysis (PCA). The procedure of the PCA was carried out in two steps—firstly with the complete dataset to identify the most representative features, whose variability was explained by the first three principal components (PC) from 60% at least. At the same time, the representative features must cover all specific technological phases of bread manufacturing as dough kneading, leavening test, baking trial, and bread color assessment. For example, information doubling was confirmed between the Mixolab features including the stability of dough consistency and the torque point C2 as heat–mechanical dough destruction (r = 0.95, p = 95%), and reversely, an irreplaceable role could be addressed to the dose of the essential oil and dough weight at the 90th min of the leavening test. After such reduction, PCA was repeated with an improved explanation rate of the scatter of the representative data.

3. Results

3.1. Mixolab Test

The non-enriched wheat flour (STD) was characterized by a standard dough development time (parameter t_C1) of 3.75 min and long stability of consistency (equal to 9.20 min;

Table 1). As supposed, the addition of both essential oils did not change values of the specific torque points C1 and C3–C5 verifiably, perhaps owing to low dosages and the absence of alpha-amylases, acting in the second pasting phase of the Mixolab test. Only in the case of dough mechanical–thermal destruction and the proper torque point C2, a somewhat higher extent of the gluten skeleton damage was observed for the dough with THY doses 20, 40, and 80 mg/100 g flour. Such a difference could be later reflected in bread volume.

Table 1. Mixolab testing of rheological behavior of non-fermented dough as affected by several doses of thyme and lemongrass essential oils (THY and LMG, respectively).

Essential Oil		Mixolab Torque Point					Mixolab Time Value	
Type	Dose (mg/100 g)	C1 (N.m)	C2 (N.m)	C3 (N.m)	C4 (N.m)	C5 (N.m)	t_C1 * (min)	STA ** (min)
Control	0	1.08 a	0.50 a	1.82 a	1.68 a	2.61 a	3.75 a	9.20 d
THY	10	1.10 a	0.51 a	1.82 a	1.67 a	2.60 a	3.95 ab	8.10 c
	20	1.12 a	0.41 a	1.79 a	1.71 a	2.82 a	3.73 a	5.90 a
	40	1.12 a	0.41 a	1.82 a	1.71 a	2.80 a	3.90 ab	5.80 a
	80	1.09 a	0.41 a	1.81 a	1.71 a	2.87 a	3.97 ab	6.90 b
LMG	10	1.08 a	0.51 a	1.85 a	1.67 a	2.66 a	3.70 a	9.40 d
	20	1.10 a	0.52 a	1.86 a	1.74 a	2.74 a	3.93 ab	9.20 d
	40	1.07 a	0.51 a	1.83 a	1.68 a	2.65 a	3.88 ab	9.30 d
	80	1.07 a	0.53 a	1.84 a	1.63 a	2.67 a	4.18 b	9.20 d
Subgroup means								
Control	0	1.08 A	0.50 AB	1.82 A	1.68 A	2.61 A	3.75 A	9.20 B
THY	10–80	1.11 A	0.43 A	1.81 A	1.70 A	2.77 A	3.89 A	6.68 A
LMG	10–80	1.08 A	0.52 B	1.84 A	1.68 A	2.68 A	3.92 A	9.28 B

* time of the occurrence of the maximal consistency C1; ** stability of dough consistency. a–d: averages in columns, signed by the same letter, are not statistically different ($p = 95\%$). A–B: In columns, capital letters mark subgroup averages and statistically significant differences ($p = 95\%$).

Only the time parameters of the Mixolab test were varied significantly—more than the time of C1 point occurrence (data variance $a - b$ only), just the stability of dough consistency was shortened by THY to approximately a half (from 9.20 min to 6.68 ± 1.07 min; data variance $a - d$). The LMG counterpart had no recognizable influence on that dough cohesivity. Pooling over the EO doses, a distinguishing of THY-dough in the stability of consistency from the STD and LMG counterparts was statistically verified ($p = 95\%$) (Table 1).

3.2. Leavening Test

In a brief overview, the leavening microtest demonstrated its ability to describe satisfyingly a course of wheat dough fermentation progress under step-by-step doubled additions of the two types of essential oil.

3.2.1. Dough Weight Loss

Both for dough weight and volume, data recorded within the leavening microtest could be processed by three-factorial ANOVA; the F1 was leavening test time, F2 essential oil type, and F3 essential oil dose. For the dough weight loss, a linear decrease for the non-fortified wheat dough (STD) as well as for both the THY- and LMG-enriched counterparts was observed. A weight loss step oscillated between 0.1 and 0.2 wt.%; to ensure the clarity of the ANOVA chart and to magnify significant differences, a dataset was reduced to the leavening times 0, 30, 60, and 90 min (Figure 1). During the entire leavening test, 1.21 g of the control wheat dough was metabolized by yeasts. For the 5THY and 10THY samples, the consumed mass portion increased nearly twice (1.97 and 2.04 g, respectively). Along with the magnifying THY doses, these total weight losses diminished surprisingly up to 1.34 g. For a row of the LMG-dough variants, the corresponding weight losses were lower—1.59 g,

1.76 g, and 1.38 g, respectively. Described differences in the effect of THY and LMG can be noticed in Figure 1, especially for specimen pairs 5THY–5LMG and 10THY–10LMG.

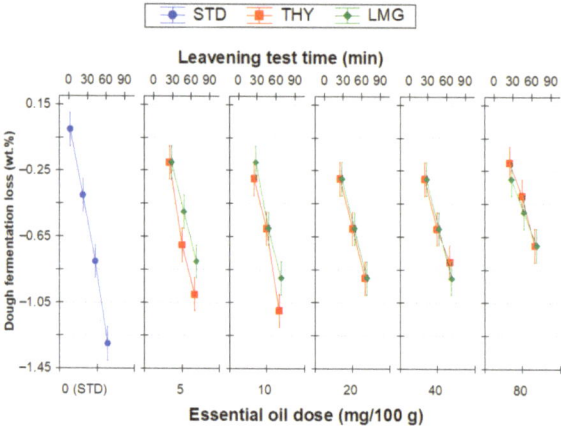

Figure 1. Comparison of the reduced results of the 90 min leavening microtest of the weight loss of the non-fortified wheat dough (STD) as influenced by the stepwise-rising doses of the thyme or lemongrass essential oil (THY and LMG). Vertical bars denote 0.95 confidence intervals.

3.2.2. Dough Volume Monitoring

In terms of dough volume monitoring, the effect of the THY and LMG was reversed—in general, the higher the oil content in the leavened dough, the harder the progress of the fermentation process. In bakeries, there is a common praxis to even double a recipe dose of yeasts in the case of the manufacturing of fat-rich Christmas or Eastern pastry. The maximal volume of the wheat control was 42 mL, which occurred at the 70th min of the leavening test (423% of the initial volume). The maxima of dough volumes of all 10 EO-enriched doughs did not level to the control (Figure 2); in the case of THY additions, observed values ranged from 31 to 38 mL (at 50th to 70th min of the leavening test; 308 and 383% of the initial volume). A stronger negative influence of the LMG oil is documented in the data pictured in Figure 2—dough volume maxima reached 29–36 mL.

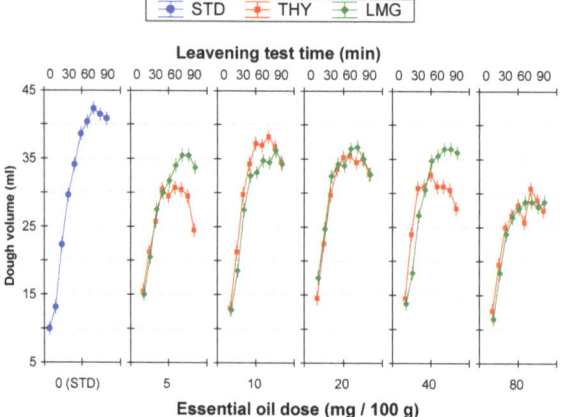

Figure 2. Comparison of the complete results of the 90 min leavening microtest of the volume rise of the non-fortified wheat dough (STD) as influenced by the stepwise-rising doses of the thyme or lemongrass essential oil (5, 10, 20, 40, and 80 mg/100 g of flour). Vertical bars denote 0.95 confidence intervals.

3.3. Baking Trial

3.3.1. Bread Yield

The control bread sample (STD) had a bread yield of 451 ± 16 mL/100 g flour. The bread yield of the samples with the addition of the THY essential oil ranged from 401 to 455 mL/100 g flour (Figure 3; average 440 ± 27 mL/100 g flour). However, data oscillation for these THY-bread variants did not demonstrate any trend (the lowest bread yield 401 mL/100 g flour found for 10THY, while the highest 478 mL/100 g flour for 20THY blend). Similarly, the additions of the LMG essential oil varied the bread volumes independently to the applied dose—the yield of the LMG-breads ranged from 405 to 482 mL/100 g flour (average 441 ± 25 mL/100 g flour; Figure 3). The bread from the 20LMG blend had the highest yield, and reversely the bread from blend 80LMG. In general, the antimicrobial activity of essential oils can have a toxic effect on yeasts or inhibit yeast activity and reduce fermentation speed. Fat may slow down the fermentation process via the greasing of starch granule surface and hardening of their accessibility to amylases.

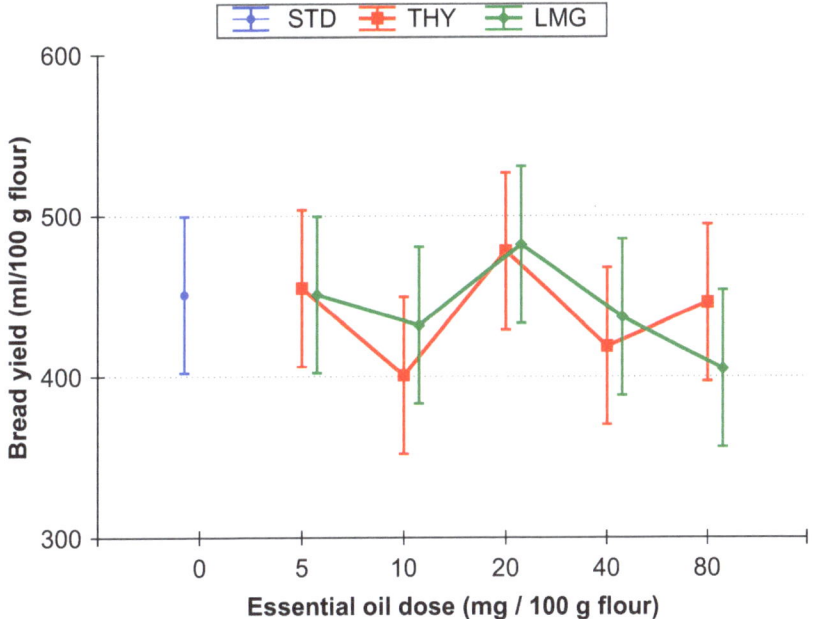

Figure 3. Comparison of the baking trial results of the non-fortified wheat bread (STD) as influenced by the stepwise-rising doses of the thyme or lemongrass essential oil (5, 10, 20, 40, and 80 mg/100 g of flour). Pictured differences were insignificant completely (two-factorial ANOVA, $p = 95\%$). Vertical bars denote 0.95 confidence intervals.

To document changes in small-bread shape and crumb porosity, bread cuts were scanned on a common office scanner at a resolution of 300 dpi. In Figure 4, wheat control and variants enriched by 80 mg THY or LMG/100 g of flour are presented. As can be noticed, the higher doses of EO affected the bread height, which corresponds with non-fermented dough elasticity (extensograph or alveograph one). The higher diameter of the breads with 80 mg THY or LMG/100 g of flour thus reflects a supporting effect of EO on dough extensibility.

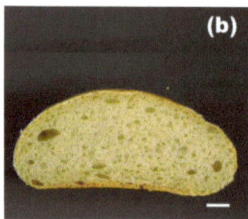
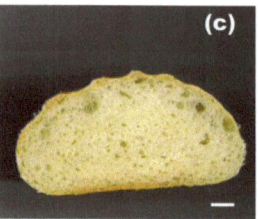

Figure 4. Comparison of cut appearance of selected small-bread variants. (**a**) unfortified wheat control, (**b**) bread with 80 mg thyme essential oil/100 g of flour, (**c**) bread with 80 mg lemongrass essential oil/100 g of flour. Note: scanned on common office scanner at resolution 300 dpi.

3.3.2. Bread Color Evaluation

The changes in the color of the bread crusts fortified by THY and LMG essential oils in increasing levels are summarized in Table 2a. By using a CM-600d spectrophotometer, the color of the crust of the control bread (STD) was characterized by lightness L^* 67.5, and the values of redness a^* 11.47 and yellowness b^* 36.42.

Table 2. (**a**) Crust color of wheat bread variants tested as affected by several doses of thyme and lemongrass essential oils (THY and LMG, respectively); (**b**) Crumb color of wheat bread variants tested as affected by several doses of thyme and lemongrass essential oils (THY and LMG, respectively).

(a)						
	Essential Oil (EO)		Crust Whiteness L^*	Crust Redness a^*	Crust Yellowness b^*	Crust ΔE^*
Type		Dose (mg/100 g Flour)				
Control		0	67.49 c	11.47 abc	36.42 ab	0.00 a
THY		5	64.74 abc	12.10 abcd	38.33 b	5.16 bc
		20	66.48 bc	12.67 abcd	34.79 ab	3.35 bc
		80	60.00 ab	15.70 cd	37.23 ab	8.82 d
LMG		5	69.83 c	9.96 ab	32.98 a	4.01 bc
		20	63.40 abc	12.36 abcd	34.50 ab	3.05 b
		80	67.17 c	12.69 abcd	36.15 ab	4.15 bc
Subgroup means						
Control		0	67.49 A	11.47 A	36.42 A	0.00 A
THY		5–80	63.70 A	13.10 A	35.94 A	6.30 C
LMG		5–80	66.00 A	12.69 A	36.70 A	4.00 B
(b)						
	Essential Oil (EO)		Crumb Whiteness L^*	Crumb Redness a^*	Crumb Yellowness b^*	Crumb ΔE^*
Type		Dose (mg/100 g Flour)				
Control		0	75.06 cd	0.17 a	19.03 ab	0.00 a
THY		5	77.49 def	0.18 a	18.45 a	2.31 bc
		20	79.43 f	0.27 abc	19.39 ab	4.39 d
		80	78.39 ef	0.35 abcd	19.66 bc	3.47 cd
LMG		5	72.56 bc	0.42 cd	19.29 ab	3.45 cd
		20	66.87 a	0.46 cd	19.23 ab	8.55 e
		80	71.64 b	0.50 d	20.65 c	3.81 cd
Subgroup means						
Control		0	75.06 B	0.17 A	19.03 A	0.00 A
THY		5–80	78.00 B	0.26 A	19.11 A	3.00 AB
LMG		5–80	71.20 A	0.42 B	19.43 A	4.20 B

L^*—luminosity, whiteness, a^*—greenness to redness, b^*—blueness to yellowness; ΔE—total color difference; in the CIE *Lab* space, Euclidean (i.e., rectangular) distance between the color of the fortified and control bread variant. a–f: averages in columns, signed by the same letter, are not statistically different (p = 95%). A–C: In columns, capital letters mark subgroup averages and statistically significant differences (p = 95%).

In the case of the crust tint of the fortified bread variants, the observed L^* values ranged between 59.0 and 69.8, while the a^* and b^* values were determined in ranges 8.79–15.70 and 32.98–38.33, respectively. There was no evidence of a trend in the change in bread crust color between both essential oil types tested, as well as among their doses in the recipe. Significant differences were observed between the 40THY and 5LMG samples, which showed significantly lower redness a^* values and at the same time higher whiteness L^* values. No higher extent of the darkening of the breads' crust was recognized along with the rising doses of both THY and LMG essential oils.

In the evaluation of bread crumb color in the CIE Lab space, the position of the control with any essential oil (STD) was $[L^*; a^*; b^*] = [75.1; 0.17; 19.03]$. The crumb color coordinates of all 11 bread samples ranged in interval 66.9–79.4 for lightness L^*, 0.17–0.50 for redness a^*, and 18.45–20.65 for yellowness b^*. Similar to the bread crust color measurement, no rising or diminishing trend was observed (Table 2b).

Both for crust and crumb color of single bread samples, any tendency connected to essential oil type or essential oil dose was not proven also for the total color difference ΔE (overlaying ranges 3.1–9.0 for the crust color, 0.7–8.6 for the crumb color). Considering an average color change as a result of THY or LMG incorporation, the tint of the crust was distinguished completely and the one of the crumb partially (Table 2a and Table 2b, respectively). As mentioned above, an average appearance of crust control and LMG-breads could be considered as just noticeably different to each other; moreover, bread counterparts with THY oil could be recognized both from the control as well as from ones containing LMG oil. In the case of the bread crumb tint, only the pairs of control—THY-breads and control—LMG-breads met the condition of a minimum difference in ΔE equal to 2.0 (Figure 5).

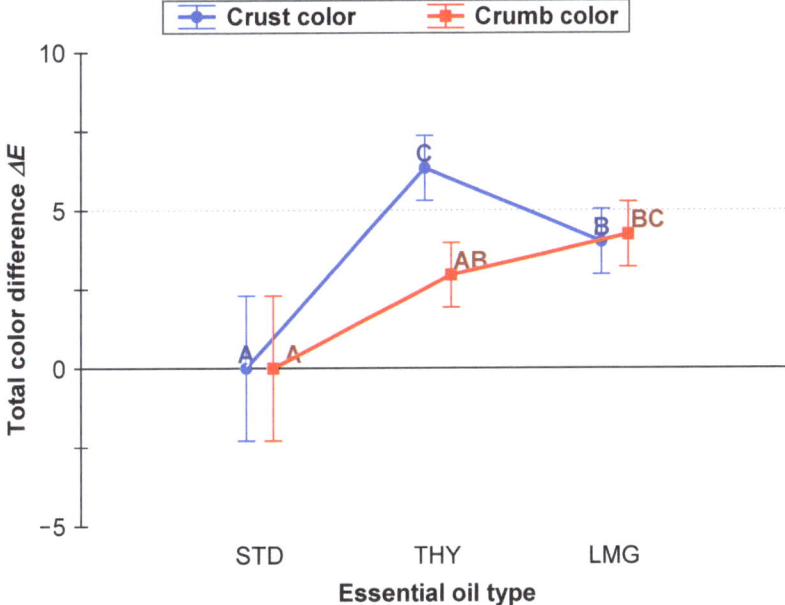

Figure 5. Comparison of the average total color difference ΔE between the crust and crumb of the non-fortified bread (STD) and its counterparts enriched by thyme and lemongrass essential oils (THY and LMG, respectively). A–C: averages signed by the same letter are not statistically different ($p = 95\%$). Note: average values pooled over all five doses of essential oil applied (5–80 mg/100 g flour). Vertical bars denote 0.95 confidence intervals.

3.4. Multivariate Statistical Analysis

After the original dataset reduction, mentioned in Section 2.2 above, 9 variables from the original 19 ones were maintained as the necessary as well as sufficient ones for a description of the recorded data variability. In correspondence with information compressed in Table 3, the first principal component (PC1) covered 46%, the PC2 22%, and the PC3 10% of the reduced dataset scatter on average. The PC1 covered mainly the Mixolab torque points of dough consistency and hot gel stability (C1 and C4) plus the crumb whiteness L^*; the PC2 is built-up from the characteristics essential oil dose, starch retrogradation rate (torque point C5), and dough weight loss in 90 min of the leavening test (Figure 6a). For the majority of the mentioned dough quality features, these PC1 and PC2 could be considered sufficient to ensure a minimal information loss (communalities sum $\geq 70\%$). For the bread yield additionally, the PC3 should also be included—as mentioned above, bread volume variance did not allow a clear distinguishing of essential oil type, or its dose used. It is a partially logical finding that the volume and shape of the final leavened bakery product depend on many factors; they begin from used dough recipe and mixer type via leavening conditions and dough pieces shaping procedure up to baking conditions. The plot of PC1 × PC2 loadings confirmed a known positive relationship between specific bread volume/bread yield and the Mixolab torque point C2—the more resistant the dough to heat–mechanical treatment, the potentially larger rise of the dough during leaving (based on stronger gluten net present in that dough).

Table 3. Communalities (%), a percentage of explained data variability by the first three principal components (PC).

Variable		PC1	PC2	PC3	Sum
Dose of essential oil		3	45	29	77
Mixolab torque point	C1	74	3	6	82
	C2	64	24	3	91
	C4	41	0	1	42
	C5	41	47	3	91
Baking trial	Dough weight$_{90\ min}$	14	53	6	73
	Bread yield	27	2	42	70
Bread color	Crumb L^*	86	1	0	87
	Crumb a^*	64	25	0	89
Average		46	22	10	78

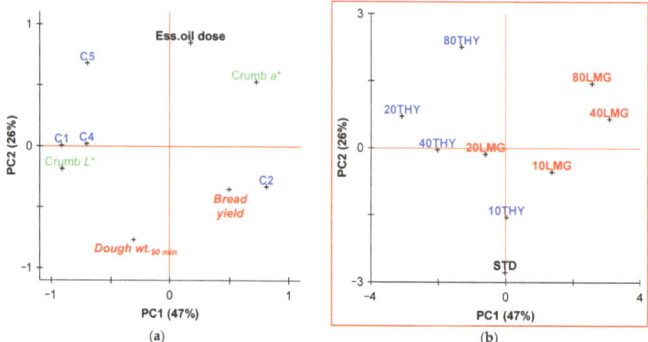

Figure 6. Principal component plot: (**a**) of variable loadings, where C1, C2, C4, and C5: Mixolab torque points; Dough wt.$_{90\ min}$—weight of the dough sample in the 90 min of the leavening test; Crumb L^*, Crumb a^*—crumb luminosity (whiteness) and redness; (**b**) of case scores, where STD—non-fortified wheat dough/bread; 10-20-40-80THY/LMG—wheat dough/bread fortified by 10 up to 80 mg/100 g four of thyme and lemongrass essential oil, respectively.

Correspondingly to the positions of the variables in Figure 6a, the tested samples are located in the same plane of the PC1 × PC2. The effects of THY and LMG on flour/dough and bread properties are listed above; in the plot Figure 6b, there is documented a statistical closeness of the samples 10THY and 10LMG to the non-enriched control (STD).

Uncovered oscillation mainly in bread yields and crumb color caused shifts in the position of the samples containing 20 or 40 mg/100 g of both essential oils (20THY, 40THY, 20LMG, and 40LMG), whilst as presumed, the most enriched specimen 80THY and 80LMG are the furthest from the control (STD).

4. Discussion

4.1. Mixolab Test

In the previous case of the Czech commercial flour testing on the Mixolab, the time of reaching the C1 torque point was prolonged by about 4.72 to a value of 8.47 min. Similarly, the dough stability as resistance to over-kneading was supported in the same direction (11.53 min) [39]. Durum wheat is known through a different proportion of gliadins and glutenins, and thus a diverse rheological behavior of non-fermented dough. In Northern Italy in the year 2014, common and durum wheat reached a shorter dough development time (1.8 and 1.3 min) than both the Czech samples, but different stability (10.0 and 3.2 min, respectively) [40]. The addition of 1 wt.% and 3 wt.% of THY into wheat flour magnified the Farinograph dough development time of wheat flour from 1.9 min to 8.3 and 15.9 min, respectively. An insignificant effect was evaluated for the same doses of wheat germ oil (from 1.9 min to 2.0 and 1.8 min) [30]. The authors also tested further essential oils from the seeds of black currant and black cumin, and the Farinograph test results ended between the mentioned ones recorded for the wheat germ and THY. The dough stability of their wheat flour was 8.3 min, and the listed plant oils varied that value both negatively and positively (5.3–15.5 min). For bakers, more important knowledge was gained in terms of dough softening, which did not overcome the value of 40 Brabender units (BU) for the control wheat flour. For the industrial production of the Central European types of leavened small-bread as rolls and buns, the recommended Farinograph dough development time oscillates around 3.0 min with dough softening degrees up to 100 BU.

4.2. Leavening Test

The leavening test does not belong among the preferred proofs of wheat flour/dough quality and owing to that, there is a lack of reference materials. In research aimed at bakery applications, moreover, essential oils are tested mainly in the aspect of their effect on a bread's quality as well as the antifungal agent. Debonne et al. [31] found a notable decrease in bread volume when thyme essential oil was added. They attributed this reduction to the impact of thyme oil on the yeast viability in the dough, which affects the optimal speed of bread volume rise. In paper [36], a leavening test was used to describe the course of semolina dough leavening under several additions of sweet lupin flour and lupin protein concentrate (3–15 wt.%). Both lupine materials increased slightly Farinograph water absorption, but significantly prolonged the dough development time and reversely the dough stability. In the leavening test, consequently, more elastic semolina–lupin dough underwent a fermentation process more quickly, reaching maximal volume in approximately half the time in contrast to the control semolina dough. These effects may be addressed to either the higher fat content in lupine flour (6–8%) or to a partial disruption of too cohesive semolina dough itself. In contrast to lupine flour, white or black chia seeds are richer in fat content (30–35%) [41]. During the dough preparation of the Farinograph, wheat flour replaced by 5 or 10 wt.% of both chia flour variants needed about 4.8 % pt. and 9.6 % pt. more water to reach the prescribed consistency of 600 units. By using the machine Fermentograph SJA, leavening proof taking 160 min brought knowledge about the one-tenth shorter optimal leavening time (time of the first reaching of the maximal dough volume). At the end of this 160 min proof, dough volumes were lessened to approx. 71% and 56% without the effect of chia seed type (seed color). Regardless of the lower portions of fermentation gases captured and maintained in the wheat–chia dough, total production of the fermentation gases was almost comparable among all five samples (687 mL for control wheat dough, and 630–615 mL vs. 645–650 mL for wheat–white chia/black chia dough variants, respectively) [42].

4.3. Bread Yield and Color

An important parameter indicating the quality of the baked product is its final volume, which is related to the entire technological process, but mainly to the amount and quality of protein, especially gluten [43]. The addition of THY and LMG essential oils to the recipe did not lead to a significant reduction in volume yield compared to the control. The concentrations tested ranged from 0 to 80 mg/100 g of flour. Debonne et al. [31] reported a negative effect of thyme essential oil on bread volume from a concentration of 137.55 mg/100 g flour (172% of own maximal dose). Lower concentrations had no damaging effect on the specific volume of the bakery product. According to Dhillon et al. [44], dried thyme positively impacted bread quality when supplemented even up to 2% of flour. This was evident in the increased specific volume values at 1% and 2% supplementation levels. On the other side, spices and herbs such as cinnamon, clove, garlic, oregano, and thyme, added at 1 or 2%, significantly decreased the bread volume [45]. Lemongrass essential oil is well known for its strong antimicrobial activity against fungi, yeasts, and bacteria [46]. There is a presumption of yeast inhibition and as a result of the reduction in fermentation, the volume of bread may be reduced [31].

The observed L^* values for the crust were measured from 59.0 to 69.8, which is a standard range determined by many other authors [47]. The values for a^* ranged from 8.79 to 15.70 and b^* from 32.98 to 38.33 indicating the crust color of elaborated breads fell within the red–yellow area, which is a favorable characteristic of baked goods for consumers. Johnston et al. [48] used white wheat all-purpose flour, produced in New Zealand, and tested wheat bread fortification by wholemeal faba bean flour and protein isolate. Their control wheat bread had similar $L^*a^*b^*$ color coordinates ([59.9; 13.1; 31.3] for crust and [70.22; 2.07; 19.18] for crumb) as the own control (STD). The addition of 20 wt.% of both faba bean forms changed the color of wheat bread crust significantly on all three axes of the color space—the evaluated coordinates were [45.5; 16.5; 29.5] and [31.2; 15.0; 17.9], respectively. The authors explained bread browning, reddening, and yellowing as just a result of the Maillard reaction, speeded up by a higher content of legume proteins.

The color of the crust and crumb are quality parameters, which are associated with the organoleptic properties of the bread [49], and these attributes hold significance from the perspective of the customers. Caramelization and Maillard reaction led to the development of non-enzymatic browning in baked goods surface (resulting in its attractive flavor and beige–brown color). Reducing sugars react with an amine, creating via glycosylamine a derivate of amino deoxy-fructose. Our results indicate that the addition of EO to the recipe did not manifest itself in a large and consumer-unfriendly way. For example, the addition of protein-rich plant materials, e.g., chickpea powder, supports crust browning. Replacing 5.0, 17.5, and 30.0 wt.% of wheat flour, the ΔE has risen from 24.9 and 30.2 up to 34.1, respectively [50]. The total color difference ΔE can be classified analytically according to Cserhalmi et al. [51] as not noticeable (0–0.5), slightly noticeable (0.5–1.5), noticeable (1.5–3.0), well visible (3.0–6.0) and great (>6.0). Thus, ΔE of the crust is categorized as well visible and, only in the case of THY doses 10 and 80 mg/100 g flour, is considered as great. However, evaluating this total color difference of crumb is more variable and is categorized as noticeable (40THY and 10LMG), great (20LMG), and well visible for the rest. Debonne et al. [31] reported the discoloration of bread after the incorporation of thyme essential oil into a recipe as a result of the oxidation process that the essential oil undergoes, causing changes in food color. Thus, it is evident that applying EO can influence the final color of the bread.

In the realm of food science research, one of the primary challenges confronting enriched bakery food production is the harmonization of fortification with optimal sensory attributes [52]. Together with product volume, color is one of the primary sensory attributes influencing consumers' behavior [53] and is strongly associated with the concept of quality [54]. The secondary sensory attributes are crust crispiness and crumb resilience (elasticity).

5. Conclusions

The aim of the study was to evaluate the impact of the use of selected essential oils on the viscoelastic properties of wheat dough, the fermentation process course, and the final quality of yeasted bakery products. In lower concentrations, the inhibition of leavening is not dramatic and does not have a negative effect on the technological process and the final quality of the small-piece bread. The results of the Mixolab rheological tests showed a possible influence of the essential oils used on the weakening of the gluten structure, which became significant during the second heating–cooling phase of the proof, potentially affecting bread volume. No effect on the water absorption, kneading, or development of the dough was detected. The comprehensive examination of rheological behavior, leavening processes, and baking outcomes provides valuable insights into the potential applications of thyme and lemongrass essential oils in bakery formulations. The clarification of the effects of essential oils on dough rheology, fermentation processes, and fermentation kinetics provides essential insights for optimizing dough performance and final product quality.

In the baking trials, both thyme and lemongrass essential oils did not significantly reduce the bread volume yield compared to the wheat control. Although the color of crust and crumb was influenced by essential oil type and dose, variations observed were maintained within acceptable ranges for consumers' preference. This finding is particularly significant for the food industry, as it suggests that essential oil enrichment can be integrated into bakery formulations without compromising product volume or quality. For all nine manufactured bread variants, an indicative sensory analysis was appended. At higher concentrations (40 or 80 mg/100 g flour), the impact on the bread taste was enormous. From the point of view of consumer's acceptability, the concentration of essential oils was too high, manifesting flavor of categories 'excessively intense' or even 'unacceptable', accompanied by feelings of bitterness, pungency, and burning with a long-lasting effect in the mouth. However, the lower concentrations were very interesting and potentially attractive to common consumers.

It is important to find the optimal level between the effective dose of the specific essential oil, consumers' preference, and formulation as well as bakery technology process optimization and the suitable method of essential oil application. In industrial bakeries, for example, ready-to-use flour-based blends are commonly processed. If the essential oils could become a part of these premixes, there is a challenge for the premix producer to consider protection against potential oxidation.

Overall, the study provided a partial insight into the complex interactions between essential oils and dough characteristics, offering opportunities for optimizing bakery processes and product quality. Given the huge potential of natural substances, including essential oils, it is necessary to continue research and find the best ways to use natural substances.

It is necessary to test different doses of thyme and lemongrass essential oils and find the ideal concentration to achieve the desired dough properties and quality of the final product. In baking recipes, it is important to strike a balance between the characteristic flavor, antioxidation effect, and potential health benefits of essential oils and functional considerations such as dough consistency and leavening, as well as bread volume. The increased knowledge of the interaction of thyme and lemongrass essential oils with yeast activity and dough fermentation can inform adjustments in proofing time, yeast dosage, and processing conditions to optimize product proofing and consistency.

Bakeries can take advantage of market trends and focus on developing innovative product lines that take advantage of the unique sensory and health benefits of essential oils to meet consumer demand for healthier and more flavorful products. Informing consumers about the use of essential oils in bakery products can increase transparency and build confidence about the safety and potential health benefits of essential oils. This educational approach can help consumers make informed choices and promote a positive perception of the bakery branch.

Author Contributions: Methodology, L.J., I.Š. and I.K.; Software, I.Š.; Formal analysis, L.J., M.B. and P.K.; Investigation, L.J., I.Š., I.K. and P.K.; Writing—original draft, L.J. and I.Š.; Writing—review & editing, M.H. and O.L. All authors have read and agreed to the published version of the manuscript.

Funding: This work was supported by the National Agency for Agricultural Research of the Ministry of Agriculture of the Czech Republic under project Biostore (QK21010064). The work used [data/tools/services/facilities] provided by the METROFOOD-CZ Research Infrastructure (https://metrofood.cz; accessed on 30 May 2024), supported by the Ministry of Education, Youth, and Sports of the Czech Republic (Project No. LM2023064).

Institutional Review Board Statement: Not applicable.

Informed Consent Statement: Not applicable.

Data Availability Statement: The data presented in this study are available on request from the corresponding author.

Conflicts of Interest: The authors declare no conflicts of interest.

References

1. Qian, M.; Liu, D.; Zhang, X.; Yin, Z.; Ismail, B.B.; Ye, X.; Guo, M. A review of active packaging in bakery products: Applications and future trends. *Trends Food Sci. Technol.* **2021**, *114*, 459–471. [CrossRef]
2. Gu, M.; Hong, T.; Ma, Y.; Xi, J.; Zhao, Q.; Xu, D.; Jin, Y.; Wu, F.; Xu, X. Effects of a commercial peptidase on rheology, microstructure, gluten properties of wheat dough and bread quality. *LWT–Food Sci. Technol.* **2022**, *160*, 113266. [CrossRef]
3. Melini, V.; Melini, F. Strategies to extend bread and GF bread shelf-life: From sourdough to antimicrobial active packaging and nanotechnology. *Fermentation* **2018**, *4*, 9. [CrossRef]
4. de Sousa, T.; Ribeiro, M.; Sabença, C.; Igrejas, G. The 10,000-year success story of wheat! *Foods* **2021**, *10*, 2124. [CrossRef] [PubMed]
5. Dewettinck, K.; van Bockstaele, F.; Kühne, B.; van de Walle, D.; Courtens, T.M.; Gellynck, X. Nutritional value of bread: Influence of processing, food interaction and consumer perception. *J. Cereal Sci.* **2008**, *48*, 243–257. [CrossRef]
6. Dziki, D.; Różyło, R.; Gawlik-Dziki, U.; Świeca, M. Current trends in the enhancement of antioxidant activity of wheat bread by the addition of plant materials rich in phenolic compounds. *Trends Food Sci. Technol.* **2014**, *40*, 48–61. [CrossRef]
7. Axel, C.; Zannini, E.; Arendt, E.K. Mold spoilage of bread and its biopreservation: A review of current strategies for bread shelf-life extension. *Crit. Rev. Food Sci. Nutr.* **2017**, *57*, 3528–3542. [CrossRef] [PubMed]
8. Mutlu-Ingok, A.; Devecioglu, D.; Dikmetas, D.N.; Karbancioglu-Guler, F.; Capanoglu, E. Antibacterial, antifungal, antimycotoxigenic, and antioxidant activities of essential oils: An updated review. *Molecules* **2020**, *25*, 4711. [CrossRef] [PubMed]
9. Dwivedi, S.; Prajapati, P.; Vyas, N.; Malviya, S.; Kharia, A. A review on food preservation: Methods, harmful effects and better alternatives. *Asian J. Pharm. Pharmacol.* **2017**, *3*, 193–199.
10. Gavahian, M.; Chu, Y.-H.; Lorenzo, J.M.; Mousavi Khaneghah, A.; Barba, F.J. Essential oils as natural preservatives for bakery products: Understanding the mechanisms of action, recent findings, and applications. *Crit. Rev. Food Sci. Nutr.* **2020**, *60*, 310–321. [CrossRef]
11. Ahmed, L.I.; Ibrahim, N.; Abdel-Salam, A.B.; Fahim, K.M. Potential application of ginger, clove and thyme essential oils to improve soft cheese microbial safety and sensory characteristics. *Food Biosci.* **2021**, *42*, 101177. [CrossRef]
12. Vergis, J.; Gokulakrishnan, P.; Agarwal, R.K.; Kumar, A. Essential oils as natural food antimicrobial agents: A review. *Crit. Rev. Food Sci. Nutr.* **2015**, *55*, 1320–1323. [CrossRef]
13. Huang, T.; Qian, Y.; Wei, J.; Zhou, C. Polymeric Antimicrobial food packaging and its applications. *Polymers* **2019**, *11*, 560. [CrossRef]
14. Bakkali, F.; Averbeck, S.; Averbeck, D.; Idaomar, M. Biological effects of essential oils—A review. *Food Chem. Toxicol.* **2008**, *46*, 446–475. [CrossRef]
15. Cardoso-Ugarte, G.A.; Sosa-Morales, M.E. Essential oils from herbs and spices as natural antioxidants: Diversity of promising food applications in the past decade. *Food Rev. Int.* **2021**, *38*, 403–433. [CrossRef]
16. Tavares, L.; Zapata Noreña, C.P.; Barros, H.L.; Smaoui, S.; Lima, P.S.; Marques de Oliveira, M. Rheological and structural trends on encapsulation of bioactive compounds of essential oils: A global systematic review of recent research. *Food Hydrocol.* **2022**, *129*, 107628. [CrossRef]
17. Li, Y.X.; Erhunmwunsee, F.; Liu, M.; Yang, K.; Zheng, W.; Tian, J. Antimicrobial mechanisms of spice essential oils and application in food industry. *Food Chem.* **2022**, *382*, 132312. [CrossRef]
18. Tajkarimi, M.M.; Ibrahim, S.A.; Cliver, D.O. Antimicrobial herb and spice compounds in food. *Food Control* **2010**, *21*, 1199–1218. [CrossRef]

19. Napoli, E.; Di Vito, M. Toward a new future for essential oils. *Antibiotics* **2021**, *10*, 207. [CrossRef] [PubMed]
20. Dong, Y.; Wei, Z.; Yang, R.; Zhang, Y.; Sun, M.; Bai, H.; Mo, M.; Yao, C.; Li, H.; Shi, L. Chemical compositions of essential oil extracted from eight thyme species and potential biological functions. *Plants* **2023**, *12*, 4164. [CrossRef] [PubMed]
21. Dimov, I.; Petkova, N.; Nakov, G.; Taneva, I.; Ivanov, I.; Stamatovska, V. Improvement of antioxidant potential of wheat flours and breads by addition of medicinal plants. *Ukr. Food J.* **2018**, *7*, 671–681. [CrossRef]
22. Sarah, M.; Ardiansyah, D.; Misran, E.; Madinah, I. Extraction of citronella oil from lemongrass (*Cymbopogon winterianus*) by sequential ultrasonic and microwave-assisted hydro-distillation. *Alex. Eng. J.* **2023**, *70*, 569–583. [CrossRef]
23. Smith, J.P.; Daifas, D.P.; El-Khoury, W.; Koukoutsis, J.; El-Khoury, A. Shelf life and safety concerns of bakery products—A review. *Crit. Rev. Food Sci. Nutr.* **2004**, *44*, 19–55. [CrossRef]
24. Murbach Teles Andrade, B.F.; Nunes Barbosa, L.; da Silva Probst, I.; Fernandes Júnior, A. Antimicrobial activity of essential oils. *J. Essent. Oil Res.* **2013**, *26*, 34–40. [CrossRef]
25. Soumya, C.; Sudha, M.L.; Vijaykrishnaraj, M.; Negi, P.S.; Prabhasankar, P. Comparative study on batter, quality characteristics and storage stability of muffins using natural ingredients (preservatives) and synthetic preservatives. *J. Food Process. Preserv.* **2017**, *41*, e13242. [CrossRef]
26. Dobraszczyk, B.J.; Morgenstern, M.P. Rheology and the breadmaking process. *J. Cereal Sci.* **2003**, *38*, 229–245. [CrossRef]
27. Xu, J.; Wang, W.; Li, Y. Dough properties, bread quality, and associated interactions with added phenolic compounds: A review. *J. Funct. Foods* **2019**, *52*, 629–639. [CrossRef]
28. Parenti, O.; Guerrini, L.; Bossa Mompin, S.; Toldrà, M.; Zanoni, B. The determination of bread dough readiness during kneading of wheat flour: A review of the available methods. *J. Food Eng.* **2021**, *309*, 110692. [CrossRef]
29. Dubat, A.; Rosell, C.M.; Gallagher, E. *Mixolab: A New Approach to Rheology*, 1st ed.; AACC International Press: St. Paul, MN, USA, 2013; 126p, ISBN 9780128104309.
30. Debonne, E.; De Leyn, I.; Verwaeren, J.; Moens, S.; Devlieghere, F.; Eeckhout, M.; Van Bockstaele, F. The influence of natural oils of blackcurrant, black cumin seed, thyme and wheat germ on dough and bread technological and microbiological quality. *LWT–Food Sci. Technol.* **2018**, *93*, 212–219. [CrossRef]
31. Debonne, E.; Van Bockstaele, F.; De Leyn, I.; Devlieghere, F.; Eeckhout, M. Validation of in-vitro antifungal activity of thyme essential oil on *Aspergillus niger* and *Penicillium paneum* through application in par-baked wheat and sourdough bread. *LWT–Food Sci. Technol.* **2018**, *87*, 368–378. [CrossRef]
32. ČSN 56 0512-8; Methods of Testing Mill Products—Determination of Ash. Czech Standards Institute: Prague, Czech Republic, 1993. (In Czech)
33. ICC 104/1; Determination of Ash in Cereals and Cereal Products. International Association for Cereal Science and Technology: Vienna, Austria, 1999.
34. Mironeasa, S.; Codina, G.G.; Mironeasa, C. The effect of wheat flour substitution with grape seed flour on the rheological parameters of the dough assessed by Mixolab. *J. Texture Stud.* **2012**, *43*, 40–48. [CrossRef]
35. Švec, I.; Hrušková, M. The Mixolab parameters of composite wheat/hemp flour and their relation to quality features. *LWT–Food Sci. Technol.* **2015**, *60*, 623–629. [CrossRef]
36. Spina, A.; Summo, C.; Timpanaro, N.; Canale, M.; Sanfilippo, R.; Amenta, M.; Strano, M.C.; Allegra, M.; Papa, M.; Pasqualone, A. Lupin as ingredient in durum wheat breadmaking: Physicochemical properties of flour blends and bread quality. *Foods* **2024**, *13*, 807. [CrossRef] [PubMed]
37. AACC International. AACCI Method 10-05.01: Guidelines for Measurement of Volume by Rapeseed. In *AACC Approved Methods of Analysis*, 11th ed.; AACC International: St. Paul, MN, USA, 2010.
38. Mokrzycki, W.S.; Tatol, M. Color difference ΔE—A survey. *Mach. Graph. Vis.* **2011**, *20*, 383–411.
39. Švec, I.; Hrušková, M.; Babiaková, B. Chia and teff as improvers of wheat-barley dough and cookies. *Czech J. Food Sci.* **2017**, *35*, 79–88. [CrossRef]
40. Blandino, M.; Marinaccio, F.; Ingegno, B.L.; Pansa, M.G.; Vaccino, P.; Tavella, L.; Reyneri, A. Evaluation of common and durum wheat rheological quality through Mixolab® analysis after field damage by cereal bugs. *Field Crops Res.* **2015**, *179*, 95–102. [CrossRef]
41. Talandová, M.; Pospiech, M.; Tremlová, B. Use of chia seeds (*Salvia hispanica* L.) and effect on human health. *Výživ. Potraviny* **2013**, *4*, 104–106. (In Czech)
42. Babiaková, B. Characteristics of Composite Flours Containing Non-Traditional Seeds. Master's Thesis, University of Chemistry and Technology Prague, Prague, Czech Republic, 2015; 90p. (In Slovak).
43. Mondal, A.; Datta, A.K. Bread baking–A review. *J. Food Eng.* **2008**, *86*, 465–474. [CrossRef]
44. Dhillon, G.K.; Kaur, A.; Bhise, S. Rheological and quality characteristics of thyme (*Thymus vulgaris*) enriched bread. *Int. J. Chem. Stud.* **2019**, *7*, 648–651.
45. Dhillon, G.K.; Kaur, A.; Bhise, S.; Ahluwalia, P. Synergistic effect of spices and herbs on rheological and bread making properties of wheat flour. *J. Pure Appl. Microbiol.* **2016**, *10*, 1099–1107.
46. Mendes, J.F.; Norcino, L.B.; Martins, H.H.A.; Manrich, A.; Otoni, C.G.; Carvalho, E.E.N.; Piccoli, R.H.; Oliviera, J.E.; Pinheiro, A.C.M.; Mattoso, L.H.C. Correlating emulsion characteristics with the properties of active starch films loaded with lemongrass essential oil. *Food Hydrocoll.* **2020**, *100*, 105428. [CrossRef]

47. Cingöz, A.; Akpinar, Ö.; Sayaslan, A. Effect of addition of wheat bran hydrolysate on bread properties. *J. Food Sci.* **2024**, *89*, 2567–2580. [CrossRef]
48. Johnston, C.; Leong, S.Y.; Teape, C.; Liesaputra, V.; Oey, I. Low-intensity pulsed electric field processing prior to germination improves in vitro digestibility of faba bean (*Vicia faba* L.) flour and its derived products: A case study on legume-enriched wheat bread. *Food Chem.* **2024**, *449*, 139321. [CrossRef]
49. Purlis, E. Browning development in bakery products—A review. *J. Food Eng.* **2010**, *99*, 239–249. [CrossRef]
50. Shrivastava, C.; Chakraborty, S. Bread from wheat flour partially replaced by fermented chickpea flour: Optimizing the formulation and fuzzy analysis of sensory data. *LWT–Food Sci. Technol.* **2018**, *90*, 215–223. [CrossRef]
51. Cserhalmi, Z.; Sass-Kiss, A.; Toth-Markus, M.; Lechner, N. Study of pulsed electric field treated citrus juices. *Innov. Food Sci. Emerg. Technol.* **2006**, *7*, 49–54. [CrossRef]
52. Rosell, C.M.; Bajerska, J.E.; El Sheikha, A.F. *Bread and Its Fortification*, 1st ed.; CRC Press: Boca Raton, FL, USA, 2015; p. 417.
53. Konieczny, P.; Stangierski, J.; Kijowski, J. Physical and chemical characteristics and acceptability of home style beef jerky. *Meat Sci.* **2007**, *76*, 253–257. [CrossRef]
54. Wibowo, S.; Vervoort, L.; Tomic, J.; Santiago, J.S.; Lemmens, L.; Panozzo, A.; Grauwet, T.; Hendrickx, M.; Van Loey, A. Colour and carotenoid changes of pasteurized orange juice during storage. *Food Chem.* **2015**, *171*, 330–340. [CrossRef]

Disclaimer/Publisher's Note: The statements, opinions and data contained in all publications are solely those of the individual author(s) and contributor(s) and not of MDPI and/or the editor(s). MDPI and/or the editor(s) disclaim responsibility for any injury to people or property resulting from any ideas, methods, instructions or products referred to in the content.

Article

The Influence of Pomace Powder of Musky Squash on the Characteristics of Foamy Confectionery Products during Storage

Natalia Netreba [1], Elena Sergheeva [1], Angela Gurev [1], Veronica Dragancea [1], Georgiana Gabriela Codină [2,*], Rodica Sturza [1] and Aliona Ghendov-Mosanu [1]

[1] Faculty of Food Technology, Technical University of Moldova, 9/9 Studentilor St., MD-2045 Chisinau, Moldova; natalia.netreba@tpa.utm.md (N.N.); elena.sergheeva1@tpa.utm.md (E.S.); angela.gurev@chim.utm.md (A.G.); veronica.dragancea@chim.utm.md (V.D.); rodica.sturza@chim.utm.md (R.S.); aliona.mosanu@tpa.utm.md (A.G.-M.)
[2] Faculty of Food Engineering, "Stefan cel Mare" University, 720229 Suceava, Romania
* Correspondence: codina@fia.usv.ro; Tel.: +40-7-4546-0727

Abstract: This paper analyzes the possibility of using pomace powder of musky squash (PPMS, 10–30% of the formulation) for the manufacture of foamy confectionery products based on Jerusalem artichoke syrup, which is a natural substitute for sugar syrup used in the food industry. The content of biologically active compounds (polyphenols, carotenoids) as well as the antimicrobial and antioxidant properties of pumpkin powder were evaluated. Sensory analysis was applied to measure the degree of product acceptance and the analysis revealed that the optimal amount of PPMS accepted by the tasters was 15%. The addition of PPMS increased pH and free water retention, color, and lightness intensification. During the storage period (40 days), the hardness and gumminess showed an essential increase and the cohesion of the samples gradually decreased. The addition of PPMS led to the improvement of textural parameters, thus contributing to the extension of the shelf life of products by 10 days, compared to the control sample. Mutual information analysis was applied to determine the influence of PPMS concentration and storage time of foamy confectionery products on mean total score, mean sensory profile score, moisture content, water activity, antioxidant activity, hardness, cohesiveness, and gumminess. The results of this research indicate that the use of pumpkin pulp in the manufacture of foamy confectionery products can significantly increase their biological value and sensory characteristics and ensure an extension of the products' shelf life.

Keywords: pumpkin pomace powder; foamy confectionery product; biologically active compounds; color analysis; quality; textural parameters; microbiological stability

1. Introduction

Color is a key commodity indicator that influences consumer evaluation of a product. Nowadays, the problem of using synthetic food colorants in production is of significant concern, as frequent consumption of these additives can negatively impact consumer health. One possible solution to this problem is the use of natural pigments extracted from plant products, such as carotenoids, anthocyanins, chlorophylls, and betalains [1]. These pigments have powerful antioxidant activity and numerous health benefits, such as slowing down aging, nervous system restoration, and anti-atherogenicity, anti-cancer, and anti-inflammatory properties [2].

More than 1100 carotenoid pigments are known today; they give yellow, orange, and red colors in plants, fruits, and vegetables [3]. Carotenoids from plant sources are important phytonutrients for animal organisms, having multiple effects, including provitamin A activity [4]. Carotenoids are effective active oxygen scavengers and can reduce oxidative stress in the human body. They have an outstanding effect on chronic diseases, preventing cardiovascular and cerebrovascular diseases, eye diseases, osteoporosis, and cancer [5].

In plant sources, carotenoids are always accompanied by other biologically active compounds (BACs) with antioxidant properties, including polyphenols. Research by Milner and Duthie confirms the role of polyphenols as antioxidants in the prevention of degenerative diseases, especially cancer and cardiovascular diseases. Polyphenols are powerful antioxidants that complement and enhance the functions of antioxidant vitamins and enzymes, protecting against oxidative stress caused by excess reactive oxygen species [6,7].

Among the vegetable sources of BACs, especially carotenoids, we can list the pumpkin (family *Curcubitaceae*), consumed widespread in the Republic of Moldova. The carotene content of pumpkin fruits is 16–17 mg/100 g of raw product, but in some forms, it reaches 35–38 mg/100 g. The brighter the color of the yellow-orange pumpkin flesh, the more carotenoids it contains [4].

When a pumpkin is processed correctly, its beneficial substances are also preserved in the powder. Pumpkin pomace powder can be used as a β-carotene supplement in food products. The potential of pumpkin pomace powder in confectionery products has been demonstrated; it can be used as a color and flavor additive [8]. Its ability to retain water allows for extending the shelf life and freshness of the product [9].

The manufacture of food products based on local raw materials is today considered a priority area for the development of the food industry. Fortifying foods with functional ingredients from pumpkin pomace powder increases their nutritional value by providing antioxidants, dietary fibers, minerals, proteins, essential fatty acids, vitamins, and phytosterols, which positively affect the body. At the same time, the use of pumpkin pulp pomace in the manufacture of foamy confectionery products can significantly increase their biological value as well as sensory characteristics.

Foamy confectionery products have a foam-like structure, made from whipped mass based on sugar syrup and a gelling agent, with or without the addition of other raw materials, food additives, and flavors [10]. The relevance of foam foamy confectionery products lies in the fact that their recipe and preparation technology are quite flexible. These products allow for the introduction of fruit or vegetable pomace powder as well as other functional ingredients, which will greatly increase its biological value without changing its basic properties.

This study aimed to develop a foamy confectionery product with pomace powder of musky squash (PPMS) and artichoke syrup to evaluate the effect of different concentrations of PPMS on the sensory characteristics, physico-chemical, and textural properties of the foamy confectionery products, including monitoring of the main parameters of the product during storage.

2. Materials and Methods

2.1. Chemical Materials

Folin–Ciocalteu phenol reagent (2.1 N) was purchased from Chem-Lab NV (Zedelgem, Belgium); 6-hydroxy-2,5,7,8-tetramethylchromane-2-carboxylic acid (Trolox) (purity $\geq$ 97%), 2,2-azinobis (3-ethylbenzothiazoline-6-sulfonic acid), diammonium salt (ABTS) ($\geq$98%), and 2,2-diphenyl-1-picrylhydrazyl-hydrate (DPPH) ($\geq$95%) were provided by Alpha Aesar (Haverhill, MA, USA). Gallic acid (GA) ($\geq$97%), quercetin (>95%), and lutein (>90%) β-carotene (>93%) were purchased from Extrasynthese (Lyon, France). Methanol, ethanol, tert-butyl methyl ether, hydrochloric acid, ethyl acetate, petroleum ether, sodium hydroxide, phenolphthalein, sodium nitrite, sodium citrate, trisodium citrate, silver nitrate, hexane, boric acid, hydrochloric acid, Kjeldahl catalyst tablets (KjTabs) VST, sulfuric acid, potassium hydroxide, n-octanol, and acetone were purchased from Chemapol (Prague, Czech Republic). Buffered peptone water and potato dextrose agar with chloramphenicol were obtained from HiMedia Laboratories Private Limited (Maharashtra, India). All reagents were of analytical or chromatographic grade. Spectrophotometric determinations were performed using the spectrophotometer UV-1900 (Shimadzu, Tokyo, Japan).

2.2. Raw Material

The musky squash pumpkin (*Cucurbita moschata* L.) was purchased from the distributor "Aliment-Ulei" LTD in the village Parata, district Dubasari, Republic of Moldova (47.1008338928222660 latitude, 29.1191673278808600 longitude and 17 m altitude). The dry matter content was 9.9 ± 0.02 g/100 g and pH = 6.0 ± 0.4.

2.3. Characterization of the Pomace Powder of Musky Squash (PPMS)

PPMS was obtained by drying and grinding musky squash pulp pomace. The musky squash pulp pomace was dried in a thermostat (Pol-Eko Aparatura ST 2, Wodzisław Śląski, Poland) with forced air circulation at a constant temperature of 60 ± 2 °C and a relative moisture content of 60–65%. Drying occurred for 20 h until the moisture content reached no more than 7%. Next, the dried mass was crushed using a grinder (Zimmer ZM-625, Rheinau, Germany) and sifted through a sieve with no more than 0.8 mm mesh size. The resulting PPMS was a homogeneous orange powder, which was stored in a dark, airtight container without access to light.

Physico-Chemical Analysis of PPMS

Moisture content (MC) was determined by the AOAC Standard (2005) gravitational method, by drying in an oven (Pol-Eko Aparatura ST 2, Wodzisław Śląski, Poland). For this purpose, 4 g ± 0.001 g of PPMS was weighed and then dried at 105 ± 1 °C to constant weight (so that the sample mass between measurements did not differ by more than 0.004 g). Each sample was analyzed in triplicate [11,12].

Ash content (AC) was determined by the AOAC Standard (2005) method. Previously dried in the hot-air oven, samples were weighed in a crucible and heated at 600 ± 10 °C in a muffle furnace (Snol, Narkūnai, Lithuania) for 5 h until resulting in a gray ash, which was cooled in the desiccator and weighed [12,13].

The pH was measured using a pH meter (TESTO 206-pH2, Pruszkow, Poland) calibrated with buffer solutions pH 4.0 and 7.0, directly immersing the electrode in the beaker containing the sample macerated with distilled water, according to the AOAC (2012) method [14].

The titratable acidity (TA) was estimated by titrating a known volume of the sample against the standard, 0.1 N NaOH, using phenolphthalein as an indicator. The results were expressed in % malic acid [13,15].

2.4. Extract Characterization

The hydroethanolic extracts from PPMS and foamy confectionery products were obtained as follows: 0.5 g of the sample was mixed with 25 mL of 70% ethyl alcohol (1:50, m/v) and left in a dark place for 24 h. The mixture was further extracted by the ultrasound-assisted method (ISOLAB 621.06.006, Eschau, Germany) at 60 ± 1 °C for 30 min and a frequency of 37 Hz. The conditions used in ultrasound extraction were optimized to achieve a high yield of bioactive compounds, including polyphenols and flavonoids, which are responsible for the antioxidant effects. These conditions were selected based on the findings from our previous research [16,17].

After that, it was centrifuged (MPW-380R, Warsaw, Poland) at 5000 rpm for 10 min at room temperature. The supernatants were analyzed by spectrophotometric methods, which are indicated below.

2.4.1. Total Polyphenol Content (TPC)

TPC in hydroethanolic extracts was determined using the Folin–Ciocalteu reagent, following the procedure as outlined by Paulpriya et al. and Waterman et al. [18,19]. The analysis involved the use of a calibration curve of gallic acid standard (0–500 mg/L, $R^2 = 0.9977$). The findings were expressed in gallic acid equivalent per 100 g dry weight of pumpkin pomace powder (mg GAE/100 g DW).

2.4.2. Total Carotenoid Content (TCC)

TCC was assessed using a modified approach as outlined by Ghendov-Mosanu et al. [20]: 3 g of the material was subjected to extraction using a mixture of methanol/ethyl acetate/petroleum ether (1:1:1, $v/v/v$). After filtration of the extract, the residue underwent two additional extraction cycles using the same solvent mixture. The TCC was then determined by the spectrophotometric method. After plotting the absorption spectrum, the TCC was measured at the wavelength of maximum absorbance (450 nm).

2.4.3. Scavenging Activity of the Free Cation-Radical ABTS (2,2-Azinobis-(3-Ethylbenzothiazoline-6-Sulfonates))

The antioxidant activity (AA) of the hydroethanolic extracts was determined by the method described in the literature [21]. ABTS solution was produced by reacting 7 mM ABTS stock solution with 2.45 mM potassium persulfate (final concentration) for 16 h in the dark at room temperature. ABTS stock solution was diluted with ethyl alcohol to an absorbance of 0.70 ± 0.02 at 734 nm. To 3.9 mL of diluted ABTS solution, 100 µL of sample or trolox standard was added and the absorbance was measured at 734 nm after 6 min of incubation at 30 °C. ABTS inhibition capacity was expressed in mg trolox equivalent per 100 g dry weight (mg TE/100 g DW) from the calibration curve constructed in the concentration range 0–500 µmol/L (curve equation $R^2 = 0.9992$) for trolox, the water-soluble analog of vitamin E (6-hydroxy-2,5,7,8-tetramethylchroman-2-carboxylic acid) [22].

2.4.4. DPPH (2,2-Diphenyl-1-Picrylhydrazyl-Hydrate) Free Radical Scavenging Activity

The scavenging activity of the DPPH free radical was determined in the hydroalcoholic extracts from PPMS obtained above (2.4.) by the spectrophotometric method, according to the method described by Paulpriya et al. [18]. Results were expressed in µmol TE/g DW from the calibration curve (0–500 µmol/L, $R^2 = 0.9992$) with trolox [22].

2.5. Preparation and Characterization of the Foamy Confectionery Products

2.5.1. Preparation of the Foamy Confectionery Products

PPMS in different concentrations (10% PPMS, 15% PPMS, 20% PPMS, 25% PPMS, and 30% PPMS) was used for the foamy confectionery products' preparation. Jerusalem artichoke syrup (carbohydrates: 69.5 g), finely chopped white gelatin (protein content: 87.2 g, carbohydrates: 0.7 g), and water were also used. The control sample (CS) was prepared without the addition of PPMS. The formula of the foamy confectionery products is presented in Table 1.

Table 1. Formulation of the foamy confectionery products.

Ingredients	CS	10% PPMS	15% PPMS	20% PPMS	25% PPMS	30% PPMS
Gelatine	2.5	2.5	2.5	2.5	2.5	2.5
Water	15	15	15	15	15	15
Jerusalem artichoke syrup	50	50	50	50	50	50
PPMS	-	7	10.5	14	17.5	21

CS—control sample without pomace powder of musky squash; PPMS—pomace powder of musky squash.

Gelatin for swelling was soaked in water at a temperature of 20 ± 1 °C for 20 min. Jerusalem artichoke syrup was heated to a temperature of 120 ± 2 °C and swollen gelatin was immediately added, stirred to homogeneity, and whipped with a mixer (Vitek, Shenzhen, China) for 5 min. To the cooled mass, sifted (through a sieve with a mesh size not more than 0.8 mm) pumpkin powder was added and whipped for another 10 ± 1 min until a white foamy mass was obtained. The obtained mass was molded and kept at a temperature of 5 ± 1 °C for 5 h. After that, the product was cut into 4 cm sized cubes and dipped in sugar powder. The samples were hermetically sealed using vacuum packaging and stored at a temperature of 4 ± 1 °C. The products were analyzed immediately after production and after 10, 20, 30, and 40 days of refrigerated storage (4 ± 1 °C).

2.5.2. Sensory Analysis of the Foamy Confectionery Products

To carry out the sensory analysis of the developed product, degustation according to ISO 6658:2017 was organized [23]. The product was evaluated according to the following criteria: consistency, smell, taste, structure, surface, and shape. A group of 9 people was formed for the tasting, which was carried out in degustation booths at room temperature using white light. Each sample was tested in duplicate in sensory analysis laboratories meeting the requirements of ISO 8589:2007 [24]. During the degustation, each participant was offered a sample of all subtypes of the developed product; each sample type was coded. Tasters rated each criterion in points from 0 to 5: 5—exceptional, ideal qualities; 4—appropriate quality; 3—with slight defects; 2—with obvious flaws; 1—with strongly pronounced defects; 0—altered, with big changes.

Each evaluation criterion was assigned an importance factor f_i, which was determined according to the degree of importance of the criterion with the overall evaluation: smell, taste, and color were the most important criteria (f_i = 0.25); for appearance and consistency, f_i = 0.2; for shape, f_i = 0.1 [25].

Calculations for this method included calculating the mean score P_m for each criterion, as well as calculating the mean fractional values P_{mp} and the overall mean score P_{mt}. Each criterion was subjected to the eigenvalue of the factor f_p, depending on the degree of importance of this criterion. The fractional values P_{mp} were calculated using the following formula:

$$P_{mp} = P_m \cdot f_i \cdot f_t \quad (1)$$

where P_m—the arithmetic means of the scores; f_i—importance factor; f_t—transformation coefficient, whereby the 5-point scale used is transformed into a 20-point scale to determine product quality and f_t = 4.

The overall average P_{mt} score was calculated by summing the P_{mp} values of all evaluation criteria. Based on this value, the final result of the analysis was determined and the product was given a score according to Table 2.

Table 2. The grade given to the product according to the overall average score.

Overall Average P_{mt}	Note
18.1–20.0	Very good
15.1–18.0	Good
11.1–15.0	Satisfactory
7.1–11.0	Unsatisfactory
0–7.0	Inadequate

2.5.3. Physico-Chemical Analysis of the Foamy Confectionery Products

AC of the foamy confectionery products was determined according to the AOAC method [12]. MC was determined according to Sudarmaji et al. by drying samples to constant mass using a Pol-Eko Aparatura ST 2, Wodzisław Śląski, Poland [11,26].

The pH was assessed utilizing a pH meter (TESTO 206-pH2, Pruszkow, Poland), according to the AOAC (2012) method [14].

The water activity (a_w) of the foamy confectionery products was measured at room temperature (25 ± 1 °C) using an electronic dew-point water activity meter, the LabMaster (Novasina AG, CH-8853 Lachen, Switzerland) [27]. Measurement of a_w was carried out until the value was concurrent.

TPC and AA were determined by the ABTS method for foamy confectionery samples (Sections 2.4.1 and 2.4.3) immediately after production. The AA of the samples was monitored at 10, 20, 30, and 40 days of storage. The physico-chemical properties, except for MC, a_w, and AA, were determined only on the first day of storage.

2.5.4. Texture Analysis

Texture indices of the finished product were measured using a TA.HDplusC texture meter (Stable Micro Systems, Godalming, UK) with a P/40 nozzle. Two samples of each sample in the form of 2 × 2 × 2 cm cubes were placed in the work area of the slide and subjected to pressure from the nozzle of the device. The samples underwent two successive cycles under pressure, each equivalent to 50% of the sample height. There was a 15 s interval between the two cycles, and the testing speed was set at 1 mm/s. The quantitative parameters determined were hardness (maximum peak force in the first cycle), cohesiveness (the ratio of the positive area under the curve during the second compression and the first compression), and gumminess (multiplication of hardness by cohesiveness) [28].

2.5.5. Color Analysis

The color analysis was performed by determination of CIELab parameters using a Chroma Meter CR-400 (Konica Minolta, Tokyo, Japan) according to the method in [29]. The L^*, a^*, and b^* values of the samples were determined, where the L^* value indicates the lightness, a^* represents the green ($-a^*$)–red ($+a^*$) values, and b^* is from blue ($-b^*$) to yellow ($+b^*$). The overall differences in color ΔE^* were calculated according to the following formula:

$$\Delta E^* = \sqrt{\left(L_i^* - L_0^*\right)^2 + \left(a_i^* - a_0^*\right)^2 + \left(b_i^* - b_0^*\right)^2} \quad (2)$$

where L_0^*, a_0^* and b_0^*—the values of the control sample; L_i^*, a_i^* and b_i^*—the values of the samples with pumpkin powder.

The color intensity, or chromaticity (C^*), represents the vividness or saturation of a color [30] and was calculated according to the following formula:

$$C^* = \sqrt{a^{*2} + b^{*2}} \quad (3)$$

The yellowing index (YI) was calculated according to the following formula:

$$YI = 142.86 \times \frac{b^*}{L^*} \quad (4)$$

The browning index (BI) is defined as brown color purity and is one of the most common indicators of browning in food products containing sugar [31]. The BI was calculated using the following expression [32]:

$$BI = \frac{(a^* + 1.75 \times L^*)}{0.17 \times (5.645 \times L^* + a^* - 3.012 \times b^*)} \times 100 \quad (5)$$

2.5.6. Microbiological Analysis

The microbial analysis of the foamy confectionery products was carried out according to ISO standard methods. The stock solution was prepared by aseptically taking 10 g of each sample and placing it in a 100 mL flask containing 90 mL of sterile 0.1% buffered peptone water, then stirring and homogenizing for two minutes [33]. The suspensions were then diluted and serial dilutions of 10^1 to 10^3 were obtained in triplicates and used for the specific medium. Aliquots of 0.1 mL and 1 mL of each dilution were used for pour plating and spread plating, respectively, into the various media. Spread plating using potato dextrose agar with chloramphenicol (2%) was used to determine yeast and molds; the plates were incubated for 7 days at 25 °C, while the total viable count was at 37 °C for 48 h [34,35]. Acceptable levels of microorganisms were based on sanitary and epidemiological rules and standards [36].

2.6. Mathematical Modeling

The MATLAB R2016a program (MathWorks, Inc., Natick, MA, USA) was applied to determine the influence of the PPMS concentration and the storage time of the foamy confectionery products on the total average score, the average score of the sensory profile, MC, a_w, AA, hardness, cohesiveness, and gumminess. The mutual information values are measured in bits. The closer the bit value is to 1, the more pronounced the influence of the pumpkin powder concentration and storage time of the foamy samples [37].

2.7. Statistical Analysis

The findings are displayed in this paper as the mean values ± standard error of the mean based on three repeated measurements. Statistical analysis was conducted using Microsoft Office Excel 2007 (Microsoft, Redmond, WA, USA). One-way analysis of variance (ANOVA) was employed along with the Tukey test at a significance level of $p \leq 0.05$. Additionally, Statgraphics software Centurion XVI 16.1.17 (Statgraphics Technologies, Inc., The Plains, VA, USA) was utilized.

3. Results and Discussion

3.1. Physico-Chemical Indications and Phytochemical Proprieties of PPMS

The physico-chemical indicators and phytochemical proprieties of PPMS are presented in Table 3.

Table 3. Physico-chemical indicators and phytochemical properties of PPMS.

Indicators	The Value of the Indicator
MC, g/100 g	6.21 ± 0.05
AC, g/100 g DW	6.27 ± 0.06
pH	7.45 ± 0.05
TA, % expressed in malic acid	1.16 ± 0.07
TCC, mg/100 g DW	27.76 ± 5.01
TPC, mg GAE/100 g DW	237.15 ± 16.8
AA (ABTS), mg TE/100 g DW	681.01 ± 21.6
AA (DPPH), mg TE/100 g DW	300.03 ± 14.4

Antioxidant activity of ABTS and DPPH was determined for hydroethanolic extracts of pomace powder of musky squash (PPMS). MC—moisture content; AC—ash content; TA—titratable acidity; TCC—total carotenoid content; TPC—total polyphenol content; AA—antioxidant activity. ABTS—2,2-azinobis-(3-ethylbenzothiazoline-6-sulfonates); DPPH—2,2-diphenyl-1-picrylhydrazyl-hydrate; GAE—gallic acid equivalent; TE—trolox equivalent; DW—dry weight. Results are presented as mean ± standard deviation.

In the result of the analysis of the physico-chemical properties and phytochemical proprieties of the raw material (Table 3), it was found that PPMS has a moisture content of 6.21% and is a rich source of ash (6.27%). The approximate PPMS compositions were more or less similar to the results of See et al., who observed lower AC (5.37%) but slightly higher MC (10.96%) in pumpkin powder [9]. On the other hand, an AC of 3.8% in PPMS was reported by Ptichkina et al. [38]. The same MC was within the safe limit, as Bothast et al. [39] noted that pumpkin powder with an MC greater than 14% was susceptible to fungal and mold growth. In PPMS, the TA was 1.16%, expressed as malic acid, close to the values recorded by Dhiman et al. of 1.03% [40]. In the research of this study, high TCC (27.76 mg/100 g DW) and TPC (237.15 mg/100 g DW) were found in PPMS, providing high AA. These values correlate with data from Bochnak et al. [41]. Hussain et al. [42] recorded lower values for TPC (134.59 mg GAE/100 g DW). Asif et al. [43] noted that the TPC determined in hot-air-dried samples was 67.6 mg/100 g DW, and that in freeze-dried samples was 63.7 mg/100 g DW.

Our study has shown excellent antioxidant potential (mg TE/100 g DW) in both assays of 681.01 (ABTS) and 300.03 (DPPH), respectively, for pumpkin powder used in foamy confectionery products (Table 3). Other studies have shown the AA of the powder pumpkin pulp to be 0.53 mmol of AAE (ascorbic acid equivalent)/100 g, as well as a TPC of 192 mg GAE/100 g and a β-carotene content of 32.87 mg/100 g DW [44].

3.2. The Sensory Properties of the Foamy Confectionery Products

The influence of terms of storage on the quality of new types of foamy confectionery products was determined by changes in sensory and physico-chemical parameters within 40 days from the date of manufacture. Foamy confectionery products with PPMS were stored at a temperature of $4 \pm 2\,°C$ and a relative humidity of no more than 75%. The samples were packaged following current requirements in polyethylene or polyethylene film and a corrugated cardboard box for confectionery products weighing 150 g. The images of foamy confectionery products with PPMS are presented in Figure 1.

Figure 1. Image of foamy confectionery products: (**a**) classic commercial marshmallow, (**b**) control sample (CS)—foamy confectionery product without PPMS, (**c**) foamy confectionery product with 10% PPMS, (**d**) foamy confectionery product with 15% PPMS, (**e**) foamy confectionery product with 20% PPMS, (**f**) foamy confectionery product with 25% PPMS, and (**g**) foamy confectionery product with 30% PPMS.

Sensory analysis is important both for the general assessment of food characteristics and for assessing the quality and safety of products based on taste, smell, texture, and appearance. This analysis is necessary to determine consumer attitudes toward food products by measuring the degree of acceptance of a new product or improvement of an existing product [45,46]. The results of the sensory analysis are presented in Table 4.

The flavor of food is a manifestation of the complex interactions between aroma, taste, and oral sensations. Aroma, in particular, is associated with volatile compounds, while taste is linked with non-volatile, high-molecular-weight components [47]. The products exhibited a pronounced pumpkin flavor profile. Slices of the foamy confectionery products showed different shades of orange color, from light to more intense depending on the concentration range of PPMS, and the CS had a milky color typical of the product without dye. The added 10% and 15% PPMS showed more consistency in the foamy confectionery products, while lower values were found in the 25% and 30% PPMS products. The taste, odor, and intensity of the appearance of the pumpkin flavoring were highly evaluated in all samples.

According to the data obtained, the CS and the sample with a PPMS content of 15% were rated "Very good", but the other processed product samples were rated "Good" and "Satisfactory".

The optimal amount of PPMS from the tasters' point of view was 15%; a higher amount of flour reduces the grade and sensory qualities of the product.

Table 4. Sensory characteristics of the foamy confectionery products with PPMS (results are presented as mean ± standard deviation).

Sensory Characteristics	CS	10% PPMS	15% PPMS	20% PPMS	25% PPMS	30% PPMS
Total average score, taking into account factors of importance	18.96 ± 0.04 [b]	19.89 ± 0.05 [c]	20.00 ± 0.0 [c]	19.91 ± 0.06 [c]	19.56 ± 0.03 [c]	17.44 ± 0.05 [a]
The average score of the sensory profile	4.78 ± 0.05 [b]	4.98 ± 0.02 [d]	5.00 ± 0.0 [d]	4.98 ± 0.02 [d]	4.87 ± 0.04 [c]	4.31 ± 0.01 [a]
Taste and odor	4.56 ± 0.01 [a]	5.00 ± 0.0 [b]	5.00 ± 0.0 [b]	5.00 ± 0.0 [b]	5.00 ± 0.0 [b]	4.56 ± 0.01 [a]
Appearance	5.00 ± 0.0 [b]	5.00 ± 0.0 [b]	5.00 ± 0.0 [b]	5.00 ± 0.0 [b]	5.00 ± 0.0 [b]	4.56 ± 0.01 [a]
Consistency	5.00 ± 0.0 [d]	5.00 ± 0.0 [d]	5.00 ± 0.0 [d]	4.89 ± 0.04 [c]	4.56 ± 0.05 [b]	4.00 ± 0.03 [a]
Shape	4.89 ± 0.05 [c]	5.00 ± 0.0 [d]	5.00 ± 0.0 [d]	5.00 ± 0.0 [d]	4.78 ± 0.03 [b,c]	4.00 ± 0.02 [a]
Color	4.44 ± 0.03 [a]	4.89 ± 0.02 [b,c]	5.00 ± 0.0 [c]	5.00 ± 0.0 [c]	5.00 ± 0.0 [c]	4.44 ± 0.04 [a]

PPMS—pomace powder of musky squash. Different letters ([a–d]) designate statistically different results ($p \leq 0.05$).

The results of changes in the average score of the sensory profile of the foamy confectionery products with PPMS during storage with consideration of importance factors are presented in Table 5.

Table 5. Changes in the average score of the sensory profile of the foamy confectionery products with PPMS during storage with consideration of important factors (results are presented as mean ± standard deviation).

Sensory Indicators/ Storage Date	CS	10% PPMS	15% PPMS	20% PPMS	25% PPMS	30% PPMS
Total average score, taking into account factors of importance						
- first day	18.96 ± 0.08 [g]	19.89 ± 0.07 [h,i]	20.00 ± 0.0 [i]	19.91 ± 0.05 [i]	19.56 ± 0.04 [h]	17.44 ± 0.04 [d]
- 10th day	18.87 ± 0.06 [g]	19.84 ± 0.06 [h,i]	20.00 ± 0.0 [i]	19.91 ± 0.03 [i]	19.51 ± 0.05 [h]	17.36 ± 0.05 [d]
- 20th day	18.64 ± 0.09 [f]	19.76 ± 0.09 [h]	19.96 ± 0.03 [i]	19.78 ± 0.04 [h]	19.38 ± 0.07 [g,h]	17.24 ± 0.03 [d]
- 30th day	18.33 ± 0.04 [f]	19.56 ± 0.05 [h]	19.87 ± 0.05 [h,i]	19.60 ± 0.05 [h]	19.16 ± 0.05 [g]	16.93 ± 0.06 [c]
- 40th day	16.71 ± 0.07 [b,c]	18.11 ± 0.06 [e]	18.96 ± 0.07 [f,g]	17.09 ± 0.08 [c]	16.22 ± 0.03 [b]	15.18 ± 0.04 [a]
The average score of the sensory profile						
- first day	4.78 ± 0.02 [i,j]	4.98 ± 0.02 [k]	5.00 ± 0.0 [k]	4.98 ± 0.01 [k]	4.87 ± 0.03 [j,k]	4.31 ± 0.03 [d,e]
- 10th day	4.76 ± 0.01 [i]	4.96 ± 0.02 [k]	5.00 ± 0.0 [k]	4.98 ± 0.01 [k]	4.84 ± 0.01 [j]	4.29 ± 0.01 [d,e]
- 20th day	4.67 ± 0.03 [h,i]	4.91 ± 0.03 [j,k]	4.98 ± 0.1 [k]	4.93 ± 0.02 [k]	4.80 ± 0.02 [i,j]	4.27 ± 0.02 [d]
- 30th day	4.58 ± 0.01 [f]	4.87 ± 0.02 [j]	4.96 ± 0.03 [k]	4.89 ± 0.03 [j,k]	4.20 ± 0.01 [c,d]	4.20 ± 0.02 [c,d]
- 40th day	4.16 ± 0.02 [c]	4.47 ± 0.01 [f]	4.71 ± 0.02 [h,i]	4.24 ± 0.02 [d]	4.04 ± 0.01 [b]	3.80 ± 0.0 [a]

PPMS—pomace powder of musky squash. Different letters ([a–k]) designate statistically different results ($p \leq 0.05$).

The effect of the storage period on the sensory attributes of the foamy confectionery products with PPMS was studied at predetermined intervals, as delineated above, to assess the acceptability of the product. The results (Table 5) indicated that the fresh product prepared with 15% added PPMS elicited a higher flavor score (20.00) than the CS without added PPMS (18.96). At 25 and 30% levels of added PPMS, the scores were somewhat lower (19.56 and 17.44).

With the advancement in the storage period, the flavor score declined consistently. On the 20th day, although the score diminished, it remained in good condition. Again, the foamy confectionery products prepared with 15% added PPMS obtained better scores than other samples. On the 30th day, the average score had decreased slightly to 4.96, indicating the acceptability of the product. On the 40th day, the products prepared with 25% and 30% levels of added PPMS were not acceptable based on flavor, as the score had declined to above 4.0. In determining the mean interaction effect between x (PPMS concentration) and y (shelf life) on the consistency and shape of foam confectionery, it was observed that the maximum score was noted for samples prepared using 15% PPMS on the 1st (fresh) and 10th days of the storage period ($x2\ y1$, and $x2\ y2$). The scores for consistency and shape decreased consistently as the period of storage elapsed. The product remained acceptable for up to 30 days as far as consistency and shape were concerned. On the 40th day, the consistency and shape of the foamy confectionery products were unacceptable. The mean interaction effects between x and y on the taste, odor, color, and appearance of the score of the foamy confectionery products revealed that the maximum average score (5.0) was noted in samples made using 15% PPMS on the 1st and 10th days of the storage period ($x2\ y2$). As the period of storage elapsed, the scores for the appearance of the product declined, but the trend remained almost the same. The 25% and 30% PPMS products elicited lower scores for appearance irrespective of the storage period. The product remained acceptable for up to 30 days.

The total average score based on taste, odor, appearance, consistency, shape, and color, recorded in Table 5, suggested that the maximum total average score, taking into account factors of importance (20.00) and the average sensory profile score (5.00), was perceived in samples prepared using 15% PPMS on the 1st and 10th days of storage ($x2\ y2$). A slightly lower total average score (taking into account factors of the importance of the foamy confectionery products) and average score of sensory profiles were observed for samples made with 20% PPMS (19.91 and 4.98, respectively). Other PPMS combinations yielded products with lower scores. The total average score, taking into account factors of the importance of the foamy confectionery products, and the average score of sensory profiles diminished with advancement in the storage period. On the 20th day, the product elicited lower scores but it was in good condition and liked by panelists. On the 30th day, the score further declined but remained in fairly acceptable condition. The product was, however, not acceptable on the 40th day as evidenced by the overall acceptability scores declining to around 4.0. The results suggested that the use of PPMS extended the shelf life of foamy confectionery products and had an additive effect.

3.3. Physico-Chemical Analysis of the Foamy Confectionery Products

The physico-chemical indicators of the foamy confectionery products fortified with PPMS on the first day and during storage are presented in Table 6.

The addition of PPMS has a significant contribution to the physico-chemical properties of the foamed confectionery products. The AC of foamy confectionery products increased significantly ($p < 0.05$) as the proportion of PPMS additive increased. Thus, at a PPMS concentration of 10%, the AC increases by 14.3 times; at a PPMS concentration consisting of 30%, the ash content increases by almost 40 times. This may be because PPMS had a high ash content (Table 3). According to Ivanova et al. and Tamashevich et al. [48–50], the use of plant raw materials in marshmallow production led to an increase in macro- and microelements by 1.1–3 times compared to the control.

The addition of pumpkin powder leads to an increase in the pH value of foamy confectionery product samples from 5.71 to 6.42 depending on the PPMS added. Darwish, A. [51] studied the effect of pumpkin powder on the quality of yogurts in quantities from 1 to 5% and also confirmed that the pH level in pumpkin probiotic yogurts with the addition of pumpkin powder increased (from 4.53 to 4.98). The increase in pH levels in foamy confectionery products with added PPMS may be due to the presence of several protein-pound polysaccharides in the pumpkin [52].

Increasing PPMS concentration in foamy confectionery products led to the retention of free water and, respectively, to an increase in the moisture content of the foamy confectionery samples. Kita et al. [53] noted that the addition of fruit powders into snacks with Jerusalem artichoke also increased the water content of obtained products. During storage, the moisture content was reduced in all analyzed samples.

Table 6. Physico-chemical indicators, a_w and AA of the foamy confectionery products with PPMS (results are presented as mean ± standard deviation).

Indicators/Storage Date	CS	10% PPMS	15% PPMS	20% PPMS	25% PPMS	30% PPMS
AC, g/100 g DW	0.04 ± 0.01 [a]	0.57 ± 0.01 [b,c]	0.79 ± 0.01 [c]	1.07 ± 0.02 [d]	1.34 ± 0.02 [e]	1.61 ± 0.01 [f]
pH	4.83 ± 0.01 [a]	5.71 ± 0.02 [b]	5.84 ± 0.02 [b,c]	5.98 ± 0.01 [c]	6.15 ± 0.01 [d]	6.42 ± 0.01 [e]
MC, g/100 g						
- first day	40.47 ± 0.03 [l]	36.63 ± 0.02 [h]	37.02 ± 0.03 [h]	38.15 ± 0.05 [j]	38.98 ± 0.03 [k]	40.85 ± 0.05 [l]
- 10th day	39.57 ± 0.03 [k]	34.08 ± 0.04 [e]	34.47 ± 0.06 [e]	36.96 ± 0.05 [h]	36.82 ± 0.07 [h]	39.02 ± 0.03 [k]
- 20th day	36.37 ± 0.06 [g,h]	32.80 ± 0.03 [c]	34.15 ± 0.05 [e]	35.59 ± 0.08 [f,g]	36.50 ± 0.02 [h]	37.85 ± 0.07 [i]
- 30th day	34.91 ± 0.07 [f]	32.42 ± 0.06 [b,c]	33.49 ± 0.05 [d]	35.27 ± 0.08 [f]	36.08 ± 0.05 [g]	37.04 ± 0.04 [h]
- 40th day	31.25 ± 0.04 [a]	31.40 ± 0.05 [a]	32.29 ± 0.06 [b]	34.03 ± 0.02 [e]	34.91 ± 0.04 [f]	36.53 ± 0.05 [h]
a_w, c. u.						
- first day	0.674 ± 0.0 [l]	0.629 ± 0.001 [d]	0.631 ± 0.001 [d,e]	0.646 ± 0.002 [g,h]	0.649 ± 0.001 [h]	0.670 ± 0.002 [l]
- 10th day	0.672 ± 0.001 [l]	0.629 ± 0.002 [d]	0.628 ± 0.002 [d]	0.640 ± 0.001 [f]	0.648 ± 0.002 [g,h]	0.665 ± 0.001 [k]
- 20th day	0.667 ± 0.002 [k,l]	0.617 ± 0.001 [b]	0.623 ± 0.002 [c]	0.636 ± 0.002 [e,f]	0.640 ± 0.003 [f,g]	0.664 ± 0.002 [j,k]
- 30th day	0.662 ± 0.001 [j]	0.615 ± 0.001 [a,b]	0.622 ± 0.001 [c]	0.635 ± 0.002 [e,f]	0.639 ± 0.002 [f]	0.664 ± 0.001 [j,k]
- 40th day	0.656 ± 0.001 [i]	0.615 ± 0.002 [a,b]	0.620 ± 0.002 [b,c]	0.634 ± 0.001 [e]	0.639 ± 0.001 [f]	0.665 ± 0.003 [k]
AA (ABTS), mg TE/100 g DW						
- first day	43.49 ± 0.16 [a]	60.20 ± 0.15 [c]	102.80 ± 0.45 [e]	116.20 ± 0.38 [f]	131.37 ± 0.46 [h]	133.89 ± 0.37 [h]
- 10th day	43.26 ± 0.11 [a]	62.78 ± 0.18 [c]	103.02 ± 0.38 [e]	116.44 ± 0.27 [f]	132.39 ± 0.39 [h]	133.95 ± 0.31 [h]
- 20th day	45.79 ± 0.18 [a]	62.82 ± 0.22 [c]	108.01 ± 0.41 [e]	123.42 ± 0.31 [g]	135.17 ± 0.41 [h]	134.63 ± 0.26 [h]
- 30th day	45.12 ± 0.10 [a]	59.02 ± 0.17 [b]	107.12 ± 0.38 [e]	125.42 ± 0.45 [g]	130.79 ± 0.26 [h]	133.28 ± 0.29 [h]
- 40th day	43.12 ± 0.21 [a]	43.80 ± 0.15 [a]	98.65 ± 0.26 [d]	110.13 ± 0.26 [e,f]	106.07 ± 0.19 [e]	99.63 ± 0.17 [d]

AC—ash content; MC—moisture content; a_w—water activity; AA—antioxidant activity; ABTS—2,2-azinobis-(3-ethylbenzothiazoline-6-sulfonates); TE—trolox equivalent; DW—dry weight. Different letters ([a–l]) designate statistically different results ($p \leq 0.05$).

Important properties characterizing the quality of products during storage are MC and the ratio of free and bound moisture in the product. It has been established that the rate of moisture loss depends on the food's chemical composition and the amount of pectinous substances, proteins, and sugars as well as and the presence of reducing substances (glucose, maltose, fructose, etc.) [54].

Pectin substances of fruit powders, including PPMS, have hydrophilic properties; they can firmly retain moisture in the product [55]. Therefore, the more PPMS in the formulation, the more firmly the molecules of the polysaccharides retain moisture in the sample, and the faster the rate of sample moisture removal decrease during storage. Thus, when the samples were stored for 40 days, the MC loss for CS was 22.8%, that for the sample with 10% PPMS was 14.3%, that with 15% PPMS was 12.8%, that with 20% PPMS was 10.8%, that with 25% PPMS was 10.4%, and that with 30% PPMS was 10.6%. Samples with PPMS can retain MC 1.1–1.3 times better after storage for 40 days, in comparison with the CS.

The a_w indicator does not exceed the norm, which indicates good resistance of the product to damage. The a_w value also decreases slightly during storage, which indicates the beneficial properties of pumpkin flour to slow down the chemical and enzymatic reactions in the product.

In the hydroethanolic extracts of foamy confectionery products, immediately after production, TPC was determined, which had average values between 35.07 and 58.23 mg GAE/100 g DW. The CS, without the addition of PPMS, had an average TPC of 10.68 mg GAE/100 g DW. This suggests that part of the polyphenols of the raw material had passed into the finished product. During storage, the TPC in the samples varied insignificantly.

The ABTS radical-cation scavenging activity of foamy confectionery samples with different concentrations of PPMS was monitored, starting on the day of manufacture and repeating every 10 days until day 40. AA in all samples was well preserved for 30 days, but on the 40th day, a decrease in values was observed. AA reduction is more insignificant during storage for samples with 15–20% PPMS. Our proximate findings were similar to the findings of Artamonova et al. [56]. A slight increase in AA values recorded on the 20th day of storage can be explained by redox reactions between the chemicals of the complete food matrix. An analysis of literature sources shows that analogous dependencies were obtained for fruit extracts [57].

3.4. Color Evaluation of the Foamy Confectionery Products

The chromatic parameters L*, a*, b*, C*, YI, and BI of the foamy confectionery products prepared without and with PPMS addition are presented in Table 7.

Table 7. Changes in color indicators of the foamy confectionery products with PPMS (results are presented as mean ± standard deviation) during storage.

CIELab Chromatic Parameters	Samples of the Foamy Confectionery Products (First Day)					
	CS	10% PPMS	15% PPMS	20% PPMS	25% PPMS	30% PPMS
L*	79.03 ± 0.11 [f]	76.1 ± 0.07 [d,e]	74.06 ± 0.06 [d]	71.89 ± 0.11 [c]	69.76 ± 0.08 [b]	65.34 ± 0.12 [a]
a*	−2.07 ± 0.03 [a]	6.08 ± 0.05 [b]	8.14 ± 0.07 [b]	13.56 ± 0.06 [c]	20.76 ± 0.11 [d]	35.01 ± 0.09 [e]
b*	23.01 ± 0.09 [a]	30.21 ± 0.12 [b]	38.02 ± 0.05 [c]	44.92 ± 0.08 [d]	52.04 ± 0.05 [e]	61.14 ± 0.11 [f]
ΔE*	-	11.27 ± 0.08 [a]	18.83 ± 0.06 [b]	27.85 ± 0.07 [c]	38.08 ± 0.08 [d]	54.93 ± 0.10 [e]
C*	23.1 ± 0.07 [a]	30.82 ± 0.08 [b]	38.88 ± 0.06 [c]	46.92 ± 0.07 [d]	56.03 ± 0.08 [e]	70.45 ± 0.09 [f]
YI	41.59 ± 0.09 [a]	56.71 ± 0.10 [b]	73.34 ± 0.07 [b]	89.27 ± 0.09 [c]	106.57 ± 0.09 [d]	133.68 ± 0.10 [e]
BI	156.1 ± 0.10 [d]	155.54 ± 0.09 [d]	149.85 ± 0.06 [c]	147.8 ± 0.05 [b]	147.07 ± 0.07 [a]	149.41 ± 0.11 [c]
	Samples of the Foamy Confectionery Products (Recommended Day of Storage—30th day)					
L*	78.58 ± 0.26 [f]	75.15 ± 0.21 [e]	73.52 ± 0.18 [d]	71.35 ± 0.22 [c]	68.99 ± 0.21 [b]	65.05 ± 0.19 [a]
a*	−1.06 ± 0.02 [a]	5.03 ± 0.05 [b]	7.45 ± 0.08 [c]	13.04 ± 0.05 [d]	20.12 ± 0.10 [e]	34.25 ± 0.11 [f]
b*	22.04 ± 0.13 [a]	30.54 ± 0.09 [b]	37.89 ± 0.11 [c]	44.18 ± 0.14 [d]	51.41 ± 0.11 [e]	60.85 ± 0.17 [f]
ΔE*	-	11.00 ± 0.14 [a]	18.69 ± 0.15 [b]	27.23 ± 0.17 [c]	37.46 ± 0.13 [d]	54.19 ± 0.15 [e]
C*	22.07 ± 0.09 [a]	30.95 ± 0.07 [b]	38.62 ± 0.10 [c]	46.06 ± 0.09 [d]	55.21 ± 0.11 [e]	69.83 ± 0.16 [f]
YI	40.07 ± 0.14 [a]	58.06 ± 0.19 [b]	73.63 ± 0.17 [b]	88.46 ± 0.21 [c]	106.46 ± 0.16 [d]	133.64 ± 0.17 [e]
BI	157.73 ± 0.15 [d]	154.09 ± 0.17 [c]	149.21 ± 0.16 [b]	147.79 ± 0.18 [b]	146.80 ± 0.13 [a]	148.97 ± 0.12 [b]

L*—lightness; a*—red–green parameter; b*—yellow–blue parameter; C*—chromaticity; ΔE*—overall difference in color, YI—yellowing index; BI—browning index. Different letters ([a–f]) designate statistically different results ($p \leq 0.05$).

The addition and increasing concentration of PPMS resulted in a color change (darkening) and yellow tint in the foamy confectionery products samples compared to the CS. The value of L* decreased while the values of a* and b* increased. Increasing the concentration of PPMS affected the intensity of the yellow color. The more PPMS was added, the more intense the yellow color of the final product. This can also be seen in the C* and YI values. An increase in the C* value indicates an increase in the brightness of the foamy confectionery product color with increasing PPMS concentration; the color of the samples also becomes more intense. A logical relationship is also characteristic of the YI and BI; a decrease in the BI and a proportional increase in the YI also indicate that the samples acquire a more saturated yellow color with an increase in the concentration of PPMS. The addition of PPMS resulted in a decrease in the lightness of the foamy confectionery products. Storage did not cause a significant change in coordinate L* values in any of the foamy confectionery products analyzed. The addition of PPMS resulted in a significant increase in a* values. It was found that for all foamy confectionery product samples, the values of L* were over 50 and were in the clear zone [58]. It was found that the L* of the foamy confectionery products showed a decreasing trend with the PPMS addition, which indicates that the experimental samples become darker compared to the CS. The value of L* decreased, and darkening was increased with an increase in the PPMS concentration from 10% to 30%. This effect was caused by the presence of natural pigments, such as carotenoids, which are naturally found in pumpkin powder [59,60]. The values of the parameters a* and b* were positive, demonstrating the predominance of red color over green and a strong predominance of yellow coloration, in disfavor of the blue, respectively. The resulting color of the foamy confectionery products was yellow. It was also found that the values of parameters a* and b* in the samples with PPMS were higher than in the case of the CS. This is probably due to the natural coloring pigments, carotenoids, in the PPMS. ΔE* represents a dimensionless parameter, resulting from the combination of the L*, a*, and b* values of the pairs of samples, which indicates whether or not there are differences in the colors perceived by the human eye, depending on the specific sensory thresholds [61]. Lo Faro et al. showed the difference between colors: if ΔE* < 0.2, there is an imperceptible difference; if 0.2 < ΔE* < 0.5, there is a very small difference; if 0.5 < ΔE* < 1.5, there is a small difference; if 2 < ΔE* < 3, there is a barely distinguishable difference; if 3 < ΔE* < 6, there is a very distinguishable difference; if 6 < ΔE* < 12, there is a large color difference; if ΔE* > 12, they are completely different colors [61]. The values of ΔE* were found to be ΔE* > 12, indicating completely different colors [62]. The storage resulted in a slight increase in coordinate a* values that was the strongest in the case of the CS. The decrease in b* values was not a significant change.

3.5. Textural Properties of the Foamy Confectionery Products

The results of the changes in the textural indicators of the foamy confectionery products with PPMS are presented in Table 8.

Table 8. Changes in textural indicators of the foamy confectionery products with PPMS (results are presented as mean ± standard deviation).

Textural Indicators/ Storage Date	CS	10% PPMS	15% PPMS	20% PPMS	25% PPMS	30% PPMS
Hardness, g						
- first day	387.2 ± 15.2 [a]	513.9 ± 18.4 [b]	587.8 ± 23.5 [b,c]	643.6 ± 11.8 [c]	695.3 ± 17.4 [c]	772.3 ± 21.4 [d]
- 10th day	590.0 ± 20.4 [b,c]	650.7 ± 26.2 [c]	735.2 ± 31.4 [c,d]	925.7 ± 22.1 [e]	989.6 ± 24.8 [f]	1138.8 ± 28.1 [g]
- 20th day	618.1 ± 31.5 [b,c]	737.8 ± 16.9 [d]	930.3 ± 27.1 [e]	934.5 ± 19.3 [e]	1246.3 ± 19.6 [h]	1363.1 ± 31.2 [i]
- 30th day	634.5 ± 29.7 [c]	823.0 ± 16.0 [d]	1080.0 ± 35.6 [f,g]	1090.8 ± 30.2 [f,g]	1264.3 ± 23.1 [h]	1586.1 ± 25.2 [j]
- 40th day	588.2 ± 31.3 [c]	835.1 ± 21.4 [d,e]	1120.3 ± 24.7 [g]	1200.3 ± 28.1 [g,h]	1280.1 ± 30.7 [h]	1602.3 ± 19.5 [j]
Cohesiveness, %						
- first day	1.38 ± 0.01 [d]	1.44 ± 0.00 [e]	1.38 ± 0.01 [d]	1.59 ± 0.02 [h,i]	1.71 ± 0.01 [k]	1.65 ± 0.02 [i,j]
- 10th day	1.33 ± 0.02 [b,c]	1.38 ± 0.02 [c,d,e]	1.35 ± 0.01 [c,d]	1.51 ± 0.02 [f,g]	1.57 ± 0.02 [g,h]	1.65 ± 0.01 [i,j]
- 20th day	1.32 ± 0.03 [a,b,c]	1.34 ± 0.02 [b,c,d]	1.33 ± 0.03 [b,c,d]	1.44 ± 0.00 [e]	1.55 ± 0.03 [f,g,h]	1.64 ± 0.02 [i,j]
- 30th day	1.29 ± 0.02 [a,b]	1.34 ± 0.01 [c]	1.33 ± 0.01 [b,c]	1.35 ± 0.01 [c,d]	1.52 ± 0.02 [f,g]	1.56 ± 0.03 [g,h]
- 40th day	1.38 ± 0.02 [a,b]	1.33 ± 0.02 [b,c]	1.32 ± 0.02 [b,c]	1.31 ± 0.02 [b,c]	1.50 ± 0.01 [f]	1.55 ± 0.02 [g,h]
Gumminess, g						
- first day	534.3 ± 30.2 [a]	740.0 ± 26.7 [b]	740.0 ± 26.7 [b]	1023.4 ± 20.6 [c,d]	1188.9 ± 23.4 [d,e]	1274.3 ± 18.3 [e]
- 10th day	784.8 ± 28.6 [b]	898.0 ± 16.7 [c,d]	898.0 ± 16.7 [c,d]	1397.8 ± 38.1 [e,f]	1553.6 ± 28.9 [f]	1879.1 ± 21.4 [g]
- 20th day	816.0 ± 34.6 [b,c]	988.6 ± 21.5 [c,d]	988.6 ± 21.5 [c,d]	1345.6 ± 33.5 [e]	1931.8 ± 38.1 [g]	2235.4 ± 33.2 [h]
- 30th day	818.5 ± 29.8 [b,c]	1102.8 ± 36.9 [d]	1102.8 ± 36.9 [d]	1472.6 ± 29.3 [f]	1921.7 ± 30.6 [g]	2474.4 ± 29.4 [i]
- 40th day	819.5 ± 35.2 [b,c]	1110.7 ± 26.1 [d]	1110.7 ± 26.1 [d]	1572.4 ± 27.2 [f]	1920.1 ± 38.2 [g]	2483.6 ± 31.5 [i]

PPMS—pomace powder of musky squash. Different letters ([a–k]) designate statistically different results ($p \leq 0.05$).

It was revealed that the hardness of samples with the addition of PPMS increases in direct proportion to the amount of pumpkin flour. However, samples containing 25% and 30% PPMS showed a too-high value of hardness (695.3 g and 772.3 g, respectively). These values are outside the normal range reported by other studies such as Mardani et al. According to the authors, the hardness of classic foamy confectionery products was 637.68 g [63].

During the 40-day storage period, the hardness and gumminess of the analyzed foamy confectionery products registered an essential increase, and the cohesiveness of the samples gradually decreased, except the CS, which, on the 40th day of storage, registered a deterioration in the parameters of texture. The improvement in the texture parameters of the fortified foamy confectionery product samples was probably due to the better water-holding capacity of the foamy confectionery products fortified with PPMS compared to the foamy confectionery products without additions. The texture parameters were influenced by increasing the PPMS concentration in the samples. During the 40-day storage period, in the case of the 15% PPMS sample, the hardness and gumminess increased from 587.8 g to 1120.3 g and from 811.2% to 1478.8%, respectively. However, the cohesiveness decreased from 1.38% to 1.32%. In general, the texture parameters of the foamy confectionery product samples with the addition of PPMS correlated with their sensory indicators (Table 4) and their MC (Table 6). The addition of 10–30% PPMS to the foamy confectionery products determined the improvement in their degree of water retention and the textural parameters of the foamy confectionery products, thus contributing to the extension of the shelf life of the foamy confectionery products by 10 days compared to the CS.

3.6. Microbiological Results

The microbiological parameters of fresh samples of foam confectionery products (on the first day of preparation) and during 40 days of storage at 4 ± 1 °C were investigated (Table 9).

Table 9. Changes in microbiological indicators of the foamy confectionery products with PPMS during storage.

Microbiological Indicators/ Storage Date	Admitted Level [36]	CS	10% PPMS	15% PPMS	20% PPMS	25% PPMS	30% PPMS
QMAFAnM, CFU, max.							
- first day		2×10^2	3×10^2	2×10^2	2×10^2	3×10^2	4×10^2
- 10th day		5×10^2	3×10^2	6×10^2	5×10^2	4×10^2	5×10^2
- 20th day	5×10^3	5×10^2	2×10^2	2×10^1	3×10^1	8×10^1	2×10^2
- 30th day		6×10^2	3×10^1	2×10^1	3×10^1	4×10^1	8×10^1
- 40th day		7×10^2	7×10^1	1×10^1	2×10^1	3×10^1	7×10^1
Mold, CFU, max.	100	Were not found during storage					
Yeast, CFU, max.							
- first day		<5	<5	<5	<5	<5	<5
- 10th day		<10	<4	<3	<4	<4	<5
- 20th day	50	<20	<2	<2	<3	<5	<4
- 30th day		<20	<2	<2	<2	<3	<3
- 40th day		<50	<7	<2	<2	<3	<3

PPMS—pomace powder of musky squash; QMAFAnM—quantity of mesophilic aerobic and facultative anaerobic microorganisms; CFU—colony forming unit.

The results of the microbiological examination of the samples showed that the product under study does not contain coliform bacteria and pathogenic microorganisms; QMAFAnM, molds, and yeasts are contained in smaller quantities than the permissible level. Most of the studied samples contain bacteria of the genus Micrococcus. The antimicrobial and antioxidant properties of pumpkin powder allow for maintaining the microbiological stability of foamy confectionery products during storage [64].

3.7. Mathematical Modeling

Mutual information analysis was applied to demonstrate the influence of the PPMS concentration (10, 15, 20, 25, and 30%) and the storage time (10th, 20th, 30th, and 40th day) of the foamy confectionery products on the total average score, average score of sensory profile, MC, a_w, AA, hardness, cohesiveness, and gumminess (Table 10).

It was shown that the greatest influence of the PPMS concentration was on the AA (mutual information, 0.999 bits), followed by the a_w (0.895 bits), the total average score (0.717 bits), and the gumminess (0.708 bits). A satisfactory influence was found on the cohesiveness (0.488 bits), the average score of the sensory profile (0.452 bits), and the hardness (0.428 bits), and the lowest mutual information value was for the MC (0.141 bits).

In the case of the storage time, its influence on the analyzed parameters was much lower than that of the PPMS concentration. It can be seen that the greatest influence of storage time was on MC (0.179 bits). The influence on other parameters was as follows: hardness: 0.023 bits; total average score: 0.021 bits; average score of sensory profile:

0.015 bits; and AA: 0.011 bits. For a_w, cohesiveness, and gumminess, the mutual information values were insignificant (0.001 bits).

Table 10. The values of mutual analysis of the influence of PPMS concentration (10, 15, 20, 25, and 30%) and the storage time (10th, 20th, 30th, and 40th day) of the foamy confectionery products on the sensory and physico-chemical quality, AA, and texture parameters.

Parameter	Foamy Confectionery Products	
	PPMS Concentration, Bits	Storage Time, Bits
Total average score	0.717	0.021
Average score of sensory profile	0.452	0.015
MC	0.141	0.179
a_w	0.895	0.001
AA (ABTS)	0.999	0.011
Hardness	0.428	0.023
Cohesiveness	0.488	0.001
Gumminess	0.708	0.001

PPMS—pomace powder of musky squash; MC—moisture content; a_w—water activity; AA—antioxidant activity; ABTS—2,2-azinobis-(3-ethylbenzothiazoline-6-sulfonates).

The informational analysis was applied regarding the research on the influence of different amounts of sea buckthorn flour and rose hip powder on the quality of wheat bread and gingerbread [20]. It was also used to investigate the influence of storage time and different amounts of apple pomace [16], microencapsulated extracts of summer savory, rosemary [65], and basil [66] on the quality of dairy products.

4. Conclusions

This study aimed to develop foamy confectionery products with PPMS and artichoke syrup. PPMS used in foamy confectionery products showed excellent antioxidant potential (mg TE/100 g DW: 681 (ABTS) and 300 (DPPH)). The antimicrobial and antioxidant properties of pumpkin powder allow for maintaining the microbiological stability of the foamy confectionery products during storage.

The effect of different concentrations of PPMS on the sensory characteristics, physico-chemical, and textural properties of foamy confectionery products was evaluated. Sensory analysis was applied to measure the degree of product acceptance (taste and odor, appearance, consistency, shape, color). The optimal amount of PPMS, accepted by the tasters, was 15%. The main parameters of the product were monitored during storage. The mean total score and the mean sensory profile score gradually decreased. On the 30th day, the score was acceptable, clearly superior to the CS. The use of PPMS extended the shelf life of foamy confectionery products.

The addition of PPMS significantly influenced the physico-chemical properties of foamy confectionery products. There was an increase in the pH value, as well as the retention of free water in foamy confectionery products, caused by the presence of reducing substances and pectin substances with hydrophilic properties, which can firmly retain the ratio between free and bound water.

The increase in the concentration of PPMS led to an intensification of the color and lightness with the increase in the concentration of PPMS, and the storage (40 days) did not cause significant changes in the color and lightness of the products. During the 40-day storage period, the hardness and gumminess of the foamed confectionery products showed an essential increase, and the cohesiveness of the samples gradually decreased. The texture parameters of the foamy confectionery products were improved due to the better water-holding capacity of the products with PPMS compared to the CS. The addition of 10–30% PPMS to foamy confectionery products led to the improvement of textural parameters, thus contributing to the extension of the shelf life of foamy confectionery products by 10 days compared to the CS.

Mutual information analysis was applied to determine the influence of PPMS concentration and storage time of foamed confectionery products on the mean total score, mean sensory profile score, MC, a_w, AA, hardness, cohesiveness, and gumminess. The greatest influence of PPMS concentration was on AA, a_w, and gumminess.

The results of the conducted research indicate that the use of pumpkin pulp in the manufacture of foamy confectionery products can significantly increase their biological value and sensory characteristics and ensure a significant shelf life of the products.

Author Contributions: Conceptualization, N.N., E.S., G.G.C., R.S. and A.G.-M.; methodology, N.N., E.S., A.G., V.D. and A.G.-M.; software, A.G.-M.; validation, N.N., E.S., A.G., V.D. and A.G.-M.; formal analysis, N.N., E.S., A.G. and V.D.; investigation, N.N., E.S., A.G. and V.D.; resources, A.G.-M.; data curation, N.N., E.S., A.G., V.D. and A.G.-M.; writing—original draft preparation, N.N., E.S., A.G., V.D. and A.G.-M.; writing—review and editing, G.G.C., R.S. and A.G.-M.; visualization, N.N. and A.G.-M.; supervision, G.G.C., R.S. and A.G.-M.; project administration, A.G.-M.; funding acquisition, G.G.C. All authors have read and agreed to the published version of the manuscript.

Funding: This research was funded by the Institutional Project, subprogram 020405, "Optimizing food processing technologies in the context of the circular bioeconomy and climate change", Bio-OpTehPAS, implemented at the Technical University of Moldova.

Institutional Review Board Statement: Not applicable.

Informed Consent Statement: Not applicable.

Data Availability Statement: The original contributions presented in the study are included in the article; further inquiries can be directed to the corresponding author.

Conflicts of Interest: The authors declare no conflicts of interest.

References

1. Delgado-Vargas, F.; Jiménez, A.; Paredes-López, O. Natural Pigments: Carotenoids, Anthocyanins, and Betalains—Characteristics, Biosynthesis, Processing, and Stability. *Crit. Rev. Food Sci. Nutr.* **2000**, *40*, 173–289. [CrossRef]
2. Lu, W.; Shi, Y.; Wang, R.; Su, D.; Tang, M.; Liu, Y.; Li, Z. Antioxidant activity and healthy benefits of natural pigments in fruits: A review. *Int. J. Mol. Sci.* **2021**, *22*, 4945. [CrossRef] [PubMed]
3. Yabuzaki, J. Carotenoids Database: Structures, Chemical Fingerprints and Distribution Among Organisms. *Database* **2017**, *1*, bax004. [CrossRef] [PubMed]
4. Cazzonelli, C.; Pogson, B. Source to sink: Regulation of carotenoid biosynthesis in plants. *Trends Plant Sci.* **2010**, *15*, 266–274. [CrossRef] [PubMed]
5. Roohbakhsh, A.; Karimi, G.; Iranshahi, M. Carotenoids in the treatment of diabetes mellitus and its complications: A mechanistic review. *Biomed. Pharmacother.* **2017**, *91*, 31–42. [CrossRef] [PubMed]
6. Milner, J. Reducing the risk of cancer. In *Functional Foods: Designer Foods, Pharmafoods, Nutraceuticals*; Goldberg, I., Ed.; Chapman & Hall: New York, NY, USA, 1994; pp. 39–70.
7. Duthie, G.; Brown, K. Reducing the risk of cardiovascular disease. In *Functional Foods: Designer Foods, Pharmafoods, Nutraceuticals*; Goldberg, I., Ed.; Chapman & Hall: New York, NY, USA, 1994; pp. 19–38.
8. Pereira, A.; Krumreich, F.; Ramos, A.; Krolow, A.; Gularte, M. Physico-chemical characterization, carotenoid content and protein digestibility of pumpkin access flours for food application. *J. Food Sci. Technol.* **2020**, *40*, 10.
9. See, E.; Nadiah, W.; Noor, A. Physico-chemical and sensory evaluation of bread supplemented with pumpkin flour. *ASEAN Food J.* **2007**, *14*, 123–130.
10. Campbell, G.; Mougeot, E. Creation and characterisation of aerated food products. *Trends Food Sci. Technol.* **1999**, *10*, 283–296. [CrossRef]
11. Yudhistira, B.; Affandi, D.; Nusantari, P. Effect of green spinach (*Amaranthus tricolor* L.) and tomato (*Solanum lycopersicum*) addition in physical, chemical, and sensory properties of the foamy confectionery products as an alternative prevention of iron deficiency anemia. *IOP Conf. Ser. Earth Environ. Sci.* **2018**, *102*, 012007. [CrossRef]
12. AOAC. *Official Methods of Analysis*, 18th ed.; Association of Official Analytical Chemist: Gaithersburg, MD, USA, 2005.
13. Ranganna, S. *Handbook of Analysis and Quality Control for Fruit and Vegetable Products*; Tata Mc-Graw Hill Publishing Company Ltd.: New Delhi, India, 2009; p. 112.
14. AOAC. *Official Methods of Analysis*, 19th ed.; Association of Offical Analytical Chemists: Washington, DC, USA, 2012.
15. Shajan, A.E. A study on physical and chemical characteristics of pumpkin (*Cucurbita* sp.) flesh. *J. Pharm. Innov.* **2023**, *12*, 2171–2175. [CrossRef]
16. Popescu, L.; Cesco, T.; Gurev, A.; Ghendov-Mosanu, A.; Sturza, R.; Tarna, R. Impact of Apple Pomace Powder on the Bioactivity, and the Sensory and Textural Characteristics of Yogurt. *Foods* **2022**, *11*, 3565. [CrossRef] [PubMed]

17. Gurev, A.; Cesko, T.; Dragancea, V.; Ghendov-Mosanu, A.; Pintea, A.; Sturza, R. Ultrasound-and Microwave-Assisted Extraction of Pectin from Apple Pomace and Its Effect on the Quality of Fruit Bars. *Foods* **2023**, *12*, 2773. [CrossRef] [PubMed]
18. Paulpriya, K.; Packia Lincy, M.; Tresina Soris, P.; Veerabahu Ramasamy, M. In vitro antioxidant activity, total phenolic and total flavonoid contents of aerial part extracts of *Daphniphyllum neilgherrense* (wt.) Rosenth. *Ethnopharm. J. Bio. Innov.* **2015**, *4*, 257–268.
19. Waterman, P.; Mole, S. *Analysis of Phenolic Plant Metabolites, Ecological Methods and Concepts*; Wiley-Blackwell: Hoboken, NJ, USA, 1994; p. 248.
20. Ghendov-Mosanu, A.; Cristea, E.; Patras, A.; Sturza, R.; Padureanu, S.; Deseatnicova, O.; Turculet, N.; Boestean, O.; Niculaua, M. Potential application of Hippophae rhamnoides in wheat bread production. *Molecules* **2020**, *25*, 1272. [CrossRef] [PubMed]
21. Arnao, M.; Cano, A.; Alcolea, J.; Acosta, M. Estimation of free radical-quenching activity of leaf pigment extracts. *Phytochem. Anal.* **2001**, *12*, 138–143. [CrossRef] [PubMed]
22. Rice-Evans, C.; Miller, N. Total antioxidant status in plasma and body fluids. *Methods Enzym.* **1994**, *234*, 279–293.
23. *ISO 6658:2017*; Sensory Analysis—Methodology—General Guidance. International Organization for Standardization: Geneva, Switzerland, 2017.
24. *ISO 8589:2007*; General Guidance for the Design of Test Rooms. The International Organization for Standardization: Geneva, Switzerland, 2007.
25. Banu, C.; Vizireanu, C.; Rasmerita, D.; Nour, V.; Musteata, G.; Rubtov, S. *Calitatea si Analiza Senzoriala a Produselor Alimentare*; AGIR: Bucuresti, Romania, 2007; p. 574.
26. Sudarmaji, S.; Suhardi, H.B. *Analysis of Food and Agricultural Materials*; Liberty Yogyakarta: Yogyakarta, Indonesia, 1989.
27. Serin, S.; Turhan, K.; Turhan, M. Correlation between Water Activity and Moisture Content of Turkish Flower and Pine Honeys. *Food Sci. Technol.* **2018**, *38*, 238–243. [CrossRef]
28. Overview of Texture Profile Analysis. Available online: https://texturetechnologies.com/resources/texture-profile-analysis#overview (accessed on 21 March 2024).
29. Loypimai, P.; Moongngarm, A.; Chottanom, P. Thermal and pH degradation kinetics of anthocyanins in natural food colorant prepared from black rice bran. *J. Food Sci. Technol.* **2016**, *53*, 461–470. [CrossRef] [PubMed]
30. Bermúdez-Aguirre, D.; Mawson, R.; Versteeg, K.; Barbosa-Cánovas, G. Composition properties, physico-chemical characteristics and shelf life of whole milk after thermal and thermo-sonication treatments. *J. Food Qual.* **2009**, *32*, 283–302. [CrossRef]
31. Octaviani, L.; Rahayuni, A. Effect of various concentrations of sugar on antioxidant activity and the level of acceptance of buni juice (*Antidesma bunius*). *J. Nutr. Coll.* **2014**, *3*, 958–965. [CrossRef]
32. Shi, J.; Marc, L. Lycopene in tomatoes: Chemical and physical properties affected by food processing. *Crit. Rev. Food Sci. Nutr.* **2000**, *40*, 2. [CrossRef]
33. *ISO 4833:2003*; Microbiology of Food and Animal Feeding Stuffs—Horizontal Method for the Enumeration of Microorganisms—Colony-Count Technique at 30 °C. International Organization for Standardization: Geneva, Switzerland, 2003.
34. Sandulachi, E.; Netreba, N.; Macari, A.; Sandu, I.; Boestean, O.; Dianu, I. Phytopathogenic microbiote of sea buckthorn and impact on storage. *J. Eng. Sci.* **2023**, *29*, 176–189. [CrossRef] [PubMed]
35. Juhaniaková, L.; Petrová, J.; Hleba, L.; Kunová, S.; Bobková, A.; Kačániová. M. Microbiological testing of selected confectionery products quality. *J. Microbiol. Biotechnol. Food Sci.* **2014**, *3*, 225–227.
36. Uniform Sanitary and Epidemiological and Hygienic Requirements for Products Subject to Sanitary and Epidemiological Supervision (Control). Approved by Decision of the Customs Union Commission No. 299. 28 May 2010. Available online: https://food.ec.europa.eu/document/download/c032397b-b4cf-4537-b657-ad0e3a04c13a_en?filename=ia_eu-ru_sps_req_req_san-epi_chap-2_1_en.pdf (accessed on 21 March 2024).
37. Paninski, L. Estimation of entropy and mutual information. *Neural Comput.* **2003**, *15*, 1191–1253. [CrossRef]
38. Ptitchkina, N.; Novokreschonova, L.; Piskunova, G.; Morris, E. Large enhancements in loaf volume and organoleptic acceptability of wheat bread by small additions of pumpkin powder: Possible role of acetylated pectin in stabilizing gas-cell structure. *Food Hydrocolloid.* **1998**, *12*, 333–337. [CrossRef]
39. Bothast, R.; Anderson, R.; Warner, K.; Kwolek, W. Effects of moisture and temperature on microbiological and sensory properties of wheat flour and corn meal during storage. *Cereal Chem.* **1981**, *58*, 309–311.
40. Dhiman, A.; Bavita, K.; Attri, S.; Ramachandran, P. Preparation of pumpkin powder and pumpkin seed kernel powder for supplementation in weaning mix and cookies. *Int. J. Chem. Stud.* **2018**, *6*, 167–175.
41. Bochnak, J.; Swieca, M. Potentially bioaccessible phenolics, antioxidant capacities, and the colour of carrot, pumpkin, and apple powders—Effect of drying temperature and sample structure. *Int. J. Food Sci. Technol.* **2020**, *55*, 136–145. [CrossRef]
42. Hussain, A.; Kausar, T.; Din, A.; Murtaza, M.; Jamil, M.; Noreen, S.; Ramzan, M. Determination of total phenolic, flavonoid, carotenoid, and mineral contents in peel, flesh, and seeds of pumpkin (*Cucurbita maxima*). *J. Food Process. Preserv.* **2021**, *45*, e15542. [CrossRef]
43. Asif, M.; Raza Naqvi, S.; Sherazi, T.; Ahmad, M.; Zahoor, A.; Shahzad, S.; Mahmood, N. Antioxidant, antibacterial and antiproliferative activities of pumpkin (cucurbit) peel and puree extracts—An in vitro study. *Pak. J. Pharm. Sci.* **2017**, *30*, 1327–1334.
44. Malkanthi, H.; Umadevi, S. Pumpkin powder (*Cucurbita maxima*)-supplemented string hoppers as a functional food. *Int. J. Food Nutr. Sci.* **2020**, *9*, 2–6. [CrossRef]
45. Meilgard, M.; Civille, G.; Carr, B. *Sensory Evaluation Techniques*, 2nd ed.; CRC Press Inc.: Boca Roton, FL, USA, 1991; p. 354.

46. Poste, L.; Mackie, D.; Butler, G.; Larmond, E. *Laboratory Methods for Sensory Analysis of Food*; Ottawa Agriculture: Ottawa, ON, Canada, 1991; p. 210.
47. Shahidi, F.; Hossain, A. Role of Lipids in Food Flavor Generation. *Molecules* **2022**, *27*, 5014. [CrossRef]
48. Ivanova, N.; Nikitin, I.; Semenkina, S. Method for Production of Marshmallows. Pat. 2749828C1 RU, A23G 3/52/, No. 2020125116. 17 June 2021.
49. Ivanova, N.; Nikitin, I.; Klokonos, M.; Berezina, N.; Bulavina, T.; Guseva, D. Marshmallow technology of increased nutritional value. *IOP Conf. Ser. Earth Environ. Sci.* **2021**, *640*, 052009. [CrossRef]
50. Tamashevich, S.; Shkolina, A. Perspectives of use of domestic fruit and vegetable half-finished products in the technology of the zephyr with increased food value. *Food Ind. Sci. Technol.* **2017**, *3*, 37–44.
51. Darwish, A. Influence of Inulin and Pumpkin Powder Addition on Sensory and Rheological Properties of Low-Fat Probiotic Yogurt. *Food Public Health* **2020**, *10*, 88–96.
52. Caili, F.; Haijun, T.; Tongyi, C.; Yi, L.; Li, Q. Some properties of an acidic protein-bound polysaccharide from the fruit of pumpkin. *Food Chem.* **2007**, *100*, 944–947. [CrossRef]
53. Kita, A.; Nowak, J.; Michalska-Ciechanowska, A. The effect of the addition of fruit powders on the quality of snacks with jerusalem artichoke during storage. *Appl. Sci.* **2020**, *10*, 5603. [CrossRef]
54. Roudaut, G.; Debeaufort, F. Moisture loss, gain and migration in foods and its impact on food quality. *Chem. Deterior. Phys. Instab. Food Beverages* **2010**, *186*, 143–185.
55. Konrade, D.; Gaidukovs, S.; Vilaplana, F.; Sivan, P. Pectin from fruit-and berry-juice production by-products: Determination of physico-chemical, antioxidant and rheological properties. *Foods* **2023**, *12*, 1615. [CrossRef]
56. Artamonova, M.; Piliugina, I.; Samokhvalova, O.; Murlykina, N.; Kravchenko, O.; Fomina, I.; Grigorenko, A. Study of the properties of marshmallow with the Sudanese rose and Black chokeberry dyes upon storage. *Eureka Life Sci.* **2017**, *3*, 15–23. [CrossRef]
57. Duan, X.; Jiang, Y.; Su, X.; Zhang, Z.; Shi, J. Antioxidant properties of anthocyanins extracted from litchi (*Litchi chinenesis* Sonn.) fruit pericarp tissues in relation to their role in the pericarp browning. *Food Chem.* **2007**, *101*, 1365–1371. [CrossRef]
58. Pereira, D.; Correia, P.; Guiné, R. Analysis of the physical-chemical and sensorial properties of Maria type cookies. *Acta Chim. Slovaca.* **2013**, *6*, 269–280. [CrossRef]
59. Ninčević Grassino, A.; Rimac Brnčić, S.; Badanjak Sabolović, M.; Šic Žlabur, J.; Marović, R.; Brnčić, M. Carotenoid Content and Profiles of Pumpkin Products and By-Products. *Molecules* **2023**, *28*, 858. [CrossRef] [PubMed]
60. Murkovic, M.; Mülleder, U.; Neunteufl, H. Carotenoid content in different varieties of pumpkins. *J. Food Compos. Anal.* **2002**, *15*, 633–638. [CrossRef]
61. Lo Faro, E.; Salerno, T.; Montevecchi, G.; Fava, P. Mitigation of Acrylamide Content in Biscuits through Combined Physical and Chemical Strategies. *Foods* **2022**, *11*, 2343. [CrossRef]
62. Batista, R.; de Morais, M.; Caliari, M.; Júnior, M. Physical, microbiological and sensory quality of gluten-free biscuits prepared from rice flour and potato pulp. *J. Food Nutr. Res.* **2016**, *55*, 101–107.
63. Mardani, M.; Yeganehzad, S.; Ptichkina, N.; Kodatsky, Y.; Kliukina, O.; Nepovinnykh, N.; Naji-Tabasi, S. Study on foaming, rheological and thermal properties of gelatin-free marshmallow. *Food Hydrocoll.* **2019**, *93*, 335–341. [CrossRef]
64. Ghendov-Mosanu, A.; Netreba, N.; Balan, G.; Cojocari, D.; Boestean, O.; Bulgaru, V.; Gurev, A.; Popescu, L.; Deseatnicova, O.; Resitca, V.; et al. Effect of Bioactive Compounds from Pumpkin Powder on the Quality and Textural Properties of Shortbread Cookies. *Foods* **2023**, *12*, 3907. [CrossRef] [PubMed]
65. Popescu, L.; Cojocari, D.; Ghendov-Mosanu, A.; Lung, I.; Soran, M.L.; Opris, O.; Kacso, I.; Ciorita, A.; Balan, G.; Pintea, A.; et al. The Effect of Aromatic Plant Extracts Encapsulated in Alginate on the Bioactivity, Textural Characteristics and Shelf Life of Yogurt. *Antioxidants* **2023**, *12*, 893. [CrossRef]
66. Popescu, L.; Cojocari, D.; Lung, I.; Kacso, I.; Ciorita, A.; Ghendov-Mosanu, A.; Balan, G.; Pintea, A.; Sturza, R. Effect of Microencapsulated Basil Extract on Cream Cheese Quality and Stability. *Molecules* **2023**, *28*, 3305. [CrossRef]

Disclaimer/Publisher's Note: The statements, opinions and data contained in all publications are solely those of the individual author(s) and contributor(s) and not of MDPI and/or the editor(s). MDPI and/or the editor(s) disclaim responsibility for any injury to people or property resulting from any ideas, methods, instructions or products referred to in the content.

MDPI AG
Grosspeteranlage 5
4052 Basel
Switzerland
Tel.: +41 61 683 77 34

Applied Sciences Editorial Office
E-mail: applsci@mdpi.com
www.mdpi.com/journal/applsci

Disclaimer/Publisher's Note: The title and front matter of this reprint are at the discretion of the Guest Editors. The publisher is not responsible for their content or any associated concerns. The statements, opinions and data contained in all individual articles are solely those of the individual Editors and contributors and not of MDPI. MDPI disclaims responsibility for any injury to people or property resulting from any ideas, methods, instructions or products referred to in the content.